A First Book in Algebra

This edition published 2023
by Living Book Press
Copyright © Living Book Press, 2023

ISBN: 978-1-922974-07-5 (hardcover)
         978-1-922974-06-8 (softcover)

Based on the 1920 edition with answer key
published by Charles E, Merrill Company.

A catalogue record for this
book is available from the
National Library of Australia

A FIRST BOOK IN ALGEBRA

BY

FLETCHER DURRELL, Ph.D.

HEAD OF MATHEMATICAL DEPARTMENT,
THE LAWRENCEVILLE SCHOOL

AND

E. E. ARNOLD, M.A.

SUPERINTENDENT, THE PUBLIC SCHOOLS OF
THE PELHAMS, NEW YORK,
FORMERLY SPECIALIST IN MATHEMATICS, THE
UNIVERSITY OF THE STATE OF NEW YORK

PREFACE

THIS book has been written to meet the changes in High School work which have developed in recent years. For instance, most pupils now entering High School are younger and more immature than such pupils were even a decade ago. A majority of them also study only first year algebra and do not take the later more advanced course. Hence a demand has arisen that the first year in Algebra be simplified and made as directly practical as possible.

The outstanding features of this book are as follows:

I. *A simplification of subject matter.* Among the omissions are the involved use of the parenthesis, the larger part of factoring, complex fractions, special devices in the solution of equations, the method of comparison in solving simultaneous equations, the square root of complicated expressions, all treatment of fractional, negative, and literal exponents, and most of radicals.

II. *Increased and systematic use of the graph.* The omissions just specified make possible, among other things, an enlarged and systematized use of the graph. At the outset, the scales to be used on the axes and also partial diagrams for graphs are supplied. The other various stages by which the pupil proceeds from the simple to the more complex cases have been carefully developed. The pupil thus acquires an organized and comprehensive grasp and a power both to construct and to interpret graphs, giving a disciplinary training equal and perhaps superior to that conferred by the more technical algebraic topics omitted, and far exceeding them in vocational and cultural values.

III. *Increased use of the formula.* A like development has been made of the formula. In particular, almost every topic in arithmetic is stated, reviewed, and further developed by aid of the formula.

iii

IV. *Improved treatment of written problems.* Particular attention is called to the new method of dealing with written problems, shown on pp. 4–7, and used throughout the book. Teachers agree that the written problem is the most difficult topic in algebra. It is believed that the method here presented meets this fundamental difficulty, just as the group method used in the authors' text on geometry has overcome the difficulty in teaching pupils to work originals in that subject.

V. *Large amount of oral exercise work in the use of algebraic language.* Each exercise in written problems is preceded by an oral exercise in the use of algebraic language. In certain of the examples of these sight drills, the more difficult written problems which follow in the next exercise are analyzed.

VI. *Organization of the chapters into Parts I and II.* In Part I of each chapter, the simple elements of the new topic are presented with as little theory as possible. It is intended that the class shall study in a first course or semester Part I of all the chapters in succession, and then return to the beginning of the book, and study the Parts II in order, reviewing each Part I, before taking up the Part II which follows it. The arrangement of material, however, is such that if the teacher prefers all the subject matter can be studied consecutively. It is strongly recommended, however, that the method of first studying all of the Parts I be followed.

It is generally understood that the most difficult problem in mathematical pedagogy is the transition from arithmetic to algebra. It is felt that the new departures in this book make a distinct contribution toward the solution of this problem.

CONTENTS

CHAPTER   PAGE

		PART I	PART II
I.	SYMBOLS; FORMULAS; PROBLEMS	1,	16
II.	NEGATIVE NUMBERS; GRAPHS	24,	33
III.	ADDITION AND SUBTRACTION	42,	53
IV.	THE EQUATION	61,	70
V.	MULTIPLICATION; DIVISION	76,	92
VI.	SHORT MULTIPLICATIONS; SIMPLIFICATIONS	101,	112
VII.	EQUATIONS (*Continued*)	118,	129
VIII.	FACTORING	140,	150
IX.	FRACTIONS	160,	176
X.	FRACTIONAL AND LITERAL EQUATIONS	188,	205
XI.	SIMULTANEOUS EQUATIONS	217,	234
XII.	GRAPHS	245,	255
XIII.	SQUARE ROOT	260,	267
XIV.	RADICALS	271,	279
XV.	QUADRATIC EQUATIONS	289,	300
XVI.	ADDITIONAL TOPICS	323	
INDEX		337	

A FIRST BOOK IN ALGEBRA

CHAPTER I

SYMBOLS; FORMULAS; PROBLEMS

PART I

1. Some Algebra Which You Already Know. In arithmetic you have learned that an expression like "5 increased by 3" may be more briefly expressed as "5 + 3"; that is, + is a short way of writing "increased by" or "added to."

Similarly, the rule, "percentage is equal to the base multiplied by the rate expressed decimally," may be stated in a short way, thus $p = b \times r$.

Give two other instances from arithmetic where symbols are useful as short substitutes for words.

2. The Purpose of Algebra is to lessen work and obtain many other useful results by a greater use of symbols than is practiced in arithmetic.

3. Short Ways of Indicating Multiplication in Algebra. In algebra if a and b each stands for a number, a multiplied by b (or a times b) may be briefly expressed in any one of three different ways, viz.: $a \times b$, $a \cdot b$, or ab. That is, multiplication is sometimes expressed by the inclined cross, $\times$; sometimes by a dot; and sometimes by no symbol between a letter and its multiplier.

Thus, "3 times x" is written $3x$. Similarly, for $b \times r$, we may write br.

1

EXERCISE 1

1. In arithmetic "diminished by" is expressed by what symbol? In the short way write "7 diminished by 3."

2. In arithmetic "divided by" is expressed by what symbol? Write in a short way "12 divided by 4."

3. Write as long a list as you can of symbols which are used in arithmetic as short substitutes for words. Opposite each symbol write the word or words it stands for.

4. Find 6 per cent of 400. In this example which number is the base? The rate per cent? The percentage?

5. Count the letters that are used in writing the following rule, "The percentage is equal to the base multiplied by the rate per cent expressed decimally." Also count the number of letters and symbols in the formula $p = br$. Hence, in this case, the work of writing the rule is how many times as great as the work of writing the formula?

By the aid of symbols write the following in the shortest way you can:

6. Five plus seven.

7. 3 plus 4.

8. x increased by 4.

9. a plus b.

10. 7 minus 5.

11. y minus 5.

12. x minus y.

13. 3 times y.

14. x times y.

15. 3 times a times b.

16. x times y times z.

17. The product of b and r.

18. The product of i, r, and t.

19. x divided by 3.

20. The quotient of p and r.

21. a equals the product of l and w.

22. v equals the product of l, w, and h.

23. v divided by the product of l and w.

In reading $5x$, instead of reading "five times x," say simply "five x." In like manner, $5ab$ is read "five ab." Using this method when necessary, express in words:

24. $5 + 2$.   **27.** $x - y$.   **30.** xy.   **32.** prt.

25. $5 + x$.   **28.** $x \div y$.   **31.** $\dfrac{ab}{p}$.   **33.** $3b + 5xy$.

26. $b + x$.   **29.** $7x$.   **34.** $5 + x - y$.

Find the value of each of the following when $a = 1$, $b = 2$, $c = 3$, $x = 4$, $y = 6$:

35. $3 + b$.   **39.** $\dfrac{x}{2}$.   **43.** abx.   **47.** $\dfrac{2x + y}{b}$.

36. $a + b$.   **40.** $\dfrac{x}{b}$.   **44.** $\dfrac{x + y}{b}$.   **48.** $\dfrac{3y - 4c}{x}$.

37. $x - a$.   **41.** $\dfrac{bc}{x}$.   **45.** $2a + 3ab$.   **49.** $\sqrt{x}$.

38. $3x$.   **42.** $3ab$.   **46.** $5x - y$.   **50.** $\sqrt{bcy}$.

51. Find the value of lw when $l = 8.5$ and $w = 4$.

52. Find the value of br when $b = \$450$ and $r = .06$.

53. Find the value of $\dfrac{p}{b}$ when $p = 60$ and $b = 120$.

54. Find the value of $\dfrac{p}{r}$ when $p = 36$ and $r = .05$.

55. If Mary has knit 6 garments and her older sister has knit 3 times as many, how many has her sister knit? How many have both knit together?

56. If Mary has knit x garments and her sister has knit 3 times as many, how many has her sister knit? How many have both knit together?

57. If Walter has earned twice as many dollars as his younger brother Harold, and Harold has earned x dollars, how many has Walter earned? How many dollars have both earned together?

4. Analysis of Problems.

Ex. 1.   Mary knit a certain number of garments for
the Red Cross and her older sister knit three times as
many.   Together they knit 48 garments.   How many did
each girl knit ?

The problem may be analyzed and made ready for solution by the use of symbols, thus :

NUMBERS DEALT WITH   SYMBOLS FOR NUMBERS

Number of garments knit by Mary $= x.$
Number of garments knit by her sister $= 3x.$
Number of garments knit by both together $= 48.$

EQUALITY FOUND AMONG THESE NUMBERS

Sum of the first two numbers = the last number.

Ex. 2.   Walter and Harold made $84 by gardening one
summer.   Walter, who was older and stronger, received a
double share of the profits.   How much did each receive?

NUMBERS DEALT WITH   SYMBOLS FOR NUMBERS

Number of dollars received by the boys together $= 84.$
Number of dollars received by Walter $= 2x.$
Number of dollars received by Harold $= x.$

EQUALITY FOUND AMONG THESE NUMBERS

Sum of the last two numbers = the first number.

EXERCISE 2

Analyze each of the following problems in the same way
in which the two problems which precede this exercise are
analyzed.   Keep your results for future use.

1. On a half holiday Robert caught a certain number
of fish and his cousin Edward caught twice as many.   Together they caught 36 fish.   How many fish did each of
the boys catch ?

2. One day Ella picked a certain number of quarts of
strawberries and her mother picked three times as many.

Together they picked 32 quarts. How many quarts did each of them pick?

3. Two boys together catch 84 fish. If the boy who owns the boat which they use receives twice as many fish as the other boy, how many fish does each boy receive?

4. A man left $12,000 to his son and daughter. To his daughter, who had taken care of him in his old age, he left a double share. What did each receive?

5. A man and boy by working a garden one summer made $128.80. If the man received a share of the profits three times as large as the share received by the boy, how much did each receive?

6. Two boys together gathered 1 bu. 4 qt. of hickory nuts. If the boy who climbed the trees received a double share, how many quarts did each receive?

7. Two girls made $18.60 by sewing. The girl who supplied the thread and machine received twice as much as the other girl. How much did each make?

8. Make up and work a similar example concerning two girls who kept a refreshment stand.

9. The total cotton crop of the world in a certain year was 15,000,000 bales, and the United States in that year produced three times as much as all the rest of the world. How many bales of cotton did the United States produce?

10. A farm is worked on shares. As the tenant supplies the tools and fertilizers, he receives twice as large a share of the profits as the owner of the farm. If the profits for one year are $6000, how much does the tenant receive? The owner?

11. If the sum of the areas of New York and Massachusetts is 57,400 sq. mi., and New York is 6 times as large as Massachusetts. what is the area of each state?

12. One number is 5 times as large as another and the sum of the numbers is 240. Find the numbers.

13. One number is twice as large as another and the sum of the numbers is 7.26. Find the numbers.

14. Separate $5\frac{1}{3}$ into two parts such that one part is 7 times as large as the other.

15. To look well, the middle part of a steeple should be twice as high as the lowest part, and the top part 8 times as high as the lowest part. If a steeple is to be 132 ft. high, how high should each part be?

16. A man wants to save $6000 in three years. If he is to save twice as much the second year as the first, and three times as much the third year as the first, how much must he save each year?

5. Solution of Problems. To illustrate the complete solution of a problem we shall now take Ex. 1, § 4, p. 4, repeat its analysis, and complete its solution.

Ex. Mary knit a certain number of garments and her older sister knit three times as many. Together they knit 48 garments. How many did each girl knit?

NUMBERS DEALT WITH	SYMBOLS FOR NUMBERS
Number of garments knit by Mary	$= x.$
Number of garments knit by her sister	$= 3\,x.$
Number of garments knit by both together	$= 48.$

EQUALITY FOUND AMONG THESE NUMBERS

Sum of the first two numbers = the last number.

Or, in symbols, $\qquad x + 3\,x = 48.$

Hence, $\qquad\qquad\qquad 4\,x = 48.$

$\qquad\qquad\qquad\qquad x = 12,$ *number of garments knit by Mary.*

$\qquad\qquad\qquad\quad 3\,x = 36,$ *number of garments knit by her sister.*

CHECK. 36 is three times 12. Also $12 + 36 = 48.$

Hence all of the conditions of the given problem are satisfied by the numbers 12 and 36.

The teacher will note that the pupil in the work of searching out all the numbers that are contained in a given problem, and in arranging them in a list, is necessarily and perhaps unconsciously analyzing the problem, and therefore advancing a long way toward its solution. Hence the teacher is urged to insist that pupils follow the above form of solution in full.

EXERCISE 3

Analyze and solve the problems in Exercise 2.

Also analyze and solve each of the following problems:

1. A girl has $66 to spend for a hat, coat, and suit. She wants to spend twice as much for her coat as for her hat, and three times as much for her suit as for her hat. How much does she spend for each?

2. A man bequeathed $84,000 to his niece, daughter, and wife. If the daughter received twice as much as the niece, and the wife four times as much as the niece, how much did each receive?

3. A certain kind of concrete contains twice as much sand as cement and 5 times as much gravel as cement. How many cubic feet of each of these materials are there in 1000 cu. yd. of concrete?

4. Make up and work a similar example for yourself where the materials in the concrete are as 1, 2, 4.

5. In a certain kind of fertilizer the weight of the nitrate of soda equals that of the ground bone, and the weight of the potash is twice that of the ground bone. How many pounds of each material are there in a ton of fertilizer?

6. If the amount of potash in a given kind of glass is 5 times as great as the amount of lime, and the amount of sand 3 times as great as the amount of potash, how many pounds of each will there be in 4000 lb. of glass?

7. The railroad fare for two adults and a boy traveling for half fare was $49.50. What was the fare for each person?

Sug. Let the smallest of the fares = x.

8. Separate 120 into three parts, such that the second part is twice as large as the first, and the third part three times as large as the first.

9. Separate 120 into three parts which are as 1, 2, 3.

10. Of three numbers whose sum is 1800, the second is twice as large as the first, and the third is three times as large as the second. Find the numbers.

11. Of four girls who rolled bandages for the Red Cross, the second rolled as many as the first, and the third and fourth each rolled twice as many as the first. Together they rolled 480 bandages. How many did each roll?

6. A **formula** is a brief statement of a rule by the aid of symbols.

Ex. 1. Rule and formula for the area of a rectangle.

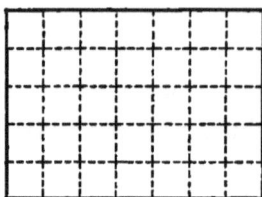

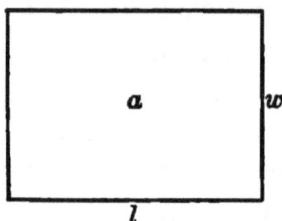

The number of square units in the area of the above rectangle $= 7 \times 5$, or 35. So in general

Rule for area of rectangle.
(Number of square units in area) = (number of linear units in length) × (number of linear units in width).

Brief rule.
Area = (length) × (width).

Formula for area of a rectangle.

$$a = lw.$$

Ex. 2. Rule and formula for percentage.

Rule. To find the per-   *Formula.* $p = br.$
centage, multiply the base
by the rate expressed decimally.

EXERCISE 4

1. Construct a rectangle 8 inches long and 5 inches
wide. Divide this rectangle into square inches, and find
the area (area here is a number of square inches) of the
rectangle by counting the number of square inches in it.
Can you find the area by a shorter method ?

2. In the formula $a = lw$, find the value of a when $l =$
32 and $w = 15$.

3. By the use of the formula $a = lw$, find the area of a
rectangle whose length is 32 in. and whose width is 15 in.

4. In the formula $a = lw$, find the value of a when $l =$
6.24 and $w = 3.25$.

5. By use of the formula $a = lw$, find the area of a rectangle whose length is 6.24 in. and whose width is 3.25 in.

6. By use of the formula, find the number of square
inches in the area of a rectangle whose length is 3 ft. 6 in.
and whose width is 2 ft. 4 in.

7. Also find the number of square inches in the area
of a rectangle whose length is 8 in. and whose width is
p inches. Also in a rectangle whose length is r inches
and whose width is s inches.

8. A boy's garden is 37 yd. long and 15 yd. wide.
By use of the formula $a = lw$, find the area of the garden
in square yards.

9. A certain floor is $23\frac{1}{2}$ ft. long and 12 ft. wide. By
use of $a = lw$, find its area.

10. A concrete walk is 20 yd. long and 4 ft. wide. By
use of $a = lw$, find its area in square yards.

11. A certain garden is to contain 140 sq. yd.   If it is to be 20 yd. long, find how wide it must be.   Use $a = lw$.

Sug.   We have given $a = 140$, and $l = 20$, to find w.   Substituting for a and l in the formula $a = lw$, we have:

$$140 = 20\,w.$$
$$\text{Hence,}\quad 7 = w,$$
$$\text{or,}\quad w = 7. \ Ans.$$

12. A certain building lot is to contain 6000 sq. ft.   If the lot is to be 250 ft. deep, how wide must it be?   Use the formula $a = lw$.

13. A certain building lot is to contain a quarter of an acre and is to be 200 ft. deep.   By use of $a = lw$, find how wide it must be.   (An acre contains 43,560 sq. ft.)

14. On squared paper mark off a rectangle 12 spaces long and 7 spaces wide.   Find the number of square units in the area of this rectangle by counting them.   Also find the area by use of the formula.   Compare the amount of work in these two methods of finding the area.   (That is, estimate how many times as great the work of one process is as that of the other process.)

15. In the formula $p = br$, find p when $b = 420$ and $r = .05$.

16. By use of $p = br$, find 5 per cent of $420.

17. By use of the formula $p = br$, find 12 per cent of 1650.

18. By use of $p = br$, find $37\frac{1}{2}$ per cent of 128.   Also $62\frac{1}{2}$ per cent.   Also $33\frac{1}{3}$ per cent.   Also $87\frac{1}{2}$ per cent.

19. An agent sold a piece of property for $3200 and received a commission of $2\frac{1}{2}$ per cent.   By use of $p = br$, find how much he received.

20. A certain macadam road cost $48,000.   The county through which the road passed paid 40 per cent of the cost.   By use of $p = br$, find how much money the county paid.

21. What per cent is 18 of 600?

Sug. We have given $p = 18$ and $b = 600$, to find r. Substituting for p and b in $p = br$, we have $18 = 600\,r$, etc.

22. If an agent received a commission of $18 for selling a property for $600, what per cent was his commission? Use the formula $p = br$.

23. A boy solves 16 out of 20 examples. What is his grade? What formula did you use?

24. If a baseball nine wins 27 games out of 35, what is its per cent of games won? Use $p = br$.

7. Algebraic Expressions. Any symbol or combination of symbols representing some number is called an algebraic expression.

Ex. $5\,x^2y - 6\,ab + 7\sqrt[3]{ax}$.

8. Terms. In order to treat algebraic expressions more efficiently it is convenient to regard the $+$ and $-$ signs in any given expression as separating the expression into terms.

Thus, $5\,a + 3\,b - 2\,c$ is considered as made up of the terms $5\,a$, $3\,b$, and $2\,c$.

9. Coefficients. In order to treat terms efficiently it is often convenient to treat part of a term separately and call it a coefficient.

Thus, in $5\,a$, 5 is called the coefficient of a. 5 shows how many times a is taken.

If the coefficient is 1, it is not written. Thus, instead of $1\,x$ we write x.

EXERCISE 5

1. How many terms are in the expression $5\,a + 2\,b$? Name these terms.

Treat in like manner:

2. $3\,x - 17\,y + 5\,z$.   **3.** $3\,ab + 5\,bc - x + 2\,y$.

4. Using numerals and the letters a, b, c, and x, write an expression containing two terms. Also one containing four terms. Read what you have written.

5. What is the coefficient of a in $7\,a$? Of y in $10\,y$? Of xy in $3\,xy$?

6. In arithmetic, in 4^3, what is the small number 3 called? How many 4's multiplied together are represented by 4^3? By 4^7? By 4^5? 4^2 is read " four square "; 4^3 is read " four cube "; 4^7 is read " four to the seventh power."

7. In like manner how many x's multiplied together are represented by x^3? How many by x^7? By x^5? By x^4?

8. In $5\,a^3b^2$, a occurs as a factor how many times? b how many times?

9. By use of an exponent write in a shorter way $yyyyyy$.

By using exponents express in the briefest way:

10. $5 \cdot 5 \cdot 5 \cdot 5 \cdot 5 \cdot 5$.

11. $3\,aabbb$.

12. $5\,cccxxxx$.

13. $3\,aaa + 5\,xxxyy$.

14. $7\,aabb - 3\,abbb$.

15. $9\,aaaxxxxyyyy$.

If $a = 1$, $b = 2$, $c = 3$, and $x = 4$, read and find the numerical value of each of the following:

16. c^2.

17. $2\,c$.

18. x^3.

19. $3\,x$.

20. $5\,c^2$.

21. a^3.

22. b^2c^3.

23. $3\,ax^2$.

24. $b^2 + a$.

25. $c^3 - b^2$.

26. $18 - x^2$.

27. $4\,ac^2 + x$.

28. $3\,x^2 - 4\,a$.

29. $c^3 - 1$.

30. $abc - x$.

31. $x^3 - 4\,abc$.

32. $\dfrac{x^2}{bc}$.

33. $\dfrac{3\,ac}{b^2}$.

34. $\dfrac{a^2 + c^3}{b^3}$.

35. $\dfrac{3\,ac - c^2}{a + b}$.

Express in symbols:

36. x plus 3. The sum of x and 3. The number which exceeds x by 3.

37. x diminished by 3.   A number 3 less than x.

38. Two times a plus three times b.

39. The sum of 4 and of 5 times x.

40. One third of the sum of a and b.

Answer the following in algebraic language:

41. If a boy has a cents and earns 10 cents, how many cents will he then have?

42. How many, if he has a cents and earns b cents? How many, if he then spends c cents?

43. Walter has x marbles and his brother has 10 more than Walter.   How many marbles has his brother?

44. Walter has b marbles and his brother has 5 more than twice Walter's marbles.   How many has his brother?

45. If Mary is a years old now, how old will she be in 3 years?   In 5 years?   In x years?

46. What is the number next larger than 5?   Than x? n? $x+1$? $x+2$? $n-1$? $x-2$?

47. Taking x as the smallest number, write two consecutive numbers.   Three consecutive numbers.   Four.   Five.

48. One number exceeds another number by 4.   If the smaller of the two numbers is denoted by x, what is the larger number denoted by?

49. Of three given numbers the second exceeds the first by 5, and the third exceeds the first by 8.   If the first of the three numbers is denoted by x, by what will each of the other two numbers be denoted?

50. Of three given numbers, the second exceeds the first by 8, and the third exceeds the second by 6.   If the first of the three numbers is denoted by x, by what will each of the other two numbers be denoted?

10. Problems.   Ex. 1. If there are 214 pupils in our
school, and the number of girls exceeds the number of
boys by 8, how many boys and how many girls are there?

NUMBERS DEALT WITH   SYMBOLS FOR NUMBERS

Number of pupils in the school $= 214.$
Number of girl pupils $= x + 8.$
Number of boy pupils $= x.$

EQUALITY FOUND AMONG THESE NUMBERS

Sum of the last two numbers $=$ the first number.
Or, in symbols $x + x + 8 = 214,$
that is, $2x + 8 = 214.$
Subtracting 8 from the equals $\underline{\quad 8 \qquad 8,}$
                        $2x \qquad = 206.$
$$x = 103, \textit{ number of boys.}$$
$$x + 8 = 111, \textit{ number of girls.}$$

CHECK. 111 exceeds 103 by 8.   Also $103 + 111 = 214.$
Hence all the conditions of the problem are met by the numbers 103 and 111.

Ex. 2. In a given test Mary, Ella, and Rachel together
solved 28 examples. Mary solved 5 more than Ella, and
Rachel 2 more than Ella. How many did each solve?

NUMBERS DEALT WITH   SYMBOLS FOR NUMBERS

Number of problems solved by Mary $= x + 5.$
Number of problems solved by Ella $= x.$
Number of problems solved by Rachel $= x + 2.$
Let the pupil complete the solution.

It cannot be too often repeated that in order to form correct
and efficient habits of analyzing and solving problems, the pupil
should strictly follow the form of solution given in § 5, p. 6.

EXERCISE 6

1. Henry has a certain number of pigeons in his loft,
and his cousin has 12 more. Together they have 86
pigeons. How many pigeons has each boy?

2. Mary has collected a certain number of postcards, and her older sister has collected 250 more than Mary. Together they have 1210 cards. How many cards has each girl?

3. Three sisters have earned money by knitting sweaters. The second knit 5 more than the first, and the third knit 8 more than the first. Together they knit 40 sweaters. How many did each knit?

4. Two boys ran a launch for a week and earned $40. The boy who owned the launch received $12 more than the other boy. How much did each boy receive?

5. A man has $3100 deposited in three banks. In the second bank he has $200 more than in the first, and in the third bank $300 more than in the second. How much has he in each bank?

6. Separate $24.80 into two parts such that one part exceeds the other by $4.60.

7. Walter and his brother together had 60 marbles, and his brother had 10 more than Walter. How many marbles had each boy?

8. A baseball nine played 62 games and won 8 more games than it lost. How many games did it win?

9. In a certain election 12,784 votes were cast. If the successful candidate had a majority of 1732, how many votes did he receive?

10. At New York on December 21, the night is 5 hr. 32 min. longer than the day. How long is the day?

11. Separate $28\frac{1}{2}$ into two parts such that one shall exceed the other by $2\frac{3}{4}$.

12. The sum of two consecutive numbers is 15. Find the numbers.

Sug.

NUMBERS DEALT WITH	SYMBOLS FOR NUMBERS
Smaller of the two numbers	$= x$.
Larger of the two numbers	$= x + 1$, etc.

13. Find three consecutive numbers whose sum is 33.

14. Separate $5000 into three parts such that the second part shall exceed the first by $300, and the third shall exceed the first by $800.

15. Separate $5000 into three parts such that the second part shall exceed the first by $300, and the third shall exceed the second by $800.

16. Four girls knit garments for the Red Cross. The second knit as many as the first. The third knit 12 more than the first. The fourth knit 15 more than the third. Together they knit 487 garments. How many did each girl knit?

17. In an athletic meet the four boys who represented a certain school together won 30 points. The first of these boys won 1 more point than the second, the third 3 more than the second, and the fourth 2 more than the second. How many points did each boy win?

18. In a competition in an algebra class, the three highest pupils together solved 26 problems in one half an hour. The highest pupil solved 6 more than the lowest, and the second in rank solved 2 more than the lowest. How many problems did each of the three pupils solve?

PART II*

11. Algebra is the study of an efficient use of numbers by the aid of letters and other symbols.

12. Utility of Algebra. We have already found that a more extended use of symbols than is practiced in arithmetic (1) shortens the work of solving problems, and (2) enables us to solve problems which we could not otherwise solve. A further study of algebra will show that it has other uses besides those just named.

* See page iv of the preface.

13. A **Term** is a part of an algebraic expression which does not contain a plus or minus sign. (Signs occurring inside a parenthesis are not considered in fixing the terms.)

Ex. 1. $5\,x^2y - 6\,ab + 7\sqrt[3]{ax}$.

This algebraic expression contains three terms, viz: $5\,x^2y$, $6\,ab$, and $7\sqrt[3]{ax}$.

Ex. 2. $5\,x + a \div b + c$.

This expression also contains three terms: $5\,x$, $a \div b$, and c.

Ex. 3. $7\,ax^2 + 5(a + b) - c^3$.

Since the parenthesis, $(a + b)$, is treated as a single quantity, three terms occur in this expression: $7\,ax^2$, $5\,(a + b)$, and c^3.

14. A **Monomial** is an algebraic expression of only one term; as $5\,x^2y$ or c.

15. A **Polynomial** is an algebraic expression containing more than one term; as $3\,ab - c + 2\,x + 5\,y^2$.

A monomial is sometimes called a *simple expression*, and a polynomial a *compound expression*.

16. A **Binomial** is an algebraic expression of two terms; as $2\,a - 3\,b$. A **Trinomial** is an expression of three terms.

17. Coefficients. A numerical factor, if it occurs in a product, is written first and is called a *coefficient*. Hence,

A **coefficient** is a number prefixed to a quantity to show how many times the given quantity is taken.

For example, in $5\,xy$, 5 is the coefficient.

The following enlarged definition of coefficient is often used. In the product of several factors, the *coefficient* of any factor, or factors, is the product of the remaining factors.

Thus, in $5\,abxy$, the coefficient of y is $5\,abx$; of xy is $5\,ab$; of ab is $5\,xy$. What is the coefficient of b? Of a? x? $5\,a$? 5?

A **numerical coefficient** is a coefficient composed only of figures; as 15 in $15\,ab$.

A **literal coefficient** is a coefficient composed only of letters ; as *ab* in *abx*.

What, then, is a *mixed coefficient?*   Give an example.

18. Power and **Exponent** are used in the same sense in algebra as in arithmetic.

A **power** is the product of equal factors.

A power is expressed briefly by the use of an exponent.

An **exponent** is a small figure or letter written above and to the right of a quantity to indicate how many times the quantity is taken as a factor.

Thus, for *xxxx*, or four *x*'s multiplied together, we write x^4, the exponent in this case being 4.   The expression is read " *x* to the fourth power."

When the exponent is unity, it is omitted.   Thus, *x* is used instead of x^1, and means *x* to the first power.

A power is composed of two parts: (1) the base (*i.e.* one of the equal factors); and (2) the exponent.

Thus, in the power a^3 the base is *a* and the exponent is 3.

EXERCISE 7

1. From the following list of expressions, $3a + 2b$; $6b^2x$; $3a + 5(x + y)$; $3a + 2b - 3c$; $4x^3$; $2a + 3x + 7y - 4z$; $6(a-b)$; $x^2 + (a+b)x + ab$; $\frac{1}{2}gt^2$; select (1) the monomials; (2) the binomials; (3) the trinomials; (4) the polynomials.

2. By making selections and combinations from the following list of terms, $3a^2$, $-2ab$, $3x^2y$, $5(a+b)$, $4b^3$, $3mpx$, form (1) three binomials; (2) three trinomials; (3) three monomials; (4) three polynomials, each containing a different number of terms from the rest.

3. In $15bx$ what is the numerical coefficient?   What is the entire coefficient of *x*?   Of *b*?

4. In $abcx$, abc is what kind of coefficient for x ?

5. Write a term containing y and a numerical coefficient. Also another term containing x and a literal coefficient. Also a term containing z and a mixed coefficient.

6. Write a power whose base is 3 and exponent 4. Also one whose base is c and exponent 5.

7. Write a power whose base is x and exponent n. Whose base is y and exponent $n-1$.

If $a = \frac{1}{2}$, $b = \frac{5}{3}$, $c = 4$, and $x = 6$, find the value of:

8. $2\,a + 6\,b$.	**13.** $2\,bx^2 - a^2c$.	**18.** $c^2 - a^2$.
9. $5\,ax - 2\,b$.	**14.** $5\,abcx$.	**19.** $x^2 - 5\,a^2c$.
10. $3\,ac^2 + bx$.	**15.** $\frac{4}{3}\,b^2cx$.	**20.** $a\sqrt{x^2 + 4\,c^2}$.
11. $\frac{1}{2}\,bx^2 + ax$.	**16.** $abx\sqrt{c}$.	**21.** $6\,ab\sqrt{9\,c}$.
12. $\frac{1}{2}\,b(c+x)$.	**17.** $4\,a - \frac{1}{2}\,c$.	**22.** $bx\sqrt{4\,x + 2\,a}$.

If $a = 4$, $b = .02$, $c = .15$, $x = 6$, $y = 5$, find the value of:

23. $3\,a - 5\,b$.	**27.** $cx - by$.	**32.** $xy - 5\,abc$.
24. $\dfrac{2\,a + c}{y}$.	**28.** $10\,ax - 3\,by$.	**33.** $\dfrac{a+b+c}{xy}$.
	29. $a^2 - b^2$.	
25. $abcy$.	**30.** $\sqrt{1000\,b + y}$.	**34.** $\dfrac{x+y}{b}$.
26. $\dfrac{a+b}{y}$.	**31.** $\dfrac{2\,a - 3\,c}{y}$.	**35.** $\dfrac{5\,ac}{b}$.

19. Order of Operations. You will have observed that in making numerical computations in algebra it is customary to follow a certain order of operations. This order is the same as that followed in arithmetic. Thus:

I. In a series of operations involving addition, subtraction, multiplication, and division (but no use of the parenthesis), *the multiplications and divisions are to be performed before any of the additions and subtractions.*

Ex. 1.   Find the value of $4 + 12 \times 3$.

$$4 + 12 \times 3 = 4 + 36 = 40. \quad \textit{Ans.}$$

(Hence $4 + 12 \times 3$ does not equal 16×3, etc.)

Ex. 2.   What is the value of $60 - 8 \div 2 + 3 \times 7$?

$$60 - 8 \div 2 + 3 \times 7 = 60 - 4 + 21 = 77. \quad \textit{Ans.}$$

Ex. 3.   Find the value of $12 \div 2 \times 6 \div 4$.

$12 \div 2$ gives 6.   Then 6×6 gives 36.   Then $36 \div 4$ gives 9.   *Ans.*

II. If a given expression contains one or more parentheses (or other signs of aggregation), *each parenthesis is to be reduced to a single number before the operations of the expression as a whole are to be performed.*

Ex. $5 + 4(6 - 2) = 5 + 4 \times 4 = 5 + 16 = 21. \quad \textit{Ans.}$

(Hence $5 + 4(6-2)$ does not equal $9(6-2)$ or 9×4, etc.)

EXERCISE 8

Find the value of the following:

1. $12 \div 3 \times 4 \div 8$.
2. $18 - 6 \div 2 \times 3$.
3. $5 + 4 \times 3 \div 2$.
4. $5 \times 6 + 8 \times 4 \div 16$.

If $a = 5$, $b = 3$, $c = 1$, and $x = 6$, find the value of:

5. $ax \div b + c \div 2$.
6. $a^3 - bx^2$.
7. $2(2a - c)$.
8. $x(a - b)$.
9. $4(a - 3c)^2$.
10. $2x(2a - 3b)^2$.
11. $3 + 2(x - a)$.
12. $5x - 3(2b + c)$.
13. $2(x^2 - a^2) + 3ac$.
14. $3x(x - 3)^2 - 9x$.
15. $(x - 1)(x - 3) + x(x - a)$.
16. $3(2x - 5c) - a(2b^2 - 3x)$.
17. $(5b + x)(x - b + a - 5c^2)$.
18. $\dfrac{a + 7c}{x}$.
19. $\dfrac{3x^2 - b}{a + 2c}$.
20. $\dfrac{5a^2}{x - 1} + \dfrac{3c}{b}$.
21. $\dfrac{(x-1)(b+1)(5c-b)}{abx}$.

EXERCISE 9

VERBAL PROBLEMS

1. What is the result of multiplying a by 3 and then adding 5 times x to the result?

2. What is the cost of 30 lb. of coffee at c cents a pound?

3. If b books cost d dollars, how much will x books cost?

4. What fractional part of a pounds is z ounces? What per cent?

5. One factor of x is c. What is the other factor?

6. Express the sum of a hours and b minutes as minutes. As hours.

7. b feet, c inches are cut off from a rope a yards long. Find in inches the length remaining. Also find the remainder in feet. Also in yards.

8. It is 143 miles from Albany to New York. After a train has traveled a miles an hour for b hours from New York toward Albany, how far is the train from Albany?

9. If y yards of dress material cost c dollars, how much will 2 yards cost?

10. How many tiles each b inches square will be required to cover a floor m feet long and n feet wide?

11. One part of m is n. What is the other part?

12. A boy had d dollars. He paid m dollars for a hat and c cents for a pencil. How many dollars did he have left?

13. Pencils are sold at the rate of a pencils for b cents. How much will a dozen pencils cost?

14. The sum of two numbers is 20. The greater of the numbers is x. The greater is what per cent of the sum? The less is what per cent of the sum? The less is what per cent of the greater?

1. How many bricks each 8″ × 4″ will be required to make a pavement 60 ft. long and 4 ft. wide ?

2. If N denotes the number of 8″ × 4″ bricks required to make a pavement l feet long and w feet wide, obtain a formula for N.

By use of the formula found in Ex. 2, solve Exs. 3 and 4.

3. Determine how many bricks will be required for a pavement 32 ft. long and 5 ft. wide.   Also for one 20.5 ft. long and 6.4 ft. wide.

4. Also for one 30 ft. 6 in. long and 4 ft. 6 in. wide. Also for one 11 yd. long and 3 ft. 6 in. wide.

5. If bricks are a inches long and b inches wide, and N denotes the number of bricks required for a pavement l feet long and w feet wide, obtain a formula for N in terms of l, w, a, and b.

6. By use of the formula obtained in Ex. 5, find the number of 10″ × 4″ bricks required for a pavement 42 ft. long and 5 ft. wide.

7. The amount of roofing on a building is often estimated in terms of the number of squares in the area of the roof.   A square of roofing is 100 sq. ft.

If a certain gable roof is 36 ft. long and 20 ft. from ridge to eaves, how many squares does the roof contain?

8. Denoting the number of squares in a gable roof by S, the length of the roof by l, and its width (from peak to eaves) by w, obtain a formula for the number of squares in the roof.

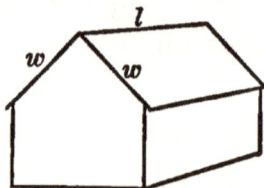

9. By use of the formula found in Ex. 8, find the number of squares in a gable roof 40 ft. long and 18 ft. from ridge to eaves.

10. Also in a gable roof 20 ft. 6 in. long and 16 ft. from ridge to eaves.

11. Also in one 24 ft. 3 in. long and 16 ft. 6 in. from ridge to eaves.

12. Also in one 9 yd. 6 in. long and 6 yd. from ridge to eaves.

13. A rectangular barn is 40 ft. long, 20 ft. wide, 18 ft. high at the eaves and 28 ft. high at the peak. Find the number of square yards in the entire outside wall surface of the barn.

14. Denoting the number of square yards in the wall surface of a barn by Y, the length of the

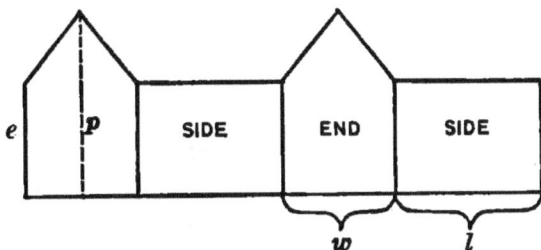

barn by l, the width by w, the height at the eaves by e, and the height at the peak by p (all dimensions being in feet), find a formula for Y in terms of the other letters.

15. By use of the formula found in Ex. 14, find how many square yards are in the walls of a barn 36 ft. long, 18 ft. wide, 16 ft. high at the eaves, and 25 ft. high at the peak.

16. Also of a barn 40 ft. long, 24 ft. wide, 16 ft. 6 in. high at the eaves, and 25 ft. 6 in. high at the peak.

17. If it costs c cents a yard to paint the walls of a barn and the total cost in dollars be denoted by D, find a formula for D. (Use the letters given in Ex. 14.)

18. Using the formula obtained in Ex. 17, find the cost of painting a barn 42 ft. long, 20 ft. wide, 18 ft. high at the eaves, and 25 ft. high at the peak, at 12 cents a square yard.

CHAPTER II

NEGATIVE NUMBERS; GRAPHS

PART I

20. Positive and Negative Numbers. A *negative number* is a number exactly opposite in some respect to another number which is taken as *positive*.

If distance east of a certain point is taken as positive, distance west of that point is called negative.

If north latitude is positive, south latitude is negative.

If temperature above zero is taken as positive, temperature below zero is negative.

If in business matters a man's assets are his positive possessions, his debts are negative numbers.

Positive and negative numbers are distinguished by the signs + and − placed before them.

Thus, $ 50 assets are denoted by + $ 50, and $ 30 debts by − $ 30. We denote 12° above zero by + 12°, and 10° below zero by − 10°.

EXERCISE 11

1. What is meant by a temperature of − 8° ? By a latitude of − 23° ? By the date − 776? (Dates after the birth of Christ are taken as positive.)

2. If the temperature was 17° at noon and − 8° at midnight, how many degrees did the temperature fall between noon and midnight?

3. If in a given time the temperature should fall from − 5° to − 12°, how many degrees would it fall?

4. If a traveler is in latitude −4° and travels north 7°, what does his latitude become? What does it become if instead he travels south 7°?

5. If a man's property is − $7000 and he saves $2000 a year for 8 years, what does his property become?

6. If a vessel, at latitude 3°, sails south 345 miles, what does her latitude become if 60 miles equal 1°?

7. Regarding each item of a boy's income as a positive number, and each item of his expenditures as a negative number, state which of the following are positive numbers and which negative :

(1) Money received for shoveling snow.

(2) Money paid for a pair of skates.

(3) Cash received for interest coupon on a liberty bond.

(4) Money spent for carfare.

(5) Cash prize for being the best scholar in a class.

(6) Money received for products of a garden.

(7) Interest on deposits in a savings bank.

Add two positive items to this list. Also two negative items.

8. Regarding an easterly direction as positive and a westerly direction as negative, name two cities which are in a negative direction from Chicago. Also two cities in a positive direction.

9. Regarding a northerly direction as positive and a southerly direction as negative, name two cities whose distances from the city of Washington are positive numbers. Also two cities whose distances are negative.

10. Name three cities whose distances from the earth's equator are negative. Also three cities whose distances from the equator are positive.

11. State which of the following changes in temperature
are positive and which negative:

(1) From 2° to 13°.   (3) From 7° to − 3°.

(2) From − 5° to − 3°.   (4) From − 8° to 4°.

(5) From 5 degrees below zero to 3 degrees below.

(6) From 7 degrees above zero to 3 degrees below zero.

(7) From 10 degrees below zero to 20 degrees above.

12. Construct a diagram like this. On your diagram,
by some letter as A, B, C, etc., placed under the point,

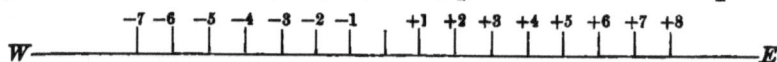

W————————————————————————E

indicate the point corresponding to + 3. To − 4. To +7.
To +2.5.   To −2.5.   To − 4½.   To zero.

13. Write " 20 degrees above zero," and also " 30 below
zero " in the shortest way by the aid of + and − signs.

14. Count how many symbols and letters there are together in "12 degrees below zero." Also how many in
" − 12°." Compare the work of writing the two ex
pressions.

15. If the thermometer stands at −12°, how much must
it rise to stand at − 5°? What must be added to − 12° to
make − 5° ?

16. If the thermometer stands at − 10°, how much must
it rise to stand at 5°? How much must be added to −10°
to make 5° ?

17. If dates before the birth of Christ are taken as negative and those after the birth of Christ as positive and
the Emperor Augustus was born in the year − 63 and
died in the year 14, how old was he when he died? What
must be added to − 63 to make 14?

18. If Alexander the Great was born in the year − 356
and died in the year − 323, how old was he when he died?
What must be added to − 356 to make − 323?

19. If a man owes $500 (that is, has —$500), how many dollars must be added to his possessions to make him worth — $200? How many to make him worth $1200?

21. Graphs. A set of numerical facts may often be combined as a geometrical picture called a *graph*. The meaning and use of negative numbers are often illustrated on a graph.

Ex. On a certain day the following were the temperatures at a given place from 8 A.M. to 8 P.M.:

8 A.M. 10°	11 A.M. 22°	2 P.M. 27°	5 P.M. 15°	7 P.M. — 2°
9 A.M. 15°	Noon 24°	3 P.M. 27°	6 P.M. 7°	8 P.M. — 6°
10 A.M. 18°	1 P.M. 25°	4 P.M. 24°		

Make a graph of the above list of temperatures.

Using paper ruled in small squares, we draw a horizontal line (called the *horizontal axis*), and on this line let each space represent 1 hour. Perpendicular to the horizontal axis, we draw the vertical axis and on it let each space represent 5°. Above (or below if the temperature is negative) each point which represents an hour, we locate

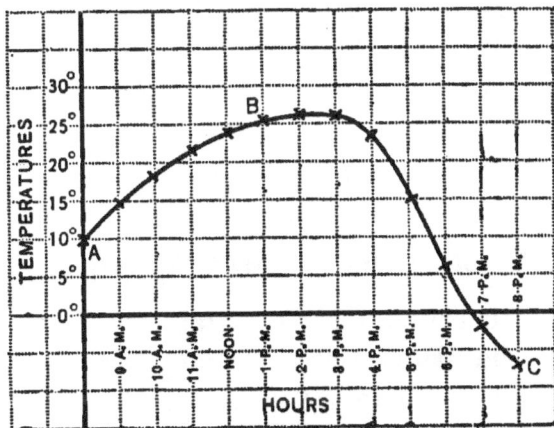

a point which represents the temperature at that hour. Through the points thus located a continuous line *ABC* is drawn. This line is the required graph.

1. By examining the graph on page 27, answer the following questions:

(1) At what hour was the temperature highest?

(2) What was the temperature at 1 : 30 P.M.?   At 4 : 30 P.M.?

(3) At what hour in the afternoon was the temperature the same as it was at 9 o'clock in the morning?

Make a graph of each of the following sets of temperatures and state three facts learned from each:

2.

8 A.M. − 3°	11 A.M. 8°	2 P.M. 18°	5 P.M. 6°	8 P.M. − 8°
9 A.M. 5°	Noon 12°	3 P.M. 17°	6 P.M. 2°	9 P.M. − 10°
10 A.M. 10°	1 P.M. 15°	4 P.M. 12°	7 P.M. − 4°	10 P.M. − 12°

3.

8 A.M. − 5°	11 A.M. 12°	2 P.M. 12°	5 P.M. 9°	8 P.M. 2ᶜ
9 A.M. 2°	Noon 10°	3 P.M. 15°	6 P.M. 7°	9 P.M. 0ᶜ
10 A.M. 7°	1 P.M. 8°	4 P.M. 13°	7 P.M. 4°	10 P.M. − 3°

4. At a certain place on a summer's day the temperatures from 6 A.M. to 8 P.M. were as follows:

6 A.M. 50°	9 A.M. 64°	Noon 80°
7 A.M. 56°	10 A.M. 66°	1 P.M. 84°
8 A.M. 60°	11 A.M. 74°	2 P.M. 85°

3 P.M. 70°	6 P.M. 68°
4 P.M. 57°	7 P.M. 60°
5 P.M. 67°	8 P.M. 48°

Make a graph of these temperatures.

The beginning of the graph is shown on the diagram at the right. Let the pupil complete the graph on a larger piece of squared paper. On the afternoon of the given day, a thunderstorm occurred. How is this' fact shown on the graph? At what time did the storm occur?

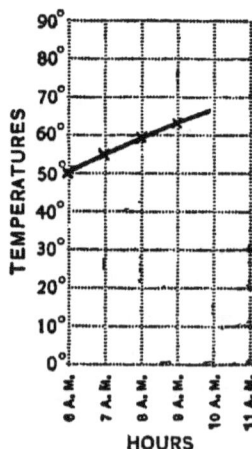

5. Graph the following set of temperatures :

Midnight — 15°	9 A.M. 2°
3 A.M. — 20°	Noon 10°
6 A.M. — 10°	3 P.M. 15°

6 P.M.   10°
9 P.M.   0°
Midnight — 10°

Sug. Let each space on the horizontal axis represent 3 hours ; and each space on the vertical axis represent 5°. See the diagram of axes to the right.

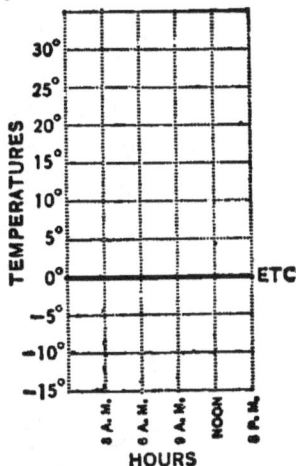

Graph each of the following sets of temperatures :

	MID-NIGHT	3 A.M.	6 A.M.	9 A.M.	12 A.M.	3 P.M.	6 P.M.	9 P.M.
6.	— 20°	— 30°	— 20°	— 10°	0°	10°	10°	0°
7.	— 10°	— 20°	— 10°	0°	10°	20°	10°	0°
8.	— 10°	— 15°	— 5°	10°	15°	25°	15°	— 5°
9.	0°	— 10°	— 5°	15°	25°	30°	15°	5°

10. The following are the mean or average temperatures for a long period of years at New York on the first day of each month of the year.

Jan. 1, 31°	Apr. 1, 42°	July 1, 71°	Oct. 1, 61°
Feb. 1, 31°	May 1, 54°	Aug. 1, 73°	Nov. 1, 49°
Mar. 1, 35°	June 1, 64°	Sept. 1, 69°	Dec. 1, 39°

Make a graph of these temperatures.

Sug. Let each space on the vertical axis represent 10°.

11. In like manner, make a graph of the average temperatures at London on the first day of each month, which are as follows :

Jan. 1, 37°	Apr. 1, 45°	July 1, 62°	Oct. 1, 54°
Feb. 1, 38°	May 1, 50°	Aug. 1, 62°	Nov. 1, 46°
Mar. 1, 40°	June 1, 57°	Sept. 1, 59°	Dec. 1, 41°

12. Obtain a set of temperatures for some day from a weather report printed in a daily newspaper and make a graph representing these temperatures.

13. At a place where a stream was 100 ft. wide, the depth of the water was measured at intervals of 10 ft. These depths were found to be in feet as follows: 5, 8, 20, 22, 15, 11, 8, 6, 4.   Make a drawing of the bottom of the river as determined by these measurements.

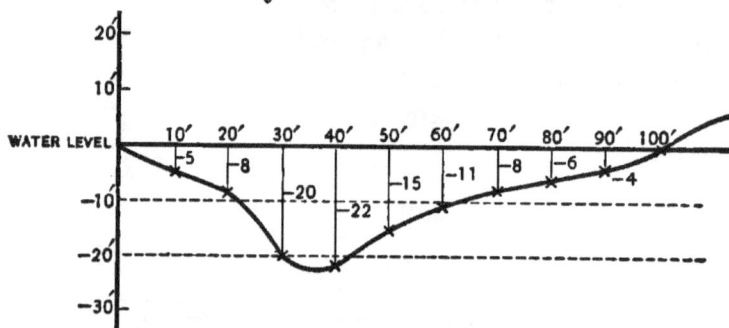

From an examination of the drawing answer the following questions :

(1) How deep is the channel of the stream ?

(2) How far from the left shore is the deepest part of the channel?

(3) How far from the right shore is the deepest part of the channel?

(4) What is the depth of the stream 20 ft. to the right of the deepest part of the channel?   20 ft. to the left ?

14. At a certain place a given stream is 240 ft. wide. At intervals of 20 ft. the depth of the water is measured in feet and found to be as follows: 2, 4, 7, 10, 9, 10, 16, 22, 35, 30, 12.   Make a graph of the bottom of the stream. State three facts which you can learn from an examination of the drawing.

15. At a given place a certain stream is 180 ft. wide. At intervals of 20 ft. the depth of the water is measured in feet and found to be as follows: 6, 12, 24, 22, 26, 18, 12, 6. The height of the east bank 20 ft. from the edge of the water is 8 ft.; 40 ft. from the edge of the water it is 12 ft. The west bank is 6 ft. high at a distance of 20 ft. from the water's edge, and 8 ft. high at a distance of 40 ft. Make a drawing of the river bottom and of the banks as far as given by the above measurements.

22. Problem. Four partners together made $47,000 in one year. The first partner received $5000 more than the second. The third received twice as much as the second; and the fourth received three times as much as the second. How much did each receive?

NUMBERS DEALT WITH   SYMBOLS FOR NUMBERS

Number of dollars received by first partner $= x + 5000.$
Number of dollars received by second partner $= x.$
Number of dollars received by third partner $= 2\,x.$
Number of dollars received by fourth partner $= 3\,x.$
Let the pupil complete the solution.

EXERCISE 13

1. A man and boy together spade up a garden containing 6000 sq. ft. If the man spades four times as much ground as the boy, how much does the boy spade?

2. Two boys earn $38 by taking passengers on a motor boat. If the boy who owns the boat receives $10 more than the other boy, how much does each receive?

3. A certain macadam road cost $18,000. The county paid twice as much as the state, and the township the same amount as the county. How much did each pay?

4. The top of the Statue of Liberty in New York Harbor is 306 ft. above the surface of the water. If the altitude of the pedestal is 4 ft. greater than the height of the statue, how high is each?

5. In a certain kind of gunpowder the weight of the charcoal equals that of the sulphur, and the amount of niter equals the combined weight of the charcoal and sulphur. How many pounds of each substance are needed to make a ton of gunpowder?

6. A man and boy together catch 320 fish, and the man receives three times as many fish as the boy. How many fish does each have?

7. A man has $3220 in two banks and the amount in one bank exceeds that in the other by $540. How much has he in each bank?

8. Two girls make $24.60 by sewing, and the younger girl receives only one half as much as the older. How much does each receive?

9. Separate $12.68 into two parts, one of which shall be smaller than the other by $5.

10. If 112,216 sq. mi. are added to 24 times the area of the British Isles, the result will be 3,025,600 sq. mi. (the area of the United States). Find the area of the British Isles.

11. Twice the height of Mt. Washington with 1567 ft. added equals the height of Pike's Peak, or 14,147 ft. Find the height of Mt. Washington.

12. The cost of a macadam road was $24,000. The county paid twice as much as the state; the township three times as much as the state. How much did each pay?

13. Three partners divided $14,000, the second partner receiving $2000 more than the first, and the third partner receiving twice as much as the first. How much did each receive?

14. Four partners together made $40,000 in one year; the second partner received the same amount as the first;

the third received twice as much as the first; and the fourth received $2000 more than the first. How much did each partner receive?

15. Find three consecutive numbers whose sum is 36.

16. Find four consecutive numbers whose sum is 106.

17. Make up and work an example concerning five consecutive numbers.

18. The area of the United States and its outlying possessions is 3,742,155 sq. mi. The area of the United States exceeds that of its outlying possessions by 2,309,045 sq. mi. What is the area of the outlying possessions?

19. Three boys together trapped 106 muskrats during the season. The first boy trapped 10 more than the second, and the third boy trapped twice as many as the second. How many did each boy trap?

20. Four partners together made $35,000 in one year. The first partner received twice as much as the third; the second received $2000 more than the third; and the fourth received $3000 more than the third. How much did each receive?

21. Four partners together made $36,000 in one year. The first and third partners received three times as much as the second. The fourth received twice as much as the second. How much did each partner receive?

PART II

23. Algebraic Numbers is a general name for both positive and negative numbers.

The **absolute value** of a number is the value of the number considered without regard to its sign.

Thus, if one man travels 5 miles east and another man travels 5 miles west, the absolute distance traveled by the two men is the same, viz. 5 miles. The two distances traveled, however,

are different *algebraic numbers,* one distance being +5 miles
and the other distance being − 5 miles.

In general the absolute value of both +5 and − 5 is 5; and
of both + a and − a is a.

1. What is the absolute value of − 4 miles? Of +4
miles? − 5 inches? − 3°? − $4200?

2. Draw a line a foot long and on it mark off inches,
half and quarter inches in the manner here indicated :

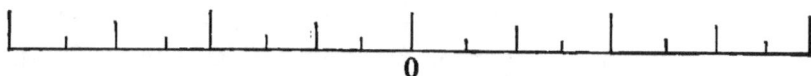

0

At the middle of the line take the inch mark as the
starting or zero point. Regard distances to the right of
the zero point as positive. On the line mark with a letter
the point which represents + 3 in. Also the point which
represents − 2 in. Also 2½ in. − 1½ in. − 2¼ in. ¾ in.
− ¾ in. 2¼ in.

3. On the above scale take the distance + 3 in. To
this add + 2 in. To this result add − 4 in. Then add
+ 1 in. Then − 7 in. Then 1½ in. Then − 2¼ in.

4. At 6 A.M. a thermometer read 57°. It then made
successive changes as follows : + 7°, − 2°, + 5°, − 3°, − 2°.
What was the final reading?

5. In a certain football game, taking a distance toward
the north goal as positive, during the first seven plays the
ball started at a certain point on the field and shifted its
position in yards as follows: + 50 − 10 − 15 − 5 + 10 − 5
− 20. Find the final position of the ball with reference
to the starting place. On squared paper show the changes
in the position of the ball, letting 5 yd. equal one space on
the paper.

Sug.   Starting at 0 the first two shifts of position are shown on the diagram.   Let the pupil complete the diagram.

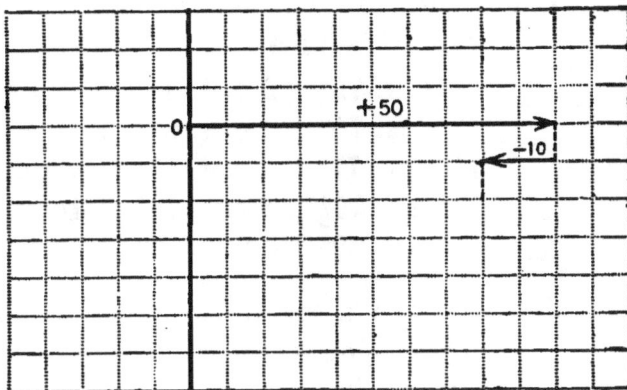

6. What is meant by saying that the altitude of a certain place in the valley of the Jordan River, with reference to sea level, is − 450 ft.?

7. If a man bought a horse for $150 and sold it for $200, what was his gain? What would his gain have been if he had sold it for $125? For $100? For $175?

8. What is meant by saving − $10? By a distance − 10 miles north?

24. The Utility of Negative Number lies in the fact that its use enables us to use two opposite or contrasted kinds of quantity in working a given problem.

Also by the use of negative numbers we are often able to choose an advantageous starting point in solving a given problem.

The full meaning of these utilities and other advantages in the use of negative quantity will appear as we advance in the study of algebra.

1. On a given day in July at different hours the following were the temperatures at a certain place :

9 A.M. 72°	2 P.M. $84\frac{1}{2}$°	7 P.M. 79°
10 A.M. 75°	3 P.M. 83°	8 P.M. $76\frac{1}{2}$°
11 A.M. 77°	4 P.M. 74°	9 P.M. $72\frac{1}{2}$°
Noon $79\frac{1}{2}$°	5 P.M. 77°	10 P.M. 67°
1 P.M. 82°	6 P.M. 80°	11 P.M. 63°

Make a graph of this list of temperatures, using scales as indicated in the adjoining diagram.

What is the probable explanation of the sudden fall in the temperature curve in the afternoon, followed by a rise?

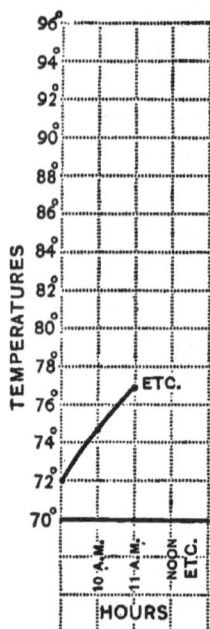

2. Using the same vertical scale as in Ex. 1, graph the following temperatures :

7 A.M. 71°	1 P.M. 84°	6 P.M. 74°
8 A.M. 74°	2 P.M. $87\frac{1}{2}$°	7 P.M. 70°
9 A.M. 78°	3 P.M. $88\frac{3}{4}$°	8 P.M. 67°
10 A.M. 83°	4 P.M. 85°	9 P.M. 62°
11 A.M. 81°	5 P.M. 80°	10 P.M. 59°
Noon 80°		

3. On page 29 (Exs. 10 and 11) are given the mean or average temperatures on the first day of each month of the year at New York and London. On the same diagram make two graphs, one showing the temperatures at New York, the other at London. Make the graph for London either a dotted or a red line. (See the adjoining diagram.)

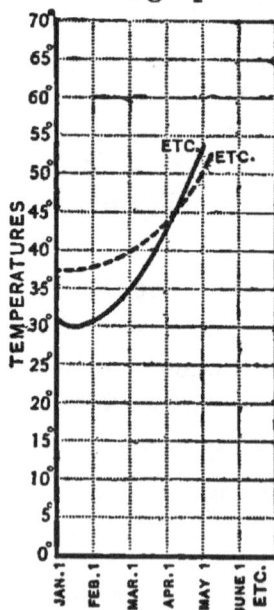

4. The populations of two neighboring villages for a series of years were as follows :

YEAR	1906	1908	1910	1912	1914	1916	1918	1920	1922	1924
Glenville	408	480	427	402	467	429	674	872	1012	1420
Moulton	612	627	582	622	550	530	480	420	356	324

On the same diagram make graphs of these two sets of facts, using axis scales as indicated on the diagram.

In the year 1917 a trolley line was built connecting Glenville with a neighboring city. How is the result shown on the graphs?

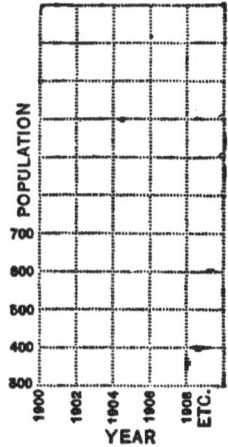

5. At a given place a certain stream is 220 ft. wide. At intervals of 20 ft. the depth of the stream was measured and the results in feet were as follows: 1.8, 2.5, 3.2, 4.8, 3.6, 5.7, 6.2, 4.5, 3.2, 1.6. Make a graph of the bottom of the stream, letting each space on the horizontal axis represent 20 ft., and each space on the vertical axis represent 1 ft. State three facts concerning the channels of the stream which you can learn from your drawing.

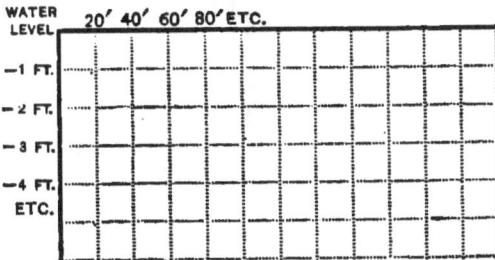

6. At a given place a certain stream is 80 ft. wide. At intervals of 5 ft. the depth of the stream was measured and the list of depths in feet was found to be as follows: 1, 1.3, 1.8, 2.3, 3.2, 4, 4.8, 4.2, 4, 3.6, 5.2, 5.8, 4, 3, 1. Make a graph of the bottom of the stream, letting each space on the horizontal axis represent 5 ft. and each space on the vertical axis represent 1 ft. State three facts which you can learn from the graph.

7. On a certain day in winter, at various hours of the day the temperatures at New York and Palm Beach were:

HOURS	8 A.M.	9 A.M.	10 A.M.	11 A.M.	NOON	1 P.M.	2 P.M.	3 P.M.	4 P.M.	5 P.M.	6 P.M.	7 P.M.	8 P.M.
New York	10°	14°	17°	19°	22°	24°	27°	26°	22°	17°	12°	8°	3°
Palm Beach	76°	78°	82°	83°	84°	86°	87°	87°	83°	78°	76°	73°	70°

Graph these two sets of temperatures on the same diagram. (Let each space on the vertical axis represent 10°.)

8. At a given place on a certain day the readings of a barometer in inches were as follows :

8 A.M. 29.35	11 A.M. 29.8	2 P.M. 29.20	5 P.M. 29.4
9 A.M. 29.45	Noon 30.24	3 P.M. 28.8	6 P.M. 29.6
10 A.M. 29.2	1 P.M. 29.65	4 P.M. 28.9	7 P.M. 30.2

Graph these readings by completing the diagram.

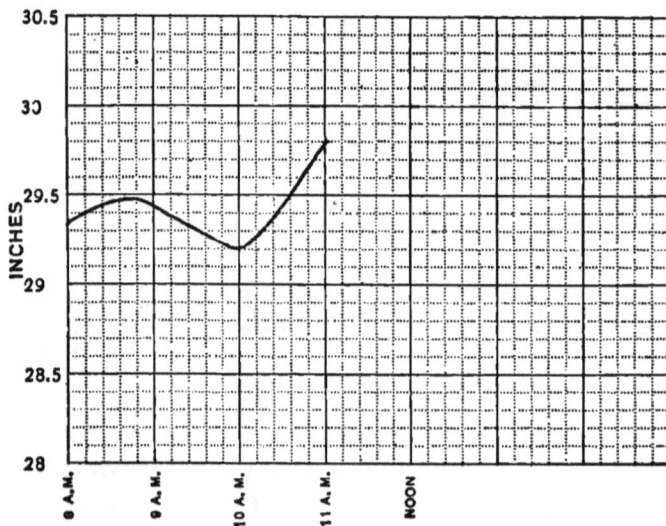

What is a possible explanation of the noticeable drop in the barometer in the afternoon?

EXERCISE 16

VERBAL PROBLEMS

1. By how much does 17 exceed 7? 17 exceed x?

2. By how much does y exceed 7? y exceed x?

3. Express c cents in dollars. y yards in inches. i inches in feet. d dollars in dimes. h hours in minutes.

4. What fractional part of a day is h hours? What per cent?

5. If a apples cost c cents, what does one apple cost?

6. If m parts of cement are mixed with n parts of sand, what fractional part of the mixture is the cement? What per cent is the cement?

7. The difference of two numbers is 5. One of the numbers is 12. What is the other number?

8. The difference of two numbers is d. One number is m. What is the other number?

9. What must be added to 8 to give the next larger even number?

10. Give three consecutive even numbers of which 8 is the smallest; of which 8 is the largest; of which 8 is the middle number.

11. What must be added to 7 to give the next larger odd number?

12. Give three consecutive odd numbers of which 7 is the smallest; of which 7 is the largest; of which 7 is the middle number.

13. If a certain even number is denoted by $2x$, what is the next larger even number?

14. If x is a whole number, can $2x$ be an odd number?

15. If x is a whole number, can $2x+1$ be an even number?

16. Write three consecutive even numbers of which $2x$ is the smallest; of which $2x$ is the largest; of which $2x$ is the middle number.

17. Write three consecutive odd numbers of which $2x+1$ is the smallest; of which $2x+1$ is the largest; of which $2x+1$ is the middle number.

EXERCISE 17

1. One fraction is three times as large as another and their sum is $\frac{1}{3}$. Find the fractions.

2. One number is 4 times as large as another and their sum is .0045. Find the numbers.

3. Separate $120 into two parts such that one part is three times as large as the other.

4. Separate .0372 into three parts which are as 1, 2, 3.

5. Separate 240 into four parts which are as 1, 1, 2, 4.

6. Separate $3800 into three parts, such that the second is three times as large as the first, and the third 5 times as large as the second.

7. In one kind of concrete the parts by weight of cement, sand, and gravel are as 1, 2, and 4 ; in another kind the three parts are as 1, 2, and 5. How many more pounds of cement are needed in a ton of one than of the other ?

8. In a certain kind of gunpowder the niter, sulphur, and charcoal are mixed in the proportions of 6, 1, 1. Find the number of pounds of each of these constituents in 1000 pounds of the gunpowder.

9. One number exceeds 4 times another number by 5, and the sum of the numbers is 100. Find the numbers.

10. One number exceeds 3 times another number by .12, and the sum of the numbers is 4.4. Find the numbers.

11. The sum of the three sides of a triangle is 24 ft. The second side is 2 ft. longer than the first, and the third side exceeds the first by 4 ft. Find the length of each of the three sides of the triangle.   Draw a diagram of the triangle using some convenient scale.

12. Find three consecutive odd numbers whose sum is 45.   Also five consecutive odd numbers whose sum is 45.

13. Find three consecutive even numbers whose sum is 60.   Also five consecutive even numbers whose sum is 60.

14. The land surface of the world is 51,240,000 square miles.   If the land area of the rest of the world is seven times that of North America, find the area of North America.

15. In a certain grade of milk the other solids equal three times the weight of the butter fat, and the liquid part of the milk weighs 7 times as much as the solids. How many pounds of butter fat in 4800 lb. of milk?

16. At a certain date the record time for the quarter-mile run was 47 seconds, and 5 times the record time for the 100 yard dash exceeded the record time for the quarter-mile by 1 second.   Find the record time for the 100 yard dash.

17. On a certain railroad in a given year the receipts per mile were $ 3085.   If the receipts per mile for freight exceeded those for passengers by $265, find the receipts per mile from each of these sources.

18. A man left $64,000 to his wife, daughter, and niece.   To his daughter he left $4000 more than to his niece, and to his wife $8000 more than to his daughter and niece together.   How much did he leave to each?

CHAPTER III

ADDITION AND SUBTRACTION

PART I

25. Examples in Addition.

Ex. 1.
$$
\begin{array}{rr}
15 \text{ dozen} & \quad \text{also} \quad 15\,a^2b \\
8 \text{ dozen} & 8\,a^2b \\
5 \text{ dozen} & 5\,a^2b \\
\hline
28 \text{ dozen} & 28\,a^2b
\end{array}
$$

These two additions are the same if $a = 2$ and $b = 3$.

Ex. 2.
$$
\begin{array}{rrr}
7 \text{ tens} \quad \text{also} & 7\,xy \quad \text{also} & 7 \\
-5 \text{ tens} & -5\,xy & -5 \\
\hline
2 \text{ tens} & 2\,xy & 2
\end{array}
$$

The first two of these additions are the same if $x = 2$ and $y = 5$.

When, for example, we add -5 and 2 and obtain -3 as the sum, -3 is called the **algebraic sum** of the given numbers.

EXERCISE 18

ORAL

At sight find the algebraic sum of each of the following:

1.	8 dozen	4.	5 dimes	7.	$15\,ab$	10.	$9\,d$
	7 dozen		3 dimes		$12\,ab$		$-3\,d$

2.	$8\,a^2b$	5.	$5\,d$	8.	-8 dimes	11.	5
	$7\,a^2b$		$3\,d$		4 dimes		-1

3.	$8\,x$	6.	$5\,x^2y^3$	9.	$-8\,ab$	12.	$8\,x^3$
	$4\,x$		$3\,x^2y^3$		$4\,ab$		$-5\,x^3$

13.	$7\,x$	17.	$-7\,p^2q^3$	21.	8 doz.	24.	$3\,x^2$
	$-\;\;x$		$7\,p^2q^3$		4 doz.		$-\;\;x^3$
					$-\;3$ doz.		$-5\,x^2$
14.	$13\,xy$	18.	-11 ft.				
	$-\;5\,xy$		$-\;6$ ft.	22.	$-\;8\,a^2b$	25.	
					$-\;4\,a^2b$		qt^2
15.	-3 doz.	19.	$-\;8\,x$		$-\;3\,a^2b$		$4\,qt^2$
	$-\;5$ doz.		$6\,x$				$-5\,qt^2$
				23.	$2\,a$		
16.	$-\;7\,ab^2$	20.	$-\;9\,a$		$5\,a$	26.	1
	$-\;3\,ab^2$		$9\,a$		$-12\,a$		-18

26. Similar Terms are terms which are alike with respect to their letters and exponents.

Thus $8\,a^2b$ and $-5\,a^2b$ are similar terms. Why are $8\,x^2y$ and $-7\,xy^2$ not similar terms?

In adding in algebra the general method is to

Arrange similar terms in the same column;

Find the algebraic sum of the numerical coefficients of each column and prefix this result to the literal factors common to the terms in the column.

Sometimes the algebraic sum of the coefficients of each group of similar terms is found without arranging the terms in columns.

Ex. Add $4\,x^2+3\,x+2$, $3\,x^2-4\,x-3$, $-2\,x^2-x-5$.

Arranging similar terms in the same column, and adding each column separately, we obtain

CHECK

$$\begin{array}{rcr}
4\,x^2+3\,x+2 = & 4+3+2 = & 9 \\
3\,x^2-4\,x-3 = & 3-4-3 = & -4 \\
-2\,x^2-\;\;x-5 = & -2-1-5 = & -8 \\
\hline
Sum\;\;5\,x^2-2\,x-6 = & 5-2-6 = & -3
\end{array}$$

To check the accuracy of the work, we let $x =$ any convenient number, as 1; find the numerical value of each row; and compare the sum of these results with the numerical value of the algebraic expression obtained as the sum.

Add and check:

1. $8\,ab$
$22\,ab$
$13\,ab$

3. $8\,ax^2$
$7\,ax^2$
$-28\,ax^2$

5. $7.2\,x^2$
$5.8\,x^2$
$-3.6\,x^2$

7. $1.4\,a^2x^3$
$1.2\,a^2x^3$
$-2.6\,a^2x^3$

2. $-15\,a^2x$
$-9\,a^2x$
$-12\,a^2x$

4. $9\,abc^2$
$-15\,abc^2$
$-4\,abc^2$

6. $\frac{1}{2}\,xy$
$\frac{3}{4}\,xy$
$-xy$

8. $.72\,ab^3$
$-.06\,ab^3$
$-.23\,ab^3$

9. $3\,x-2\,y$
$-2\,x+3\,y$
$x-\quad y$

10. $5\,x^2+\ 7$
x^2-10
$-7\,x^2+\ 1$

11. $a^2-\quad ax+4\,x^2$
$3\,a^2+2\,ax-5\,x^2$
$-a^2-\quad ax-\quad x^2$

12. $a-2\,b,\ 3\,a+4\,b,\ a+5\,b,\ -5\,a-b,\ a-5\,b.$

13. $3\,x^2+y^2,\ 2\,x^2-7\,y^2,\ -4\,x^2-5\,y^2,\ x^2+3\,y^2,\ -3\,y^2.$

14. $3\,ax^3-5\,by^3,\ 2\,ax^3+4\,by^3,\ 2\,by^3-4\,ax^3,\ by^3-ax^3.$

Reduce each of the following to its simplest form:

15. $x^2-xy+3\,y^2+2\,x^2+2\,xy-2\,y^2+x^2+y^2+3\,x^2-xy.$

16. $mn-3\,n^2+m^2+m^2+2\,n^2-3\,mn+m^2-n^2+mn-2\,m^2.$

17. $x^2+y^2-2\,z^2+3\,x^2-y^2+2\,z^2+z^2-2\,x^2+x^2-z^2.$

18. $2\,x^2-xy+3\,xy-5\,y^2+3\,y^2-3\,x^2+x^2+2\,y^2-2\,xy.$

Collect similar terms in the following and check each result:

19. $2\,x-3\,y-5\,x+4\,z+4\,y+z-2\,y-x-3\,z+2\,x-3\,y.$

20. $3\,xy-5\,ax+3\,y^2-2\,xy-3\,x^2+4\,ax-2\,y^2+3\,ax-2\,xy.$

21. $x-3\,y+2\,z+2\,y-2\,x-z-3\,x-4\,z-2\,x+z+2\,x.$

22. $2\,x-1+5\,y-2+3\,x+2+3\,y-3-2\,x+1-x-3\,y.$

27. Addition of Unlike Terms. Similar terms can be combined into one term. Thus, $4\,a^2b+5\,a^2b=9\,a^2b.$

Dissimilar (or unlike) terms cannot be combined into a single term, but they can be combined into an algebraic sum. Thus $2\,a$ added to $3\,b$ gives $2\,a+3\,b$ as the sum.

Ex. Add $2a + b^2$, $3a - c$, and $b^2 + 3d$.

Arranging similar terms in the same column, we obtain

$$
\begin{array}{llll}
2a+ & b^2 & = 2+1 & = 3 \\
3a & -c & = 3 \quad -1 & = 2 \\
& b^2 \quad +3d = & +1 \quad +3 = 4 \\
\hline
Sum \; 5a + 2b^2 - c + 3d & = 5 + 2 - 1 + 3 = 9
\end{array}
$$

EXERCISE 20

Add and check:

1. $2a+b$, $3a+c$.
2. $3a-b$, $2b-5$.
3. $3a+b$, $c+d$.
4. $5x-7$, $2-y$.
5. 3 yd. 2 in., 1 ft. 5 in., 2 yd. 1 ft.
6. 8 gal. 1 pt., 5 gal. 1 qt., 3 gal. 1 qt.
7. $3ab+5$, -7, $12+x$.
8. -7, $6a-3b+5$, $8-3a$.
9. $7a^2$, $3ab-4b^2$, $10ab + 9b^2$.
10. $3xy-5x^2$, $-5xy$, $-x^2 + 2y^2$.
11. $3a+b$, $5a-c$, $2a+b+4c$, $2c-3b-2a$.
12. Add 2 to $n-1$.
13. Add $n-2$ and $3n$.
14. Add $2a+b$ and $a+c$.
15. Add 1 to $n-1$.

16. If we denote tens by t and units by u, we have the following corresponding forms for the addition of 24 and 35.

Arithmetical Addition	Algebraic Addition
24	$2t + 4u$
35	$3t + 5u$
59	$5t + 9u$ ·

17. In like manner express both in arithmetical and algebraic form the addition of 32 and 45.

18. Of 51 and 27.
19. Of 213 and 325.

20. Reduce $3xxxyy + 8xxxyy - 5xxxyy - 2xxxyy$ to its simplest form. About how much briefer is the form you obtain than the given form?

SUBTRACTION

28. Examples in Subtraction.

Ex. 1.   How many dozen must be
added to 8 dozen to make 12 dozen?

12 dozen	$12\,a^2b$
8 dozen	$8\,a^2b$
4 dozen	$4\,a^2b$

Ex. 2.   What must be added to
$8\,a^2b$ to make $12\,a^2b$?

Note that the above two subtractions are the same if
$a = 2$ and $b = 3$.

Ex. 3.   What must be added to
-6 dimes to make 3 dimes?

3 dimes	$3xy$
-6 dimes	$-6xy$
9 dimes	$9xy$

Ex. 4.   What must be added to
$-6xy$ to make $3xy$?

Note that the two subtractions in Exs. 3 and 4 are the
same if $x = 2$ and $y = 5$.

EXERCISE 21

ORAL

In each of the following, state what must be added to the
lower number to make the upper number.

1. 7 dozen 5 dozen	**7.** 8 4	**13.** $7x$ $-4x$	**19.** -3 yd. 4 yd.
2. $7\,a^2b$ $5\,a^2b$	**8.** $8.2\,x^3$ $5.1\,x^3$	**14.** $+12\,x^2$ $-5\,x^2$	**20.** $-3xy$ $4xy$
3. 12 dimes 3 dimes	**9.** 7 doz. -4 doz.	**15.** 16 -3	**21.** $-5a$ $2a$
4. 12 xy 3 xy	**10.** $7\,a^2b$ $-4\,a^2b$	**16.** $7.5\,x^3$ $-1.3\,x^3$	**22.** $-6\,a^2b^3$ a^2b^3
5. $8x$ $5x$	**11.** 9 nickels -2 nickels	**17.** -7 doz. 2 doz.	**23.** -2 5
6. $9\,x^2$ x^2	**12.** $9n$ $-2n$	**18.** $-7\,a^2b$ $2\,a^2b$	**24.** $8.6\,m^4$ $-2.2\,m^4$

25. $-$ 3 doz.	**29.** $-\quad x$	**33.** $-$ 5 doz.	**37.** $-$ 4 in.
$\underline{\ -\ 2\ \text{doz.}}$	$\underline{-5\,x}$	$\underline{\ \ 2\ \text{doz.}}$	$\underline{-2\ \text{in.}}$
26. $-$ 7 dimes	**30.** $-\quad 5$	**34.** $7\,\cancel{c}$	**38.** -9
$\underline{-\ 4\ \text{dimes}}$	$\underline{-12}$	$\underline{-4\,\cancel{c}}$	$\underline{+3}$
27. $-5\,ab$	**31.** -4.5	**35.** $8\,ab$	**39.** -7.2 ft.
$\underline{-2\,ab}$	$\underline{-2.2}$	$\underline{\ 2\,ab}$	$\underline{-3.6\ \text{ft.}}$
28. $-\ 6\,x^3$	**32.** $-16\,x^3$	**36.** $\ \ 6\,a^2b^3$	**40.** $-\frac{3}{4}\,x$
$\underline{-\quad x^3}$	$\underline{-\ 2\,x^3}$	$\underline{-\ 5\,a^2b^3}$	$\underline{\ \frac{1}{2}\,x}$

29. Another Method of Subtraction. Notice that in each example in § 28, the result might have been obtained by *changing the sign of the number to be subtracted and then adding it.*

The sign of number to be subtracted, however, should be *changed only in the mind*, not on paper.

Ex. From $5x^3-2x^2+x-3$, subtract $2x^3-3x^2-x+2$. Check the work by letting $x=1$.

CHECK

$$5\,x^3-2\,x^2+\quad x-3=5-2+1-3=1$$
$$\underline{2\,x^3-3\,x^2-\quad x+2=2-3-1+2=0}$$
Difference $\ 3\,x^3+\quad x^2+2\,x-5=3+1+2-5=1$

The coefficient of x^3 is $5-2$, or 3, of x^2 is $-2+3$, or 1, etc.

EXERCISE 23

Subtract and check :

1.	22 tens	**5.**	$-51\,x$	**9.**	$-x^2y^2$
	$\underline{-15\ \text{tens}}$		$\underline{-26\,x}$		$\underline{\ x^2y^2}$
2.	$-35\,t$	**6.**	$41\,ab$	**10.**	$-8.5\,x$
	$\underline{\ 12\,t}$		$\underline{23\,ab}$		$\underline{\ 2.3\,x}$
3.	$-8\,x^2y^3$	**7.**	$-8\,abc$	**11.**	$-12.7\,ab$
	$\underline{\ x^2y^3}$		$\underline{-3\,abc}$		$\underline{-\ 3.9\,ab}$
4.	-12 in.	**8.**	$-\ 9\,ab^3$	**12.**	$-2\frac{1}{2}\,m$
	$\underline{\ 45\ \text{in.}}$		$\underline{14\,ab^3}$		$\underline{\ 1\frac{3}{4}\,m}$

13. 7 ft. 4 in.
3 ft. 2 in.

16. $3x - 9$
$-5x - 1$

19. $5x^2 + 4x - 3$
$x^2 - 3x + 5$

14. $7f + 4i$
$3f + 2i$

17. 7 gal. 3 qt.
5 gal. 1 qt.

20. $3a^2 + 5ab - 5b^2$
$4a^2 + 2ab + b^2$

15. $3x^2 - 4x$
$2x^2 + x$

18. $2x^3 - 5$
$x^3 + 2$

21. $3x^3 - 2x + 5$
$5x^3 - 3x - 1$

22. From $3a + 2b - 3c - d$ take $2a - 2b + c - 2d$.

23. From $7 - 3x + 2x^2$ take $15 - 4x - 5x^2$.

24. From $x^2 - y^2 - z^2 + 8$ take $2x^2 + y^2 - 2z^2 + 10$.

25. From $2 - x + x^2 + x^4$ take $3 + x - x^2 - x^3 - 2x^4$.

26. Subtract $10x^2y + 3x^2y^2 - 13xy^2$ from $x^2y - xy^2 + 2x^2y^2$.

27. Subtract $3 + 3ac - 4cd$ from $5 - ac + 8cd$.

28. Subtract $3x^3 - 2x^2 + 5x - 7$ from $3x^3 + 2x^2 - x - 7$.

29. From 5 yd. 2 ft. 9 in. take 3 yd. 1 ft. 4 in.

30. From $5a + b$ take $2a + c$.

Arranging similar terms in the same column, we obtain

CHECK

$5a + b \qquad = 5 + 1 \qquad = 6$
$2a \qquad + c = 2 \qquad + 1 = 3$

Ans. $\overline{3a + b - c = 3 + 1 - 1 = 3}$

31. From $3a + 2b$ take $a - 2c$.

32. From $n - 1$ subtract 2.   **33.** From 3 subtract $n - 2$.

34. From $5x^3 - 2x^2 + 2x$ take $4x^3 - 4$.

35. Subtract $-3x + z$ from $6x + 2y$.

36. From $5xy - 3xz + 5yz + x^2$ take $4xz - 2xy - x^2$.

37. Subtract $1 + x - x^2 + x^3 - x^4$ from $2 - x - x^2 - x^3 + x^5$.

38. Subtract $a + 2b - 3c + 4d$ from $m + 2b + d - x + a$.

39. Subtract $-x^5 - 2x^4 + x^2 + 5$ from $x^5 - x^3 + x^2 - 2x + 5$.

40. If we denote hundreds by t^2 and tens by t, we have the following corresponding forms of the subtraction of 342 from 875.

Arithmetical Subtraction	Algebraic Subtraction
875	$8 t^2 + 7 t + 5$
342	$3 t^2 + 4 t + 2$
533	$5 t^2 + 3 t + 3$

41. In like manner express both in arithmetical and algebraic form the subtraction of 452 from 895.

42. Of 217 from 589.

43. Reduce $7\,aaabb + 5\,aaabb - 3\,aaabb$ to its simplest form.

THE PARENTHESIS

30. The **parenthesis**, ( ), is used, as in arithmetic, to indicate that all the quantities inclosed by it are to be treated as a single quantity; that is, subjected to the same operation.

Thus, in arithmetic $5 \times (3 + 4)$ means that the sum of 3 and 4 is to be multiplied by 5. Similarly in algebra $5(2\,a - b + c)$ means that the quantities inside the parenthesis, viz. $2\,a$, $-b$ and $+c$, are each to be multiplied by 5.

Again, in arithmetic $18 - (5 + 2)$ means that the sum of 5 and 2 is to be subtracted from 18. So in algebra, $15\,b - (2\,a + b + 3)$ means that $2\,a + b + 3$ is to be subtracted from $15\,b$.

The best way to perform an addition or subtraction in dicated by a parenthesis, is to remove the parenthesis, changing every sign inside the parenthesis if the parenthesis is preceded by a minus sign; and then collect terms. (See § 29, p. 47.)

Thus $15\,b - (2\,a + b + 3) = 15\,b - 2\,a - b - 3$
$$= 14\,b - 2\,a - 3. \quad \textit{Ans.}$$

Instead of the parenthesis, to prevent confusion, the following signs are sometimes used: the brackets [ ], the braces { }, and the vinculum ———.

These forms are of especial value when one parenthesis is to be used inside another.

$$\text{Thus } 8\,x + y - [3\,x + (2\,x - y)] = 8\,x + y - [3\,x + 2\,x - y]$$
$$= 8\,x + y - 3\,x - 2\,x + y$$
$$= 3\,x + 2\,y. \quad Ans.$$

When one parenthesis is given inside another, in simplifying it is customary to remove the inner parenthesis first.

EXERCISE 23

Remove parentheses and collect similar terms. Check each result either by substitution of numerical values, or by reversing the order in which the parentheses are removed.

1. $a + (b + c)$.
5. $3a + (2x - 3y)$.
9. $2x - (x - 1)$.

2. $a - (b + c)$.
6. $3a - (2x - 3y)$.
10. $x + (1 - 2x)$.

3. $a + (x - y)$.
7. $a^2 + (3a - 1)$.
11. $3x - (1 + 3x)$.

4. $a - (x - y)$.
8. $7x^2 - (1 - 3x^2)$.
12. $x - (-x - 1)$.

13. $3p^3 - (2 - 5p^2)$.
18. $5x + (1 - [2 - 4x])$.

14. $a^3 + [a^3 + (a^2 - 1)]$.
19. $2 - \{1 - (3 - a) - a\}$.

15. $3a + (2a - b)$.
20. $2x - [-x - (x - 1)]$.

16. $x + 2y - (2x - y)$.
21. $2y + \{-x - (2y - x)\}$.

17. $x - [2x + (x - 1)]$.
22. $a - \{-a - (-a - 1)\}$.

By use of a parenthesis indicate

23. That $3a + 2b + 5c$ is to be subtracted from $7a - 6b$.

24. That $x^2 + 2x + 1$ is to be subtracted from $3x^2 - 2x - 5$.

25. That the sum of x and y is to be subtracted from their difference.

26. That $a^2 + 2ab + c^2$ is to be subtracted from $4x^2$.

27. That $x^2 - 4xy + 4y^2$ is to be subtracted from $x^2 + 4xy + 4y^2$.

1. What is the complete rule for finding the volume of a box-shaped solid? What is the brief form of this rule? What is the formula for the volume of a box-shaped solid?

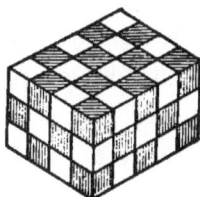

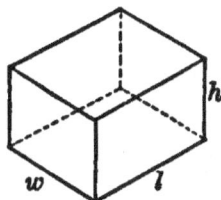

2. By use of $v = lwh$, find the number of cubic yards in a cellar 7 yd. long, 5 yd. wide, and 2 yd. deep.

If inch blocks are available, build up a representation of the cellar, letting each block stand for a cubic yard. Verify your answer by counting the number of inch blocks used.

3. By use of $v = lwh$, find the number of cubic inches in a brick whose dimensions are 8, 4, and 2 inches.

4. A rectangular tank is to contain 280 cu. ft. of water. If the tank is to be 8 ft. long and 7 ft. deep, how wide must it be? Use $v = lwh$.

5. State the rule for finding the interest on a given sum of money, for a given number of years, at a given rate.

6. In the interest formula, $i = prt$, what does the i stand for? What does each of the other letters stand for?

By use of the formula i = prt, *solve Exs.* 7–10.

7. Find the interest on $400 for $3\frac{1}{2}$ yr. at 6 per cent.

8. If $i = \$150$, $p = \$500$, and $t = 6$ yr., find r.

9. At what rate must $500 be at interest for 6 years, to produce $150 interest?

10. If $p = \$700$, $i = \$210$, and $r = .06$, find t.

11. For how many years must $700 be at interest at 6 per cent to produce $210 interest? Use $i = prt$.

12. If a train travels at the rate of 40 miles an hour, how far will it go in 5 hours? In t hours?

13. If a motorcyclist travels at the rate of v miles per hour, how far will he go in t hours?

14. For a body moving at a uniform rate, we have the brief rule that "distance equals the rate (or velocity) multiplied by the time." Convert this rule into a formula by denoting distance by d, rate or velocity by v, and time by t.

15. If an automobile travels at the rate of 20 miles an hour, how far will it go in $7\frac{3}{4}$ hours? Use $d = vt$.

16. A cyclist must travel from New York to Philadelphia, a distance of 90 miles, in $7\frac{1}{2}$ hours. At what rate per hour must he go? Use the formula $d = vt$.

17. Take a coin and measure its diameter. By rolling the coin along a straight line on a piece of paper, measure the circumference of the coin. Then divide the length of the circumference of the coin by the diameter and thus determine how many times the diameter is contained in the circumference.

18. What is the radius of the coin? How many times is the radius contained in the circumference? How does your result compare with the formula $c = 2\pi r$, where $c =$ the circumference of a circle, $r =$ the radius, and $\pi = \frac{22}{7}$?

19. By use of the formula $c = 2\pi r$, find the circumference of a circle whose radius is 28.

20. If the circumference of a wheel is 88 in., find the radius, by use of $c = 2\pi r$.

21. If the diameter of a wheel is 21 in., find the circumference by use of $c = 2\pi r$. Draw a diagram of the wheel using a scale of 1 to 7.

PART II

31. Addition, in algebra, is the combination of several algebraic expressions into a single equivalent expression.

Addition is sometimes described as *collecting terms* in an expression.

The **Algebraic Sum** of two or more algebraic numbers is the result of combining the given algebraic numbers into a single number.

Thus, the algebraic sum of 4 and -7 is -3.

32. The Utility of Addition in Algebra.

Ex.   Find the value of $3\ ab^2 + 5\ ab^2 + 2\ ab^2$, when $a = 2$ and $b = 3$.

PROCESS WITHOUT ALGEBRAIC ADDITION

If we substitute directly in the given expression, we obtain
$$3\ ab^2 + 5\ ab^2 + 2\ ab^2 = 3 \times 2 \times 3^2 + 5 \times 2 \times 3^2 + 2 \times 2 \times 3^2$$
$$= 54 + 90 + 36$$
$$= 180. \quad Ans.$$

PROCESS AIDED BY ALGEBRAIC ADDITION
$$3\ ab^2 + 5\ ab^2 + 2\ ab^2 = 10\ ab^2$$
$$= 10 \times 2 \times 3^2$$
$$= 180. \quad Ans.$$

In solving the above example, algebraic addition enables us to save more than half the work.   Algebraic addition also has many other uses, some of them direct, others indirect, which will appear later.

EXERCISE 25

Find the sum of the following:

1.	2.	3.	4.
6.42	$-\ 5\frac{1}{2}$	$3.2\ x^2 + 4.8\ x - 3$	$\frac{1}{2}a + \frac{1}{3}b + \frac{1}{4}c$
-3.25	$-16\frac{2}{3}$	$-1.6\ x^2 \qquad\qquad +5$	$-\frac{1}{6}a - \frac{1}{6}b - \frac{1}{8}c$
4.36	$20\frac{3}{4}$	$-\ .8\ x^2 + 7.2\ x$	$a -\quad b +\quad c$
-2.5	$8\frac{1}{4}$	$-2.5\ x^2 \qquad\qquad -2.2$	$\frac{1}{4}a \qquad\qquad +\frac{1}{2}c$

5. $-7.1 + 16.3 - 2.09.$ **6.** $-5\frac{1}{2} - 9\frac{3}{4} + 1 - \frac{1}{4}.$

7. $2 - 1\frac{3}{4} + 1\frac{1}{2} - 1\frac{3}{4} + 2\frac{1}{4} - 3 + 1.$

8. $3.2 - 1.5 - .8 + 1.7 - 3.62 + 5.24.$

9. The level of the water in a reservoir underwent the following changes in succession: a rise of 3 in., a fall of 2.25 in., a fall of .3 in., a rise of .45 in., a rise of 2.2 in., a fall of 1.8 in. Write these as + or − numbers, and then determine the net rise or fall.

10. Add $3\,x^2 - 2\,x + 5$; $5\,x^2 + 10$; $22\,x - 10 - 3\,x^2$; $20\,x^2 - 20\,x + 4 - 8\,x^3 + 4\,x^4.$

11. Add $\frac{1}{2}\,a^3 - b^2c + 2\,abc$; $5\,b^2c - \frac{1}{5}\,a^3$; $7\,bc - \frac{1}{2}\,abc + \frac{1}{8}\,a^3$; $-3\,a^3 - \frac{1}{2}\,bc + 5\,abc.$

12.	**13.**	**14.**	**15.**
$3(a+b)$	$-6(x-y)$	$5\sqrt{a+x}$	$4\,\pi r^2$
$4(a+b)$	$4(x-y)$	$-6\sqrt{a+x}$	$-2\,\pi r^2$
$-4(a+b)$	$-5(x-y)$	$2\sqrt{a+x}$	$\frac{1}{2}\,\pi r^2$

16.	**17.**	**18.**	**19.**
$(2\,a + 3\,b)x$	$(a+3)x^3$	$(a+b)xy$	7×25
$(3\,a - 2\,b)x$	$(a-5)x^2$	$(a+c)xy$	9×25
$(-a + 2\,b)x$	$(a+c)x^2$	$(b+c)xy$	-11×25

20.	**21.**	**22.**	**23.**
$a(x-y)$	$5(a^2 - b^2)$	$cx + ax^3$	$ax + bx^2$
$3\,a(x-y)$	$3\,a(a^2 - b^2)$	$-bx + bx^3$	$xy - 2\,x^2$
$-2\,a(x-y)$	$-4\,b(a^2 - b^2)$	$cx - 3\,cx^3$	$xz + cx^2$

24. $7\,x + y + 5\,z - 10\,xy + 2\,y - 3\,z + 13\,xy - 4\,xz + 5\,z - 6\,x - 4\,xz + 2\,xy - 3\,y + 9\,z + 7\,x - xz + 21\,xz - 16\,z + x - 5\,xy.$

25. $x^3 + 3\,x^2y + 3\,xy^2 + y^3 + x^3 - 3\,x^2y + 3\,xy^2 - y^3 + 2\,x^2y - 2\,xy^2 + y^3 + x^3 - y^3 + x^2y - 4\,x^3 - xy^2 - y^3 + y^3 + x^3 - x^2y + xy^2.$

26. $3\,a^2b - 2\,a^2c + 3\,a^2 - 5\,a^2b - a^2 - 3\,a^2c + a^2b + 6\,a^2c - 2a^2.$

27. $5\,x^3 - 3\,x + 4 - 2\,x^2 - 6\,x^3 + 4\,x - 7 - x^2 + x^3 + 3\,x^2 - x + 5 + 3\,x^2 - 6\,x - x^2 + 4\,x - 2\,x^2 + 2\,x.$

28. Write a single term equivalent to $5\,ab + 4\,ab.$

29. Can $\frac{3}{4}x + \frac{1}{4}y$ be expressed as a single term?

30. When $x = 3$ and $y = 4$, find the value of $7\,x^2y + 3\,x^2y - 5\,x^2y$, (1) without first adding the terms; (2) after adding the terms.

Compare the amount of work in the two processes.

31. In the examples of this Exercise, among the expressions added, point out five binomials. Also five monomials. Also five trinomials. Also two polynomials, each of which contains more terms than a trinomial.

32. Make up and work an example in which four monomials are added.

33. Also one in which three binomials are added.

34. Also one in which two trinomials and one binomial are added.

35. Also one in which three binomials, two monomials, and two trinomials are added.

33. Subtraction, in algebra, is the process of finding a quantity which, added to a given quantity (the subtrahend), will produce another given quantity (the minuend).

Thus, if we subtract $3\,ab$ from $10\,ab$, we obtain $7\,ab$, for $7\,ab$ added to $3\,ab$ (subtrahend) gives $10\,ab$ (minuend).

EXERCISE 26

Subtract and check :

1. 8.4	**5.**	$2\frac{1}{4}\,ab$	**9.**	-2.46	**13.**	ay
1.32		$-1\frac{1}{2}\,ab$		1		$-by$
2. 7.2	**6.**	-8.47	**10.**	$-8.27\,x^3$	**14.**	$-a^2x$
-1.32		-3.2		$4.23\,x^3$		b^2x
3. -5	**7.**	$9.$	**11.**	$-7.24\,a$	**15.**	$-\,px^2$
7.2		3.72		$-3.8\,a$		$-3\,cx^2$
4. $5.3\,x$	**8.**	12	**12.**	ax	**16.**	$-ay^3$
$1.2\,x$		$-\ 5.25$		bx		$-by^3$

17.	**18.**	**19.**
$-2\,x^3-\ .5$	$5\,x^2+4\,x-3$	$8.5\,a^2+\ \ 7\,a-3.06$
$\underline{\ -\ x^3+2.4}$	$\underline{-x^2-3\,x+5}$	$\underline{-1.7\,a^2-1.3\,a-2.4}$

20. $\quad 5(a+b)$
$\qquad \underline{3(a+b)}$

23. $\quad -4\sqrt{b-y}$
$\qquad \underline{2\sqrt{b-y}}$

26. $\quad a(x+y)$
$\qquad \underline{b(x+y)}$

21. $\quad 7(x+y)$
$\qquad \underline{-3(x+y)}$

24. $\quad \frac{3}{2}\,x^2-\frac{1}{4}\,y^2$
$\qquad \underline{x^2-\frac{1}{2}\,y^2}$

27. $\quad 7(x+y)$
$\qquad \underline{-5(x+y)}$

22. $\quad 2\sqrt{a+x}$
$\qquad \underline{-3\sqrt{a+x}}$

25. $\quad (5\,a-b)x$
$\qquad \underline{(3\,a-2\,b)x}$

28. $\quad (-3\,a+b)x$
$\qquad \underline{(-2\,a-c)x}$

29. From $3\,a^2-5\,ab+7\,b^2$ subtract $-a^2-3\,ab-10\,c^2$.

30. From $3\,x^2+7\,y^2-3\,xy+5\,bc$ take $3\,x^2-5\,y^2-10\,xy+bc^2$.

31. Subtract $4\,ab+5\,bc-6\,ac$ from $ab+bc+ac$.

32. What quantity must be added to $4\,a^2-4\,ac+4\,c^2$ in order that the sum shall be $2\,a^2+4\,ac+3\,c^2$?

33. From $3\,a+2\,b-3\,c-d$ take $2\,a-2\,b+c-2\,d$.

34. From $7\,b-3\,x+2\,x^2$ take $15\,a-4\,x-5\,x^2$.

35. From $x^2-y^2-z^2+8$ take $2\,x^2+y^2-2\,p^2+10$.

36. Find the value of $3-10.2$.

37. Take $a+c$ from $2\,a+b$.

38. Write a single term equivalent to $7\,a^2b-5\,a^2b$.

39. Can $6.6\,x-2.4\,y$ be expressed as a single term?

40. From tne sum of $2\,x$ and $3\,y$ subtract their difference.

41. From 0 subtract $-3\,x$. From 0 subtract $x-y$. From zero subtract $3\,a^2-2\,ab+b^2$.

If $A=x^3-3\,x^2+1$, $B=2\,x^2-5\,x-3$, $C=3\,x^3+x^2+3\,x$, find the value of each of the following:

42. $A+B+C.$

44. $A+B-C.$

43. $B-A+C.$

45. $A-B+C.$

46. If $a = 3$ and $b = 2$, find the numerical value of $17\,a^2b^3 - 15\,a^2b^3$, (1) without subtracting the last term from the first, (2) after subtracting. Compare the amount of work in the two processes.

Make up and work an example in which

47. A binomial is subtracted from a trinomial.

48. A binomial is subtracted from a monomial.

34. Double Use of + and — Signs. The signs + and — are employed for two purposes: first, to indicate the operations of addition and subtraction; and second, to express positive and negative numbers. We are able to make this double use of these signs because, in each use, the signs are governed by the same laws.

Thus in $5 - (x - 3)$ the minus sign to the left of the parenthesis means that the quantity inside the parenthesis is to be subtracted from 5, and the minus sign is therefore a sign of operation. On the other hand, the minus sign next to 3 may be regarded either as a sign of operation or as showing that 3 is a negative number to be added to x and therefore as a sign of quality.

EXERCISE 27

Simplify:

1. $10 - (-7.2)$.

2. $-5\frac{1}{2} + (-9\frac{3}{4})$.

3. $-52 - (-16)$.

4. $-8.3 - (-2.6)$.

5. $10.2 - (-7.2) - 3.27$.

6. $-16.52 + 3.7 - (-8.26)$.

7. $18.37 - 5.2 + (-7.23)$.

8. $9.37 - (-18.2) - (-7.2)$.

9. $2.76 - (-3.56) + 10.78 - (-3.06) - (-.07)$.

10. $.5 - (2\,a - .4)$.

11. $2.4 - (x - .7)$.

12. $5 + (-4 + b)$.

13. $1.8 - (.4 - 2\,x)$.

14. $8 - (-x - 1.5)$.

15. $1.8 - [2\,x + (x - 7)]$.

16. $5\,x + (1.7 - [2\,x - 4.8])$.

17. $1.2 - \{4 - (.3 - a) - a\}$.

18. $2.9 - [-.8 - (x - 1)]$.

19. $.28 + \{-x - (6.4 - x)\}$.

20. $5\,a - \{-a - (-a - 2)\}$.

21. $3\,x^2 - [2\,x^2 + (x^2 - x)]$.

22. $x^2 - [(x^2y - z^2) - z^2] + (x^2y - x^2)$.

23. $5.4 \, m^2 - [7.2 \, m^2 - (m^2 + 3) + 7]$.

24. $3.7 \, a^2 - \{4 \, a^2 - (2.7 \, a^2 - a) - 5.2\}$.

25. In $3 \, x - (5 \, a - 2 \, b + c)$ what is the sign of $5 \, a$ as the example stands?

35. Insertion of a Parenthesis. It is clear that the process of removing a parenthesis may be reversed; that is, that terms may be inclosed in a parenthesis. Thus,

Terms may be inclosed in a parenthesis preceded by the plus sign, provided the signs of the terms remain unchanged.

Terms may be inclosed in a parenthesis preceded by the minus sign, provided the signs of the terms are changed.

Ex. $a - b + c + d - e = a - b + (c + d - e)$,

or, $= a - b - (- c - d + e)$. *Ans.*

EXERCISE 28

In each of the following, insert a parenthesis inclosing the last three terms, each parenthesis to be preceded by a minus sign. Check the work either by removing the parenthesis in the answer, or by numerical substitution.

1. $x^3 - 3 \, x^2 + 3 \, x - 1$.   **5.** $x^4 + 4 - x^2 - 4$.

2. $a - b + c + d$.   **6.** $a^2b^2 - 2 \, cd - c^2 - d^2$.

3. $1 + 2 \, a - a^2 - 1$.   **7.** $4 \, x^4 - 9 \, x^2 + 12 \, xy - 4 \, y^2$.

4. $1 - a^2 - 2 \, ab - b^2$.   **8.** $x^4 - 4 \, x^3 + 4 \, x^2 + 4 \, x - 4$.

In the following inclose the last two terms in a parenthesis preceded by the minus sign:

9. $ax + bx - ay - by$.   **11.** $x^3 + 2 \, x^2 + 5 \, x - 1$.

10. $x^3 - y^3 - x^2 + y$.   **12.** $am - bn - ap + bq$.

EXERCISE 29

1. If 38 cu. ft. of a certain kind of coal make a ton, and the length in feet of a rectangular bin be denoted by l, the width by w, the height by h, and the number of tons in the bin by T, obtain a formula for T.

2. By use of the formula for T found in Ex. 1, find the number of tons of coal that will be contained by a bin 20 ft. long, $3\frac{1}{2}$ ft. wide, and 4 ft. deep.

3. Also by a box 9 ft. 6 in. long, 3 ft. 4 in. wide, and 2 ft. 9 in. deep.

4. Also by a bin 12.4' by 3.8' by 5.6'.

5. A man wishes to build a bin to contain 10 tons of coal. If the bin is to be 15 ft. long and $6\frac{1}{3}$ ft. wide, how deep must it be? Use the formula for T found in Ex. 1.

6. If $2\frac{1}{5}$ cu. ft. of corn on the cob will yield one bushel of shelled corn, find the number of bushels of shelled corn that can be obtained from a bin 20 ft. long, 9 ft. high, and 5 ft. wide, filled by corn on the cob.

7. If $2\frac{1}{5}$ cu. ft. of corn on the cob is taken as equivalent to one bushel of shelled corn and the dimensions in feet of a bin are denoted by l, w, h, and the number of bushels of shelled corn by B, find a formula for B, when the bin is filled with corn on the cob.

8. A bin 18 ft. long, $5\frac{1}{2}$ ft. wide, and 8 ft. deep is filled with corn on the cob. This corn will make how many bushels of shelled corn? Use formula for B found in Ex. 7.

9. A farmer wishes to build a bin of such a size that it will contain enough corn on the cob to yield 100 bushels of shelled corn. If the bin is to be 22 ft. long and 5 ft. wide, how deep must it be? Use formula found in Ex. 7.

10. A lumber or board foot is a piece of board 1 ft. square and 1 in. thick. Find the number of lumber feet in a plank 14 ft. long, 8 in. wide, and 3 in. thick.

11. Obtain a formula for the number of board feet (L). in a board l feet long, w inches wide, and t inches thick.

12. How many board feet are there in a beam 6 in. by $4\frac{1}{2}$ in. and 16 ft. long? Use formula for L found in Ex. 11.

13. In computing lumber, if the thickness of a board is less than one inch, the computation is made as if the board were one inch thick. Find the number of lumber feet in a board 16 ft. long, 10 in. wide, and $\frac{5}{8}$ in. thick. Use formula for L found in Ex. 11.

14. A cranberry grower wishes to obtain a formula for estimating the number of bushels (B) per acre of a crop when there are n berries on one square foot of the meadow. By counting them he finds that there are 1010 berries in a quart. Show that $B = \dfrac{4n}{3}$ would be a convenient approximate formula for his purpose.

15. If he counts the berries on a number of different square feet, and finds that the average is 60, at this rate how many bushels per acre are there?

16. In the interest formula $i = prt$, t denotes the number of years. By use of the formula find the interest on $450 at 6 per cent for 82 days, regarding 360 days as a year (bank's method).

Sug. Number of years or $t = \frac{82}{360}$, etc.

17. Show that if the number of days be denoted by n and 360 days be taken as a year, the interest formula becomes $i = pr\left(\dfrac{n}{360}\right)$.

By use of this formula find the interest on $675 for 75 days at 5 per cent.

18. If 365 days is regarded as making a year (used by the United States government in computing interest) and n denotes the number of days, show that $i = pr\left(\dfrac{n}{365}\right)$.

By use of this formula compute the exact interest on $675 for 75 days at 5 per cent.

CHAPTER IV

THE EQUATION

PART I

36. Equations; Root of an Equation.

Ex. 1. If $x = 2$, does $x + 4 = 7$?

We find that it does not, since, on substituting 2 for x in $x + 4$ we get $2 + 4$ or 6, which does not equal 7.

Ex. 2. If $x = 3$, does $x + 4 = 7$?

We find that it does, since $3 + 4 = 7$.

A relation like $x + 4 = 7$, where x is an unknown number to be found, is called an *equation*.

In the above equation, the value of x (that is 3) which makes $x + 4 = 7$ is called the *root* of the equation.

37. The Solution of an Equation is the process of finding the value of the root of the equation.

Ex. 1. Find the value of x in the equation $x - 5 = 7$.

We have given, $x - 5 = \ 7.$

Adding 5 to each of the equals, $\underline{5 \qquad 5}$

                         $x = 12.$ *Root.*

CHECK. Substituting 12 for x in the given equation,

$$12 - 5 = 7,$$
$$\text{or,} \qquad 7 = 7.$$

Ex. 2. Solve $5x = 28 + 3x$.

We have given, $5x = 28 + 3x.$

Subtracting $3x$, $\underline{3x \qquad\quad 3x}$

Hence, $2x = 28,$

                       $x = 14.$ *Root.*

CHECK. $70 = 28 + 42.$

                       $70 = 70.$

Ex. 3. Find the root of $5x - (x - 4) = 20 + 2x$.

Removing the parenthesis,

$$5x - x + 4 = 20 + 2x$$
$$-2x \qquad\qquad -2x$$
$$\overline{3x - x + 4 = 20}$$
$$\qquad -4 \quad -4$$
$$\overline{3x - x \quad\; = 16}$$
$$2x = 16,$$

Let the pupil check the work. $\qquad x = 8.$ *Root.*

EXERCISE 30

Find the root of each of the following equations and check:

1. $x + 4 \text{ in.} = 9 \text{ in.}$ 3. $x + 3 \text{ ft.} = 10 \text{ ft.}$
2. $3 \text{ in.} + x = 8 \text{ in.}$ 4. $x + 4 = 9.$

5. $3 + x = 8.$ 9. $3x = \$12.$ 13. $x + 4 = 7.$
6. $x - 3 = 10.$ 10. $5x = 15 \text{ in.}$ 14. $2x - 5 = 9.$
7. $x - 4 = 24.$ 11. $8x = 16.$ 15. $5x - 7 = 23.$
8. $x - 5 = 12.$ 12. $x - 4 = 3.$ 16. $8x = 20 + 3x$

17. $7x = 20 + 3x.$ 25. $5x - 1 = 3x + 7.$
18. $4x - 3 = x + 9.$ 26. $3x - 2 = 2x + .74.$
19. $7x + 5 = 3x + 11.$ 27. $8x - 7.2 = 6x - 2.8.$
20. $3x - 2 = x + 8.$ 28. $7x - (x + 2) = 22.$
21. $x + 2x - 3 = 6.$ Sug. $7x - x - 2 = 22$, etc.
22. $4x - 3 = 12 - x.$ 29. $7x - (5x + 4) = -2.$
23. $13x + 7 + 3x = 71.$ 30. $9x = 10 - (x + 5).$
24. $6x - 25 + 3x = 2.$ 31. $8x + (3x - 4) = 25.$

32. When $x = 3$, does $x + 3 = 7$?
33. When $x = 4$, does $x + 3 = 7$?
34. Is 5 the root of $2x + 3 = 13$?
35. Is 4 the root of $2x - 3 = 7$?

Find the value of the letter in each of the following:

36. $y + 3 = 6.$ 39. $3l - 5 = l - 9.$
37. $5 + s = 12.$ 40. $5w - (w + 6) = 20.$
38. $7 + 2p = 13.$ 41. $3y - 9 = 7 - (2 + 5y).$

42. Find whether $x + 3 = 5$, when $x = 1$, when $x = 4$, when $x = 3$. For what value of x does $x + 3 = 5$?

43. Does $2p - 3 = 7$ when $p = 8$? when $p = 6$? For what value of p does $2p - 3 = 7$?

EXERCISE 31
VERBAL PROBLEMS

1. A has x marbles and B has twice as many. How many has B? How many do both have together?

2. A boy had m marbles and lost 3. How many were left?

3. A man bought a horse for $120 and sold it so as to gain $35. For how much did he sell it? If he had sold it so as to gain l dollars, for what would he have sold it?

4. A man bought a horse for x dollars, and sold it so as to gain a dollars. What did he receive for it?

5. A man by selling a horse for $200 lost $40. What did he pay for the horse?

6. A man sold a horse for $200, and thus lost x dollars. What did the horse cost?

7. If I am x years old now, how old was I ten years ago? a years ago? How old will I be in c years?

8. A boy rides a miles an hour; how far will he ride in c hours?

9. A bicyclist rides x yards in y seconds. How far will he ride in one second? In n seconds?

10. John has x dollars, and James has seven dollars less than three times as many. How many has James?

11. If I have x dollars, and you have three dollars more than twice as many, how many have you? How many have we together?

12. The difference between two numbers is 15, and the less is x. What is the greater? What is their sum?

Using algebraic symbols express the following statements as equations:

13. x diminished by 19 equals 17.

14. x increased by 7 equals 18.

15. x increased by 10 equals 20 diminished by x.

16. 2 x increased by 14 and then diminished by x equals 10.

17. Five times x equals 9 diminished by twice x.

18. If x is multiplied by 3, and the result diminished by 40, the result is 140.

19. If Walter has x marbles and his brother has 10 less, how many has the brother? How many have both boys together?

20. A certain distance is x miles and another distance is 20 miles less than four times the first distance. State the second distance in terms of x.

21. A certain number is x. If 210 be added to 24 times this number, what is the result?

22. A boy paid x dollars for a baseball bat and 25 cents more than twice as many dollars for a mitt. How many dollars did he pay for the mitt? How many cents?

EXERCISE 32

1. Find the number which, when diminished by 19, equals 37.

2. What number increased by 19 equals 37?

3. What number diminished by 1.3 equals 4.5?

4. What number increased by twice itself and then by 24 equals 144?

5. What number increased by twice itself and then diminished by 24 equals 144?

6. A number increased by 3 times itself and then by 40 equals 180. Find the number.

7. If a number is multiplied by 3 and then diminished by 40, the result is 140. Find the number.

8. If 5 times a certain number is increased by 20.5, the result is 870. Find the number.

9. If five times a certain number is increased by 20.4, the result is equal to three times the number increased by 160. Find the number.

10. Walter has a certain number of marbles and his brother has 10 less. Together they have 90 marbles. How many marbles has each boy?

Sug.	NUMBERS DEALT WITH	SYMBOLS FOR NUMBERS
	Number of marbles Walter has	$= x.$
	Number of marbles his brother has	$= x - 10.$
	etc.	etc.

11. Mary earned a certain number of dollars one month by picking berries and her younger sister earned $5 less. Together they earned $21. How much did each earn?

12. In a certain school the number of boys is 12 less than the number of girls. All together there are 72 pupils in the school. Find the number of each in the school.

13. A basketball team has played 27 games and has lost 3 less than it has won. How many games has it won?

14. In a certain election 12,420 votes were cast, and the defeated candidate had 210 less votes than the winning candidate. How many votes had each candidate?

15. Ernest solved a certain number of problems and Robert solved 8 less than twice as many. Together they solved 40 problems. How many did each of the boys solve?

Sug.	NUMBERS DEALT WITH	SYMBOLS FOR NUMBERS
	Number of problems solved by Ernest	$= x.$
	Number of problems solved by Robert	$= 2x - 8.$
	Number of problems solved by both	$= 40$, etc.

16. A boy paid a certain sum for a baseball and 25 cents more than twice as much for a mitt. He expended $4.75 in all. How much did he pay for the baseball?

17. A certain part of the Panama Canal passes through the hills. The part through the lowlands is 4 miles longer than 8 times the length of the part through the hills. The total length of the canal is 49 miles. Find the length of each part.

18. Margaret has a certain number of postcards. Louise has 7 less than twice as many. Together they have 83 postcards. How many has each girl?

19. The sum of two numbers is 76. The second number exceeds twice the first by 13. Find the numbers.

20. Two partners in one year earned $8500. The second partner received $1000 more than twice what the first partner received. How much did each receive?

21. $x + 2$ oz. $= 2$ lb. From this equation, find x, which represents the weight of sugar in the picture.

22. If a package of coffee and a $\frac{1}{4}$ lb. weight together weigh 3 lb., find the weight of the coffee.

Construct a diagram for Ex. 22 similar to that of Ex. 21.

EXERCISE 33
REVIEW

1. Find the value of $a + 3b - 3x$, when $a = 5$, $b = 2$, and $x = 1$.

2. If $s = vt + \frac{1}{2}gt^2$, find the value of s when $v = 10$, $g = 32.16$, and $t = 4$.

3. If $x = 3$, find the value of $4x^2$. Also of $(4x)^2$.

4. David caught a certain number of fish one day and the next day caught 8 more than on the first day. All together he caught 30 fish. How many did he catch the first day?

Express in simplest form:

5. $3 a^2 - 5 ab + 2 b^2 - c^2 + 4 b^2 + 2 ab - 2 a^2 - 4 ab - 5 c^2 + 6 a^2.$

6. $2 x^4 - 5 x^2 - 3 x^2 + 2 x - 5 + 2 x^3 - 3 x^4 - 2 x + 2 x^2 - 2 x + 2 x^2 - 6 + 3 x^2 + x^4 - 3 x^3 + 7 - x + 2 + 3 x^3 + 2 x^4 - 4 x - 2 x^2.$

7. Subtract $3 x^3 - 2 x^2 + 5 x - 3$ from $8 x^3 - x^2 - 1$. Also $5 x^3 - 3 x^2y + y^3$ from $3 x^3 + xy^2 - y^3$.

8. Two girls take in $21.60 at a refreshment stand. One girl furnishes the materials sold and receives four times as large a share of the profits as the other girl. How much does each girl receive?

9. Three boys caught 84 fish together. The boy who furnished the boat received twice as large a share as each of the other two boys. How many fish did each boy receive?

10. Simplify $3 x - (5 x - 2) + 7.$

11. Simplify $3 a - [7 a - (4 a + 5) - 4] - 2.$

12. Solve $3 x - (5 x - 1) = 9.$

13. Find four consecutive numbers whose sum is 106.

14. Reduce to its simplest form $a + a + a + a + b + b + b.$ Also $3 a \cdot a \cdot a \cdot a \cdot b \cdot b \cdot b.$

15. Alexander the Great was born in the year $- 356$ and he lived 33 years. In what year did he die?

16. What is the formula for the area of a rectangle whose length is l and width w? By use of this formula find the number of square feet in a boy's garden which is $17\frac{1}{2}$ yd. long and 24 ft. wide.

17. If b denotes the base, r the rate expressed decimally, and p the percentage, what is the formula for p in terms of b and r? Find r if $p = 4$ and $b = 64.$

18. What is the rate of income from a 4 per cent liberty bond which cost $94? Solve by use of the formula $p = br.$

19. Graph the growth of the population of the United States using the following table:

Year	1790	1800	1810	1820	1830	1840	1850
Millions . . .	4	5	7	10	13	17	23

Year	1860	1870	1880	1890	1900	1910
Millions	31	39	50	63	76	92

Mark the axes as indicated at the right.

From your graph determine as accurately as you can the population in 1815. In 1835. In 1895. In 1905.

From your graph determine as nearly as you can in what year the population was 35 millions. 70 millions. 80 millions.

20. Four boys together earned $182 on a farm one summer. The first boy earned the same amount as the third. The second earned $12 more than the third. The fourth earned twice as much as the third. How much did each boy earn?

21. A school contains 162 boys and 198 girls. By use of the formula, $p = br$, find what per cent of the whole are boys.

Solve and check:

22. $3 - (x - 2) = 7 - 5x.$ **23.** $4x + (x - 1) = 3x - (x + 2).$

24. Subtract $5x^2 - 3ax - 2a^2$ from $-3x^2 + 2ax^2 - a^4.$

25. Find the value of $5x^2 - 3(a - 2x) + 5a^2$, when $a = 4$ and $x = 1.$

26. Add $5x^2 - 3ax + 4a^2$, $5ax - 3x^2 + a^2$, and $3ax - x^2 - 2ax.$

27. Simplify $x^2 - [5ax + (a^2 - 2x^2 - ax) - 3x^2] - 5a^2.$ Test the accuracy of your work by letting $a = 1$ and $x = 2.$

28. By use of the formula $i = prt$, find the interest on $450 for 2 yr. 6 mo. at 5 per cent.

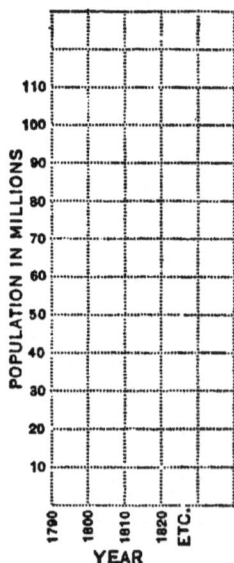

29. By use of the formula $v = lwh$, find the volume of a wagon body $10\frac{1}{2}$ ft. long, 4 ft. wide, and $1\frac{3}{4}$ ft. deep.

30. Copy the following tabulation and fill in each vacant place with the proper rule or formula:

Subject	Brief Rule	Formula
Volume of box-shaped solid	*Volume equals length × width × height*	
Area of rectangle		$a = lw$
Interest on money		$i = prt$
Percentage	*Percentage equals base × rate*	
Circumference of a circle		$c = \frac{44}{7} r$
Motion of a body		$d = vt$

31. The following table gives the normal or average height of a boy and girl at different ages:

Age in Years	3	6	9	12	15	18	21
Height of boy	$2'\,11''$	$3'\,8''$	$4'\,2''$	$4'\,7\frac{1}{2}''$	$5'\,2\frac{1}{4}''$	$5'\,6\frac{3}{4}''$	$5'\,8\frac{1}{4}''$
Height of girl	$2'\,11''$	$3'\,7''$	$4'\,2''$	$4'\,9''$	$5'\,1\frac{1}{2}''$	$5'\,3\frac{1}{4}''$	$5'\,3\frac{3}{4}''$

Graph the above facts on two separate diagrams.

From these graphs determine as accurately as you can the normal height of a boy and girl at 10 years of age.   At 14 years.

32. The distance of the sun from the earth is 92,800,000 mi. This distance exceeds 107 times the diameter of the sun by 95,200 mi.   Find the diameter of the sun.

33. A man bequeathed $ 20,000 to his wife, daughter, and son. To his daughter he left $ 2000 more than to his son, and to his wife three times as much as to his son.   How much did he leave to each?

34. The distance of the moon from the earth is 238,850 mi. This exceeds 110 times the moon's diameter by 1030 mi.   Find the diameter of the moon.

PART II

38. **A Root** of an equation is a number which, when substituted for the unknown quantity, satisfies the equation; that is, reduces the two members of the equation to the same number.

Ex. If in the equation, $3x - 1 = 2x + 3$, we substitute 4 in the place of x in each member,

we obtain,
$$3x - 1 = 12 - 1 = 11,$$
$$2x + 3 = 8 + 3 = 11.$$

The equation is satisfied. Hence, 4 is the root of the given equation.

EXERCISE 34

Solve for x and check:

1. $x - .7 = 0.$
2. $x + 1.8 = 0.$
3. $3x - 1.5 = 0.$
4. $x - a = 0.$
5. $4x - 2.4 = 2x.$
6. $9x - 3.69 = 6x.$
7. $x - b = 2b.$
8. $x - 2 = a.$
9. $-4 = 3x + 8.$
10. $5x - .2 = 4x - .7.$
11. $9x + .5 = 5x + 8.$
12. $12x - .5 = 11x - .6.$
13. $x + b = 5 + b.$
14. $3x - a = 12 - a.$
15. $5x - (2x + .3) = 1.2.$
16. $3x = .61 + (1.3 - 'x).$
17. $2x - (x - 2.82) = 4.6.$
18. $x + a = b.$
19. $3x - (x - a) = 5a.$
20. $5a - (x - a) = -5x.$
21. $-3x + (12x - 1.74) = 3x.$
22. $5x - (10p - x) = 2p.$

EXERCISE 35

Verbal Problems

1. The sum of two numbers is c and one of the numbers is n. What is the other number?

2. If a boy is x years old now, how old was he 7 years ago? How old will he be 10 years hence? If his sister is twice as old as he is now, how old will she be 5 years hence?

3. $\frac{a}{b}$ is a fraction. Express this fraction when each term is increased by 2.

4. The product of two numbers is p. One of the numbers is n. Find the sum of the numbers.

5. How much less is b than a?

6. How many times 7 is 21? How many times a is b?

7. What fractional part of a is b?

8. Mary has x post cards. Sarah has 8 less than twice as many. How many has Sarah?

9. A given canal is x miles long. Another canal is 5 miles more than three times as long. How long is the second canal?

10. Let d be the distance traveled in h hours at the rate of m miles an hour. Then express in words $d = hm$. Also $m = \frac{d}{h}$. Also $h = \frac{d}{m}$.

Express each of the following in algebraic symbols as an equation:

11. x when diminished by 7.2 becomes equal to 15.8.

12. x exceeds 5 by 4.2.   **14.** x exceeds b by c.

13. x exceeds a by 10.   **15.** Three times x exceeds 20 by 4.

16. x diminished by the sum of $3x$ and 4 equals 28.

EXERCISE 36

1. Find the number which when diminished by 2.7 equals 3.72.

2. What number increased by 4.62 equals 16.5?

3. What number increased by twice itself and then by 3.6 equals 18.69?

4. If a certain number is multiplied by 5 and then diminished by 1.75, the result is 13.6. Find the number.

5. If four times a certain number is increased by 20.5, the result is equal to three times the number increased by 160. Find the number.

6. One number exceeds 4 times another number by .12, and the sum of the numbers is 4.4. Find the numbers.

7. One fraction exceeds 5 times another fraction by $\frac{1}{3}$, and the sum of the fractions is $1\frac{3}{9}$. Find the fractions.

8. The sum of three numbers is 50. The first is twice the second, and the third is 16 less than three times the second. Find the numbers.

9. Three girls together had 111 post cards. The second had 10 more than the first. The number which the third had exceeded twice the number which the first had by 17. How many did each of the girls have?

10. Three boys together earned $98. If the second earned $11 more than the first, and the third $28 less than the other two together, how much did each earn?

11. The record time for the 100 yd. swim at a certain date was $55\frac{2}{5}$ sec. This was $7\frac{2}{5}$ sec. more than 5 times that for the 100 yd. dash. What was the record time for the latter?

12. The length of the Suez canal is 2 miles longer than 8 times the length of the Simplon tunnel. If the canal is 100 miles long, find the length of the tunnel.

13. The temperature of the electric arc is 5400° F. This is 464° more than 8 times the temperature at which lead fuses. Find the temperature at which lead fuses.

14. The velocity of sound in the air is 1090 ft. per second. This rate is 10 ft. more than 9 times the rate at which sensation travels along a nerve. Find the rate at which sensation travels. How does this compare with the velocity of an express train going 60 miles per hour?

15. A running horse with a rider has gone 1 mi. in 1 min. $35\frac{1}{2}$ sec., which is $13\frac{1}{2}$ sec. more than three times the time in which an automobile has gone one mile. Find the latter time.

16. The distance from New York to San Francisco by way of Cape Horn is 13,800 mi. This is 1920 mi. less than three times the distance from New York to San Francisco by way of Panama. Find the latter distance.

17. Make up and work an example similar to Ex. 16, using the fact that the distance from London to Bombay by the way of the Cape of Good Hope is 11,220 mi., but by way of the Suez Canal is 6332 mi.

18. Find five consecutive numbers whose sum is 3 less than 6 times the least of the numbers.

19. In a recent year, the record for the baseball throw was 426 ft. 6 in., which was 8 ft. $2\frac{3}{4}$ in. more than 17 times the record for the running long jump. What was the latter record?

20. The sum of three numbers is 73. The second is twice the third, and the first exceeds three times the third by 1. Find the numbers.

21. In a recent year, the record time for 1 mile traveled on a bicycle was 1 min. 7 sec., which was $12\frac{1}{8}$ sec. more than twice the record for 1 mile traveled by an automobile. Find the latter record.

22. The distance from New York to Chicago is 912 miles. If this is 24 miles less than 4 times the distance from New York to Boston, what is the latter distance?

23. The Eiffel Tower is 984 ft. high. If this is 126 ft. less than twice the height of the Washington Monument, what is the height of the Washington Monument?

EXERCISE 37
REVIEW

1. Find the value of $3\,a^2 - 5\,ab + 3\,b^2 - x^3$ if $a = .4$, $b = .05$, and $x = 1.2$.

2. Add $\frac{1}{2}x - \frac{3}{4}x + \frac{1}{8}$, $\frac{1}{4}x^2 + \frac{1}{8}x - \frac{1}{2}$, and $\frac{3}{4}x^2 - \frac{1}{2}x + \frac{3}{8}$.

3. Subtract $\frac{1}{8}x^2 - \frac{1}{2}x + \frac{3}{8}$ from $\frac{1}{2}x^2 - \frac{3}{4}x - \frac{3}{8}$.

4. Add $.5\,a^2 - .15\,a + 2.5$, $1.2\,a^2 + .3\,a - 1.5$, and $-.75\,a^2 + .3\,a - .7$.

5. Subtract $.27\,a^2 - .12\,a - 2.3$ from $1.5\,a^2 + 2\,a - 1.7$.

6. Add $2(x+y) - 3(x+z) + 2(y+z)$, $4(x+z) - 3(x+y) - 5(y+z)$, and $4(x+y) - (x+z) + 4(y+z)$.

7. From the sum of $a^2 - 7\,ab + 3\,b^2$ and $2\,a^2 - 6\,b^2 + 7\,a^2b^2$ take the sum of $4\,a^2b^2 - 3\,a^3 + 2\,a^2 - b^2$ and $3\,ab - 2\,b^2 + a^2$.

8. What must be added to $x^2 - x + 1$ that the sum may be x^3? That the sum may be $3\,x$? 15? 0?

9. What must be subtracted from $2\,x^2 - 3\,x + 1$ that the remainder may be x^3? May be $x^2 + 10$? 7? $a - x + 1$?

10. Four times a certain number diminished by 12.07 equals twice the number increased by 1.13. Find the number.

11. Separate 1000 into three parts such that the second part is three times as large as the first part, and the third part exceeds the first part by 100.

12. Simplify $4.7 + 3\,x - [8.24 - (x - .27)]$.

13. Inclose in a parenthesis preceded by a minus sign the last three terms of the expression $4\,x^2 - a^2 + 2\,ab - b^2$.

14. Solve the equation $3\,x - (x - 7.2) = 4.8$.

15. Solve for p, $.78 - 2\,p = 5\,p + .01$.

16. In a certain year the greatest mountain height climbed by man was 24,853 ft., which was 3441 ft. less than twice the height of Pike's Peak. Find the height of Pike's Peak.

17. If 5 mi. exceeds 8 kilometers by 153 ft. 4 in., how many yards are there in a kilometer?

18. In estimating freight space it is customary to allow 40 cu. ft. for one ton of general merchandise. Using this estimate, find a formula for the number of tons (T) contained in a box car l feet long, w feet wide, and h feet high.

19. By use of this formula determine how many tons of general freight will be held by a box car 32 ft. long, $8\frac{1}{2}$ ft. wide, and 8 ft. high.

20. The following table shows the number of years which a person having attained a certain age may expect to live. Construct a graph of life expectancy from the data.

Age in Years	0	2	4	6	8	10	20	30
Life Expectancy in Years	38.7	47.6	50.8	51.2	50.2	48.8	41.5	34.3

Age in Years	40	50	60	70	80	90	100
Life Expectancy in Years . .	27.6	21.1	14.3	9.2	5.2	3.2	2.3

From this graph determine your life expectancy at the present time, and also that of several acquaintances of various ages.

21. Express the following in as few terms as possible:
$$3 x^3 - .5 x^2y + .6 y^3 - 7 xy^2 - y^2 - .5 x^3 - 4 xy^2 - 3 x^2y + 5 z^2$$
$$+ 8 x^2y + .06 x^2 - 9 y^2.$$

22. If $x = 5$, fill in the second member for each of the following: (1) $3 x - 2 = ?$ (2) $7 x - 2 + 3 x = ?$

23. Draw on one diagram the graphs for the average heights of a boy and girl at different ages (see Ex. 31, p. 69).

If $A = 4 x^3 - 2 x^2y + 3 xy^2 + y^3,$ $C = 3 x^3 - x^2y + 2 y^3,$
   $B = 4 x^3 - x^2y - xy^2 - 3 y^3,$ $D = x^3 - 2 xy^2 + y^3,$
find the value of

24. $A - B + C - D.$ **26.** $A - (B + C) + D.$

25. $A - [B - (D + C)].$ **27.** $B + \{A - [C - D]\}.$

28. Allowing 22 bricks for one cubic foot of wall, find a formula for the number of bricks that will be required in building a wall l feet long, h feet high, and w inches thick.

29. By use of this formula, determine the number of bricks that will be needed in building a wall 120 yd. long, 9 in thick, and 6 ft. high.

CHAPTER V

MULTIPLICATION; DIVISION

PART I

39. Multiplication of Coefficients. To multiply $4\,a$ by $3\,b$, we evidently take the product of all the factors of the multiplier and the multiplicand, and thus get $4 \times a \times 3 \times b$. Rearranging factors, we obtain as the product,

$$4 \times 3 \times a \times b \text{ or } 12\,ab.$$

Hence, in multiplying two terms,

Multiply the coefficients to produce the coefficient of the product.

40. Multiplication of Literal Factors, or Law of Exponents.

Ex.   Multiply a^3 by a^2.

$$\text{Since } a^3 = a \times a \times a,$$
$$\text{and } a^2 = a \times a,$$
$$\therefore a^3 \times a^2 = a \times a \times a \times a \times a = a^5.$$

This may be expressed in the form

$$a^3 \times a^2 = a^{3+2} = a^5,$$

or, in general, $a^m \times a^n = a^{m+n}$,

where m and n are positive whole numbers.

Hence, in multiplying one term by another,

Add the exponents of each letter that occurs in both multiplier and multiplicand.

Ex. $4\,a^2bc^3 \times 3\,a^3b^2x = 12\,a^5b^3c^3x.$

41. Law of Signs.

(1) Place four -3's in a column as shown
at the right, and add them. The result
is -12.

$$\begin{array}{r} -3 \\ -3 \\ -3 \\ -3 \\ \hline -12 \end{array}$$

Hence $4 \times (-3) = -12$.

(2) Similarly place three -5's in a column and add.
The result will be -15.

Hence $(-5) \times 3 = -15$.

(3) By adding it is also easily shown that $4 \times 5 = 20$.

(4) To determine the sign of the product of two negative numbers, we proceed thus:

Ex. Multiply $9-5$ by $6-3$ term by term.

$$\begin{array}{l} 9-5 \\ 6-3 \\ \hline 54-30 \\ -27\,(?)\,15 \\ \hline 54-57+15 \end{array}$$

Since $9-5$ equals 4 and $6-3$ equals 3, the
value of $(9-5)(6-3)$ must be the same as
that of 4×3 or 12. Hence in the multiplication of $9-5$ by $6-3$ term by term, the product of -5 and -3 must be $+15$ in order
that the net result of the whole process shall
be 12.

Summing up results, we see from (3) and (4) that

both $+ \times +$, and $- \times -$, give $+$,

and from (1) and (2), that

both $- \times +$, and $+ \times -$, give $-$.

In brief, in multiplication,

Two like signs give plus; two unlike signs give minus.

MULTIPLICATION BY A SINGLE TERM

42. Examples in Multiplication.

Ex. 1. Multiply $4\,a^2x^3$ by $5\,ax^4$.
The product is $20\,a^3x^7$.

Ex. 2. Multiply $-3\,a^2bx$ by $5\,ac^3y$.
The product is $-15\,a^3bc^3xy$.

EXERCISE 38

At sight give the product of:

ORAL

1. 3
2

2. 3 doz.
2

8. $3 a^2 b$
2

4. $4 a^2 b$
$5 b$

5. $5 x^2 y^3$
$4 xy^2$

6. $- 5$ dimes
2

7. $- 5 ab$
2

8. $- 6$ doz.
4

9. $- 6 a^2 b$
4

10. $- 5 a^2 b^5$
$2 a^3 b^2$

11. 4 dimes
$- 3$

12. $4 ab$
$- 3$

13. 6 doz.
$- 5$

14. $8 ab^3$
$- 5 ab$

15. $9 a^2 x^3$
$- 3 ax$

16. $- 5$ dimes
$- 3$

.17. $- 5 ab$
$- 3$

18. $- 7$ doz.
$- 4$

19. $- 8 a^2 x^5$
$- 4 ax^3$

20. $- 10 a^2 b^3$
$- \frac{1}{2} ab$

21. $- 7 ab$
$3 a^2 b^3$

22. $5 abx$
$- 4$

23. $- 7 x$
$- 4 x$

24. $- 2 xy^2$
$3 xy^2$

25. $- 15 x^3 y^3$
$- \frac{1}{3} x$

26. $- 7 xy$
$- 1$

27. $3 xy$
$- 1$

28. $- \frac{1}{2} ab^3$
$- 2 ax$

29. $- 2$
$- 5 xy$

30. $- \frac{3}{4}$
$8 x^3$

EXERCISE 39

Multiply:

1. -15
14

2. $13 a$
-20

3. $23 ab$
-5

4. $30 x^2 y^2$
-12

5. $.4 x$
$-2 x$

6. $-.5 x$
$-3 x$

7. $13 ax^3$
$-4 ax$

8. $-6 xy^2$
$-7 x^2 y^2$

9. $-7\,a^2x$ $-3\,ay^2$	**13.** $12\,ab$ $5\,ac$	**17.** ab xy	**21.** $.5\,x$ $.03\,x$
10. $-5\,a^2b$ $-4\,cd^2$	**14.** $23\,ax^2$ $-15\,c$	**18.** $-13\,ax^2$ $-5\,y^2$	**22.** $2.1\,y^3$ $.5\,y^2$
11. $-6\,c^2d$ $3\,cd^2$	**15.** ab x	**19.** $.4\,x^2$ $.2\,x^3$	**23.** $2\frac{1}{2}\,x^3$ $\frac{3}{4}\,x$
12. $-2\,x^2yz$ $-8\,xy^2z^3$	**16.** $3\,abc$ $-2\,cd$	**20.** $\frac{1}{2}\,ax^3$ $\frac{2}{3}\,a^2x^3$	**24.** $\frac{3}{4}\,x^2$ $.5\,x$

Find the value of following indicated products:

25. $ab \cdot ax \cdot ay$.

26. $-3\,ax \cdot 5\,ay \cdot 2\,xy$.

27. $(3\,ab^2)(-2\,ab^3)(3\,bc)$.

28. $(ab)(2\,ax)(-3\,b^2y)(.4\,xy)$.

29. In $2^3 \cdot 2^4$ how many two's are to be multiplied together?

30. In $x^4 \cdot x^5$ how many x's are to be multiplied together?

31. In $a^2x^3 \cdot a^3x^4$ how many a's are to be multiplied together? How many x's?

32. Express $7\,aaaxx \cdot 5\,aaaxxxx$ in the briefest form.

33. What number multiplied by itself makes 25? Makes 36? Makes $36\,a^2$?

34. What is the square root of 25? Of $36\,a^2$? Of $4\,x^2$? Of $64\,a^2b^2$? Of $25\,a^2b^4$?

35. Find the value of $\sqrt{9}$. Also of $\sqrt{16\,a^2}$. Of $\sqrt{121\,x^4}$. Of $\sqrt{36\,a^2b^2}$.

36. How much money do five empty pocketbooks contain? $5 \times 0 = ?$

37. Find the value of 7 times 0. Of $5\,a \times 0$. Of $0 \times 6\,x^2y^3$. Of $3(x + y) \times 0$.

If $a = 4$, $b = \frac{3}{2}$, $c = 0$, $x = 1$, and $y = 9$, find the value of

38. abc.

39. a^2c.

40. $5\,cxy^2$.

41. $a^2 + 3\,cy$.

42. $4\,c^3 + a$.

43. $(3\,c + x)^3$.

43. Example.   Multiply $2\,a^3 + 3\,a^2b + 5\,b^2c$ by $-3\,a^2b$.
Multiply each term of the first expression separately by $-3\,a^2b$.

$$
\begin{array}{llll}
2\,a^3 + 3\,a^2b\ + 5\,b^2c & = & 16 + 24 + 20 = & 60 \\
-3\,a^2b & = & & = - \quad 24 \\
\hline
-6\,a^5b - 9\,a^4b^2 - 15\,a^2b^3c & = -384 - 576 - 480 & = -1440
\end{array}
$$

The check is obtained by letting $a = 2$, $b = 2$, $c = 1$.

EXERCISE 40

Multiply and check :

1. $2\,a + 3\,x$
    $3\,ax$

2. $3\,x - 2\,y$
    $-5\,xy$

3. $4\,x^2y - xy^2$
    $2\,xy$

4. $7\,ax + 4\,by$
    $-3\,abxy$

5. $3\,x + 5\,y$
    -4

6. $7\,a - 3\,b$
    -5

7. $6\,m + 5\,n$
    $-2\,x$

8. $-7\,ab^2 + 3\,ax^2$
    $5\,by$

9. $2.5\,x^2 - 3.7\,x + .51$
    $.4\,x$

10. $\frac{3}{4}\,x - \frac{1}{2}\,x^2 - \frac{4}{5}$
    $\frac{1}{2}\,x$

11. $\frac{2}{5}\,ax^2 - \frac{1}{5}\,ax - \frac{3}{5}\,a$
    $-\frac{1}{3}\,ax$

12. $.4\,l - .5\,m - 2\,n$
    -1.2

Multiply :

13. $8\,ac^2 - 3\,m^2n$ by $5\,an$.
14. $m - m^2 - 3\,m^3$ by $-7\,m^2n$.
15. $8\,x^2y - 5\,xy^2 - y^3$ by $3\,xy$.
16. $.4\,a^2 - 1.2\,ab - .3\,b^2$ by $.5\,ab$.

17. By setting down 3 ft. + 2 in. 5 times and adding, show that 5(3 ft. + 2 in.) equals 15 ft. + 10 in.

18. By an addition show that 4(2 sq. ft. + 9 sq. in.) = 8 sq. ft. + 36 sq. in.

Perform the following indicated multiplications:

19. $3\,ab(2\,a^2 - 3\,b^2)$.
20. $(a + b + c)a^2b^2$.
21. $3\,ax(3\,a^2 - 5\,ax + 2\,x^2)$.
22. $3(x - a)$.
23. $3\,xy(x - y)$.
24. $\frac{2}{3}(3\,x + 5)$.

25. Write $(a+2b)+(a+2b)+(a+2b)+(a+2b)$ as a product of $(a+2b)$ by a numerical coefficient.

26. Reduce $(7\,aaabb - 5\,aaabb) \times 6\,aabb$ to its simplest form. Compare the size of the result with that of the original expression.

MULTIPLICATION BY A TWO-TERMED EXPRESSION

44. Examples in Multiplication.

Ex. 1. Multiply $3x+5$ by $2x+7$.

We proceed by multiplying each term of the first expression by each term of the second, thus:

$$3x+5 \qquad = 6+5 \qquad = 11$$
$$2x+7 \qquad = 4+7 \qquad = 11$$
$$\overline{6x^2 + 10x}$$
$$+21x+35$$
$$\overline{6x^2+31x+35} = 24+62+35 = 121$$

Ex. 2. Multiply $3x^2 - 4xy - y^2$ by $2x - 5y$.

$$3x^2 - 4xy - y^2 \qquad\qquad = -2$$
$$2x - 5y \qquad\qquad = -3$$
$$\overline{6x^3 - 8x^2y - 2xy^2}$$
$$-15x^2y + 20xy^2 + 5y^3$$
$$\overline{6x^3 - 23x^2y + 18xy^2 + 5y^3} = \quad 6$$

The *check* is obtained by letting $x=1$ and $y=1$ (or $x=y=1$). Note that this method checks only the signs and coefficients, not the letters or their exponents. Mistakes in letters and exponents, however, are rare in comparison with mistakes in signs and coefficients. A convenient check for both the letters and their exponents as well is obtained by letting $x=y=2$.

Multiply term by term and check:

1. $7-2$ by $5-4$.
2. $8-3$ by $6-2$.
3. $15-8$ by $9-4$.
4. $x-2$ by $x-4$.
5. $y-3$ by $y-2$.
6. $3x-5$ by $x-4$.

7. $8\,a - 5$ by $2\,a - 7$.

8. $a - b$ by $2\,a - b$.

9. $3\,x - 5\,y$ by $4\,x - 3\,y$.

10. $x + 2$ by $x + 5$.

11. $3\,x + 5$ by $2\,x + 3$.

12. $3\,x + y$ by $x + 3\,y$.

13. $a^2 + 3\,b^2$ by $3\,a^2 + 2\,b^2$.

14. $x + 5$ by $x - 7$.

15. $2\,x + 5$ by $x - 4$.

16. $7\,x + 4\,y$ by $3\,x - 5\,y$.

17. $x - 9$ by $x + 5$.

18. $3\,x - 5$ by $2\,x + 1$.

19. $3\,x^2 - 5\,y^2$ by $4\,x^2 + 3\,y^2$.

20. $3\,x - 2\,y$ by $2\,x - 5\,y$.

21. $5\,x^2 + 6\,y^2$ by $2\,x^2 + 9\,y^2$.

22. $3\,m - 4\,p^2$ by $4\,m - 7\,p^2$.

23. $8\,a - 9\,bc$ by $5\,a + 11\,bc$.

24. $5\,xy + 6$ by $6\,xy - 7$.

25. $3\,x^2 + 2\,x + 4$ by $2\,x + 5$.

26. $5\,x^2 + 2\,x + 3$ by $7\,x + 2$.

27. $3\,a^2 + 2\,ab + b^2$ by $3\,a + 5\,b$.

28. $4\,x^2 - 3\,x + 5$ by $2\,x + 1$.

29. $3\,x^2 + 7\,x + 1$ by $2\,x - 5$.

30. $7\,x^2 - 3\,xy + y^2$ by $3\,x + 2\,y$.

31. $x^2 + xy + y^2$ by $x - y$.

32. $3\,x^2 - 5\,x + 4$ by $2\,x - 5$.

33. $3\,a^2 - 5\,ab - 6\,b^2$ by $4\,a - 5\,b$.

34. $\frac{1}{2}\,a + \frac{1}{2}\,b$ by $\frac{1}{2}\,a - \frac{1}{2}\,b$.

35. $\frac{2}{3}\,x^2 - 4\,x + \frac{4}{9}$ by $\frac{3}{4}\,x + \frac{9}{2}$.

36. $.5\,a - .4\,b$ by $.2\,a - .3\,b$.

37. $1.8\,x^2 - 3.2\,x + .48$ by $2.5\,x + .5$.

38. $x^3 + 2\,x^2 + x + 3$ by $x + 2$.

39. $a^3 + 3\,a^2b + 5\,ab^2 + 4\,b^3$ by $3\,a + 2\,b$.

40. $4\,x^3 - 3\,x^2 + 2\,x - 1$ by $2\,x + 1$.

41. $5\,m^3 - 4\,m^2p - 2\,mp^2 - 6\,p^3$ by $3\,m - 4\,p$.

42. $4\,r^3 - 8\,r^2s - 3\,rs^2 + 8\,s^3$ by $2\,r - 7\,s$.

43. $x^3 + x^2y + xy^2 + y^3$ by $x - y$.

44. $a^3 - a^2b + ab^2 - b^3$ by $a + b$.

Perform the following indicated multiplications:

45. $(3\,x - 5)(2\,x + 7)(x - 5)$.

46. $(3\,x^3 + 8)(2\,x^3 + 5)(x^3 - 2)$.

47. $(4\,a^2 - b^2)(a^2 - 3\,b^2)(a^2 + b^2)$.

48. $(3\,a^2 - 5\,a - 2)(2\,a + 3)$.

49. $(3\,x - 4)(3\,x - 4)(2\,x - 7)$.

50. What is the shortest way of indicating the multiplication of $x + 3$ by itself?

51. Express in the shortest way the following product:
$$(3x + 2)(3x + 2)(3x + 2).$$

DIVISION

45. Coefficients, Exponents, and Signs in Division. In arithmetic, from the fact that $2 \times 3 = 6$, it follows that $\frac{6}{2} = 3$, and also $\frac{6}{3} = 2$.

So in algebra, since $(-2a^2)(-3a^3) = 6a^5$, it follows that $\frac{6a^5}{-2a^2} = -3a^3$.

Similarly, since $(4a^2)(-5a^3) = -20a^5$, it follows that $\frac{-20a^5}{4a^2} = -5a^3$; and also that $\frac{-20a^5}{-5a^3} = 4a^2$.

Hence in division the law of signs is the same as the law of signs in multiplication, viz.: like signs give plus, and unlike signs give minus.

DIVISION BY A SINGLE TERM

46. Rule. Hence in general to divide one term by another term,

Divide the coefficient of the dividend by the coefficient of the divisor;

Obtain the exponent of each literal factor in the quotient by subtracting the exponent of each letter in the divisor from the exponent of the same letter in the dividend;

Determine the sign of the result by the rule that like signs give plus, and unlike signs give minus.

Ex.   Divide $27 a^3b^4x^3$ by $-9 a^2bx^3$.

Since the factor x^3 in the divisor cancels x^3 in the dividend,

$$\frac{+27 a^3b^4x^3}{-9 a^2bx^3} = -3 ab^3. \quad \textit{Ans.}$$

EXERCISE 42

ORAL

Give the quotient for each of the following:

1. $3 \text{ tens})\overline{15 \text{ tens}}$.
2. $3x)\overline{15x}$.
3. $4)\overline{12 \text{ dozen}}$.
4. $4)\overline{12 a^2b}$.
5. $6x^2y)\overline{18x^4y}$.
6. $3x^3)\overline{18x^5y^3}$.

7. $3a^3x^5)\overline{15a^5x^5}$.
8. $-3)\overline{15a^2b}$.
9. $-5x)\overline{15x^2y}$.
10. $-6ax^3)\overline{12a^2x^3}$.
11. $7)\overline{-21xy}$.
12. $9x)\overline{-36x^4}$.

13. $3)\overline{-15a^3b}$.
14. $-3)\overline{-15x}$.
15. $-7r)\overline{-21r^2}$.
16. $-8a^3)\overline{-8a^5}$.
17. $-9x)\overline{18x^3}$.
18. $9x)\overline{-18x^3}$.

19. Divide $-18x^3$ by $-9x$.
20. Divide $8a^3b^2$ by $4a^2b^2$.
21. Divide p^3q^2 by $-pq^2$.
22. Divide $-12a^2b^2$ by $-6ab$.

EXERCISE 43

Divide:

1. $15a$ by $-5a$.
2. $-3x^3$ by x.
3. $8a^2x^2$ by $-4ax^2$.
4. $-30x^3y^2$ by $-6x^2y$.
5. $-7xz^3$ by $7z^3$.
6. $21x^2yz$ by $-3xz$.
7. $18bc^3d^3$ by $-9c^3d$.
8. $-33x^5y^6z^7$ by $11xy^3z^5$.
9. $28x^2y^2z^3$ by $-14xy^2z^3$.

10. $-m^3n$ by $-m^3$.
11. $-3x^2$ by -1.
12. $-8ax$ by $\frac{1}{2}x$.
13. $16by^2$ by $-\frac{2}{3}by$.
14. $8mx$ by $.2x$.
15. $.4ax^2$ by $.8x^2$.
16. $.04ax$ by $.5ax$.
17. $2\frac{1}{2}x^3$ by $\frac{3}{4}x^2$.
18. $-\frac{1}{4}x^2$ by $.5x$.

19. $4\pi r^2$ by 2π. By r^2. By πr.
20. $\frac{1}{2}gt^2$ by gt. By $\frac{1}{4}g$. By $\frac{1}{2}t$.
21. $\frac{1}{2}mv^2$ by $\frac{2}{3}m$. By $.5v^2$. By $.25v$.

22. How many 2's are multiplied together in 2^{10}? In 2^4? In the quotient of $2^{10} \div 2^4$?

23. How many x's in x^{10}? In x^4? In the quotient of $x^{10} \div x^4$?

24. If an empty box is divided by partitions into 5 equal parts, will each compartment of the box be empty?

25. What is the value of $0 \div 5$? Of $0 \div 7$? State the meaning of the latter in a manner similar to that used in Ex. 24.

26. Give the value of $0 \div 10$. Of $0 \div a$. Of $0 \div 2x$. Of $\dfrac{0}{7\,a}$, $\dfrac{0}{7\,ab}$, $\dfrac{0}{7\,abx}$, $\dfrac{0}{7\,a^2b^2x^3}$.

27. What is the value of $\dfrac{3\,ax}{7\,y}$ when $a = 0$? When $x = 0$?

If $a = 2$, $b = 3$, $c = 0$, $x = 1$, find the value of each of the following:

28. $\dfrac{bc}{a}$.

29. $\dfrac{c(a+x)}{b}$.

30. $\dfrac{c(2\,b - x)}{4\,a}$.

31. $\dfrac{5\,ac}{b+x}$.

32. $\dfrac{3\,a + 5\,c}{7\,x + b}$.

33. $\dfrac{ax - 3\,c + bc}{a + b + c}$.

47. Example. Divide $12\,a^3x - 10\,a^2y + 6\,a^4z^2$ by $2\,a^2$.

$$2\,a^2 \overline{)12\,a^3x - 10\,a^2y + 6\,a^4z^2}$$
$$6\,ax - \quad 5\,y + 3\,a^2z^2. \quad \textit{Quotient.}$$

EXERCISE 44

Divide:

1. $x^3 - 3\,x^2$ by $-x$.

2. $20\,x^2 - 8\,xy$ by $4\,x$.

3. $4\,ab^2 - 6\,a^2bc$ by $-2\,ab$.

4. $-6\,x^3 + 7\,x^2 - x$ by $-x$.

5. $15\,x^3y - 10\,x^2y^2 - 5\,xy^3$ by $5\,xy$.

6. $-m - m^2 + m^3 - m^4$ by $-m$.

7. $14\,x^3y^2z - 21\,xy^2z^3 + xyz$ by $-xyz$.

8. $-3\,x^2 - 2\,x + 5$ by -1.

9. $.6\,x^2 - .12\,x + 9$ by $-.3$.

10. $.02\,a^2 - .04\,ab - .8\,b^2$ by $.5$.

11. $\frac{1}{2}x^2 - \frac{2}{3}x - \frac{5}{2}$ by $-\frac{2}{3}$.

12. $\frac{2}{3}a^4b^2 - \frac{1}{6}a^3b^3 - \frac{1}{3}a^2b^4$ by $\frac{2}{3}ab$.

Perform the following indicated divisions:

13. $(6 x^3 + 12 x^2 + 3 x) \div 3 x.$

14. $(12 a^3 - 14 a^2 b) \div (- 2 a^2).$

15. $(.4 a^3 b^3 + .8 a^2 b^2 + 1.8 ab) \div 2 a.$

16. $(.9 x^2 y - .6 xy^2 - 1.2 xy) \div (.3 xy).$

17. $3 a^2) \overline{9 a^3 - 6 a^2 b - 12 a^2}.$

18. $- 4 xy) \overline{8 x^4 y^2 - 12 x^3 y^3 - 16 x^3 y}.$

19. $.6 x^3) \overline{.6 x^3 - 1.2 x^4 y^4 - 1.8 x^6}.$

20. $- 3 pq) \overline{6 p^3 q - 1.2 p^2 q^2 - 3 pq^3}.$

DIVISION BY TWO-TERMED EXPRESSIONS

48. Examples in Division.

Ex. 1.   Divide $6 x^3 + 17 x^2 + 16 x + 6$ by $2 x + 3.$

$$
\begin{array}{r|l}
6 x^3 + 17 x^2 + 16 x + 6 & \underline{2 x + 3} \qquad = 45 \div 5 \\
\underline{6 x^3 + 9 x^2} & 3 x^2 + 4 x + 2 = 9 \\
\end{array}
$$

$$
\begin{array}{r}
8 x^2 + 16 x \\
8 x^2 + 12 x \\
\hline
4 x + 6 \\
4 x + 6 \\
\hline
\end{array}
$$

In the above process we divide the first term of the dividend, $6 x^3$, by the first term of the divisor, $2 x$, and thus obtain the first term of the quotient, $3 x^2$.   We then multiply the entire divisor by $3 x^2$, subtract the result from the dividend, and so on.

Ex. 2.   Divide $12 a^3 - 25 a^2 b - 3 ab^2 + 20 b^3$ by $3 a - 4 b$

$$
\begin{array}{r|l}
12 a^3 - 25 a^2 b - 3 ab^2 + 20 b^3 & \underline{3 a - 4 b} \qquad = 4 \div (-1) \\
\underline{12 a^3 - 16 a^2 b} & 4 a^2 - 3 ab - 5 b^2 = -4 \\
\end{array}
$$

$$
\begin{array}{r}
- 9 a^2 b - 3 ab^2 \\
- 9 a^2 b + 12 ab^2 \\
\hline
- 15 ab^2 + 20 b^3 \\
- 15 ab^2 + 20 b^3 \\
\hline
\end{array}
$$

Divide and check :

1. $x^2 + 7x + 10$ by $x + 2$.
2. $x^2 + 9x + 20$ by $x + 5$.
3. $x^2 + 6xy + 8y^2$ by $x + 2y$.
4. $a^2 + 7ab + 12b^2$ by $a + 3b$.
5. $2x^2 + 3x + 1$ by $x + 1$.
6. $3x^2 + 10xy + 8y^2$ by $3x + 4y$.
7. $3p^2 + 10pq + 3q^2$ by $3p + q$.
8. $x^2 - 3x - 10$ by $x + 2$.
9. $x^2 + 2xy - 8y^2$ by $x + 4y$.
10. $a^2 - a - 12$ by $a - 4$.
11. $12x^2 - 5x - 2$ by $3x - 2$.
12. $2x^2 + xy - 10y^2$ by $2x + 5y$.
13. $3a^2 + 4ab - 4b^2$ by $3a - 2b$.
14. $32m^2 + 4mn - 45n^2$ by $4m + 5n$.
15. $6x^4 - 7x^2 - 5$ by $3x^2 - 5$.
16. $x^6 - 5x^3 + 6$ by $x^3 - 2$.
17. $a^2 - 13ab + 12b^2$ by $a - b$.
18. $14x^8 - 43x^4 + 20$ by $2x^4 - 5$.
19. $6x^3 + 13x^2 + 8x + 3$ by $2x + 3$.
20. $12x^3 + 19x^2 + 11x + 2$ by $3x + 1$.
21. $15a^3 + 22a^2b + 17ab^2 + 6b^3$ by $3a + 2b$.
22. $7p^3 + 39p^2 + 17p - 15$ by $p + 5$.
23. $6x^3 - 17x^2 + 2x + 15$ by $2x - 3$.
24. $10m^3 - 37m^2n + 15mn^2 - 28n^3$ by $2m - 7n$.
25. $2x^4 + 5x^3 - 8x^2 + 11x - 20$ by $x + 4$.

EXERCISE 46

VERBAL PROBLEMS

1. If a house rents for a dollars a month, what is the annual rental?

2. Write an expression which represents the product of any two numbers (as of a and b).

3. Write an expression which represents the quotient of any two numbers.

4. If a man is now x years old, how many years ago was he half as old?

5. How many feet are there in a yards, b feet, c inches?

6. One number is a, and another number is b. Express one ninth of the first plus two thirds of the second.

7. Express the sum of 4 dollars, d dimes, and c cents as dollars. Also as cents.

8. A pipe can fill a cistern containing g gallons in m minutes. How many gallons per minute flow in?

9. Give three consecutive numbers, the largest of which is $3x$.

10. A boy walks c miles per hour for h hours. How far does he walk?

11. If post cards are sold 2 for a cents, what will n cards cost?

12. If the perimeter of a square is i inches, what is the length of one side?

13. A rectangular lot is m rods long and n rods wide. Find the perimeter of the lot.

14. Express the sum of a yards, b feet, and i inches in inches. In feet. In yards.

15. How many cubic feet in a box a feet long, b feet wide, and c feet deep?

16. How many square feet in s square yards?

17. How many cubic yards in c cubic feet?

EXERCISE 47

1. Find the number of square yards in the area of a floor 22 ft. long and 15 ft. wide.

2. If a rectangular floor is l feet long and w feet wide, and the number of square yards in the area of the floor be denoted by y, find the formula for y in terms of l and w.

3. By use of the formula for y obtained in Ex. 2 find the number of square yards in a boy's garden which is 120 ft. long and 45 ft. wide.

4. Also find the number of square yards in a floor 24.3 ft. long and 18.6 ft. wide.

5. Also find what will be the width in feet of a floor that is to contain 48 sq. yd. and be 24 ft. long.

6. If 32 sq. ft. of wall require one roll of paper, how many rolls will be required in papering the walls (not the ceiling) of a room 20 ft. long, 14 ft. wide, and 9 ft. high?

7. If a room is l feet long, w feet wide, and h feet high, find a formula for the number of rolls (R) of wall paper the walls of the room will require if 32 sq. ft. require one roll.

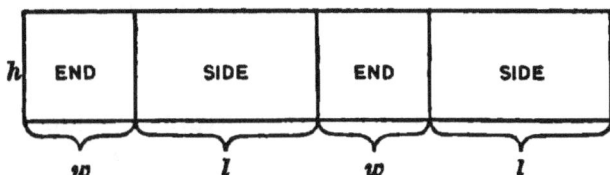

SUG. The ends and sides of the room together form a rectangle whose length is $2\,l + 2\,w$ and whose width is h.

Hence, we obtain $R = \dfrac{h(l + w)}{16}.$

8. By use of the formula for R found in Ex. 7, compute the number of rolls of wall paper required for a room 24 ft. long, 16 ft. wide, and 9 ft. high.

9. Also for a room 20 ft. long, 15 ft. wide, and 10.4 ft. high.

10. Also for a room 22 ft. 6 in. long, 14 ft. wide, and 8 ft. 8 in. high.

11. Find the number of cubic yards in a cellar 20 ft. long, 18 ft. wide, and 7 ft. deep.

12. Find a formula for the number of cubic yards (Y) in a cellar l feet long, w feet wide, and h feet deep.

13. By use of the formula for Y obtained in Ex. 12, find the number of cubic yards in a cellar 24 ft. long, 18 ft. wide, and 6 ft. deep.

14. Also find the number of cartloads of dirt (one cart-load $= 1$ cu. yd.) that must be excavated in digging a cellar 20 ft. long, 15 ft. wide, and $4\frac{1}{2}$ ft. deep.

15. If a rectangular tank is to contain 40 cu. yd., and be 24 ft. long, and 6 ft. wide, how many feet deep must it be? Use the formula for Y found in Ex. 12.

16. Find a formula for the number of gallons (G) held by a tank a feet long, b feet wide, and c feet deep, if 1 cu. ft. equals $7\frac{1}{2}$ gal.

17. Make up and solve two numerical problems by use of the formula obtained in Ex. 16.

49. Broken-Line Graphs. The graphs which we have studied thus far have been curved lines. Another important kind of graph much used in business is the broken line; that is, a line composed of a set of small straight lines pieced together.

For instance, when the income (or outlay) for a given period of time, as a month, is summed up as a single number, and a series of these numbers is obtained, the graph used in representing them is a broken line.

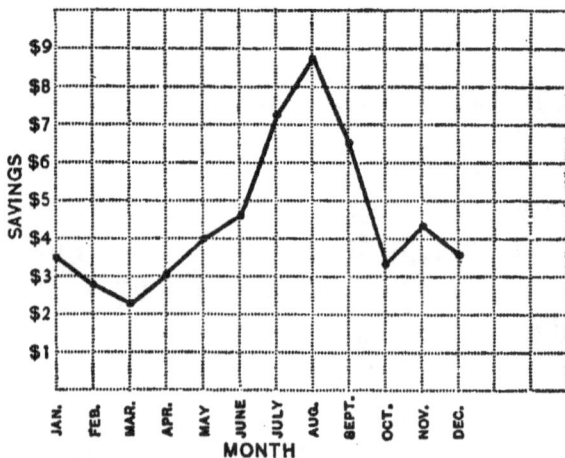

Ex. During a certain year a boy saved and deposited in the savings bank at the end of each month, the following sums: Jan., $3.44; Feb., $2.80; Mar., $2.40; Apr., $3.05; May, $4.00; June, $4.60; July, $7.20; Aug., $8.70; Sept., $6.50; Oct., $3.20; Nov., $4.12; Dec., $3.60. Make a graph of his savings during the year.

During the summer the boy raised vegetables in a garden, and sold the same. How is this fact shown in the graph?

EXERCISE 48

1. During a certain year a boy saved and deposited in a savings bank at the end of each month, sums as follows: Jan., $4.20; Feb., $3.60; Mar., $2.10; Apr., $1.20; May, $2.15; June, $3.20; July, $9.20; Aug., $8.75; Sept., $3.15; Oct., $2.10; Nov., $1.60; Dec., $3.60. Make a graph of his savings during the year.

During the winter he earned money by shoveling snow and tending a furnace; during the summer vacation by working on a farm. How are these facts shown on the graph?

2. During a certain year a girl saved money during the various months as follows: Jan., $2.10; Feb., $1.60; Mar., $2.40; Apr., $1.60; May, $1.70; June, $2.75; July, $6.10; Aug., $5.80; Sept., $4.20; Oct., $8.20; Nov., $3.20; Dec., $1.25. Make a graph of these savings.

During the summer she earned money by canning tomatoes raised by herself. Many of the tomatoes canned during August and September were sold during October. How are these facts shown on the graph?

3. During the different months of a year a business firm spent money in advertising as follows: Jan., $350; Feb., $260; Mar., $200; Apr., $275; May, $310; June, $180; July, $175; Aug., $150; Sept., $300; Oct., $400; Nov., $600; Dec., $830. Make a graph of these expenditures.

On September 1, the firm employed a special advertising agent and organized a system of aggressive advertising. How is this fact shown on the graph?

4. A boy's grades in mathematics for the eight half years of his high school course were as follows: 72, 70, 76, 82, 80, 87, 88, 92.

Make a graph of these grades (use axes marked as shown to the right). What does the graph show as to his improvement in this branch of study?

5. A girl's grades in English for the eight half years of her high school course were as follows: 86, 84, 74, 77, 72, 85, 88, 94. What can be learned from this graph?

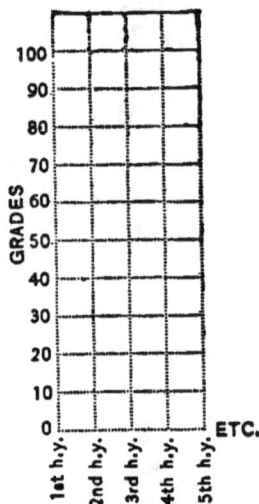

6. At a certain boys' boarding school, the number of cases of illness reported each day for two weeks were as follows: Sun., 3; Mon., 2; Tues., 12; Wed., 8; Thurs., 4; Fri., 5; Sat., 2; Sun., 4; Mon., 3; Tues., 13; Wed., 7; Thurs., 3; Fri., 1; Sat., 1. Make a graph of these facts (use axes marked as at the right).

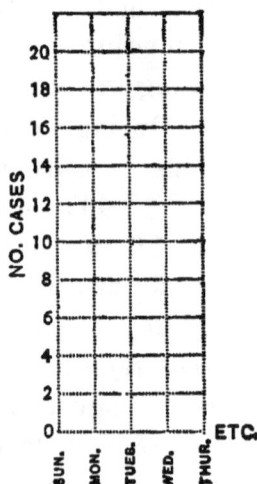

The boys received their allowances of spending money on Monday. What inference can be made by the aid of the graph?

PART II

50. Multiplication, for present purposes, may be regarded as the process of finding the result (called the *product*) of taking one quantity (the *multiplicand*) as many times as there are units in another quantity (the *multiplier*).

51. Examples in Multiplication.

Ex. 1.   In each of the expressions $5x^2 - 3x + 1 + x^3$ and $2 + 4x^2 - 3x$ arrange the terms in descending order of powers of x and then multiply the two expressions.

$$
\begin{array}{ll}
x^3 + 5x^2 - 3x + 1 & = 4 \\
4x^2 - 3x + 2 & = 3 \\
\hline
4x^5 + 20x^4 - 12x^3 + 4x^2 & \\
\quad\ - 3x^4 - 15x^3 + 9x^2 - 3x & \\
\quad\qquad\qquad 2x^3 + 10x^2 - 6x + 2 & \\
\hline
4x^5 + 17x^4 - 25x^3 + 23x^2 - 9x + 2 = 12. \quad \textit{Ans.}
\end{array}
$$

Ex. 2.   Multiply $x + a + b$ by $x - a - b$.

$$
\begin{array}{l}
x + a + b \\
x - a - b \\
\hline
x^2 + ax + bx \\
\quad\ - ax \qquad - a^2 - ab \\
\quad\qquad - bx \qquad - ab - b^2 \\
\hline
x^2 \qquad\qquad - a^2 - 2ab - b^2. \quad \textit{Ans.}
\end{array}
$$

Let the pupil check by letting $x = 4$, $a = 1$, and $b = 1$.

EXERCISE 49

Multiply :

1. $\ \ 3.2x^2$	3. $\ \ 2\frac{1}{5}x^3$	5. $3(x+y)^2$	7. $\ \ 4(2a+b)$
$\underline{-1.5x}$	$\underline{-3.5p^2}$	$\underline{2(x+y)^3}$	$\underline{-8(2a+b)}$
2. $-4.6ab$	4. $-1.45xy$	6. $-4(a-b)^3$	8. $\ \ .5(x-y)$
$\underline{.05}$	$\underline{2a}$	$\underline{-2(a-b)}$	$\underline{-.2(x-y)^2}$

Perform the following indicated multiplications :

9. $\left(\dfrac{3x}{4y}\right)^2$.

10. $\left(\dfrac{3x}{4y}\right)^3$.

11. $-5ab(a^2 - 3ab)$.

12. $-2.7(.4x - 1.5)$.

13. $-3a(\frac{1}{2}x^2 - \frac{1}{6}y^2)$.

14. $(3x - 2y)(4x - 3y)$.

15. $(4a^2 - b^2)(a^2 - 3b^2)$.

16. $(.3a^2 - .2b)^2$.

Multiply and check:

17. $3x^2 - 4x + 2$ by $2x^2 - 5x - 3$.

18. $5a^2 - ab - 2b^2$ by $2a^2 + 3ab - 4b^2$.

19. $a^2 + ax + x^2$ by $a^2 - ax + x^2$.

Arrange the terms of the following in descending order of some letter, and multiply:

20. $4x - 3x^2 - 5 + 2x^3$ by $x + 4$.

21. $3x^2 - 5 - x$ by $x + 4x^2 - 2$.

22. $2x^3 + y^3 - 4xy^2 + 3x^2y$ by $y^2 + 3x^2 - 2xy$.

23. $3x^2 + x^3 - x - 5$ by $5 + x^2 - 4x$.

24. $4x^2 + 9y^2 - 6xy$ by $4x^2 + 9y^2 - 6xy$.

Multiply and check:

25. $x + a$ by $x + b$.   **27.** $x + y + a$ by $x + y - b$.

26. $x + y$ by $a + b$.   **28.** $x + a - b$ by $x - a + b$.

29. $x^3 - 3x^2 + 2x - 1$ by $2x^2 + x - 3$.

30. $3x^2y - 4xy^2 - y^3$ by $x^2 - 2xy - y^2$.

31. $x^3 - 3x^2y + 3xy^2 - y^3$ by $x^2 - 2xy + y^2$.

32. $4x^3 - 3x^2 + 5x - 2$ by $x^2 + 3x - 3$.

33. $x^4 - 3x^2 + 5$ by $x^2 - x - 4$.

34. $x^3 - 3xy + y^3$ by $x^3 - 3xy - y^3$.

35. $a^2 - ab + b^2$ by $a^2 + ab + b^2$.

36. $4x^2 + 9y^2 - 6xy$ by $4x^2 + 9y^2 + 6xy$.

37. $x^4 - 7x^2y^2 + 6xy^3 - y^4$ by $x^3 - 2xy^2 + y^3$.

38. $x^3 - 6ax^2 + 12a^2x - 8a^3$ by $-x^2 - 4ax - 4a^2$.

39. $a^2 + b^2 + x^2 + 2ab - ax - bx$ by $a + b + x$.

40. $ab + cd + ac + bd$ by $ab + cd - ac - bd$.

41. Multiply 32 by 43.   Also $3t + 2u$ by $4t + 3u$.

$$
\begin{array}{l}
32 \\
43 \\
\hline
96 \\
128 \\
\hline
1376
\end{array}
\qquad
\begin{array}{l}
3t + 2u \\
4t + 3u \\
\hline
12t^2 + 8tu \\
\quad\ + 9tu + 6u^2 \\
\hline
12t^2 + 17tu + 6u^2
\end{array}
$$

If we denote tens by t and units by u, show that the two processes are the same

42. Multiply 46 by 23 in both the arithmetical and algebraic ways as in the preceding example.

43. In a similar manner multiply 342 by 23.

44. Make up and work an example in which a trinomial is multiplied by a binomial expression, at least one term in each expression being negative.

52. Division is the process of finding one factor when the product and the other factor are given.

The **dividend** is the product of the two factors, and hence it is the quantity to be divided by the given factor.

The **divisor** is the given factor.

The **quotient** is the required factor.

Thus, to divide $10\,xy$ by $5\,x$, we must find a quantity which, multiplied by $5\,x$, will produce $10\,xy$. The factor $5\,x$ is the divisor, $10\,xy$ is the dividend, and the other factor, or required quotient, is evidently $2\,y$.

The division of a by b may be indicated in each of the following ways :

$$b\overline{)a}, \quad a \div b, \quad \frac{a}{b}, \quad \text{or } a/b.$$

53. Examples in Division.

Ex. 1.   Change to descending order of terms and divide:

$$2\,x - 17\,x^2 + 15 + 6\,x^3 \text{ by } -4\,x - 5 + 3\,x^2.$$

$$
\begin{array}{l}
6x^3 - 17x^2 + 2x + 15 \\ \hline
6x^3 - 8x^2 - 10x \\ \hline
 - 9x^2 + 12x + 15 \\
 - 9x^2 + 12x + 15
\end{array}
\quad
\begin{array}{|l}
3x^2 - 4x - 5 \\ \hline
2x - 3
\end{array}
\quad
\begin{array}{l}
= 6 + (-6) \\
= -1
\end{array}
$$

The check is obtained by letting $x = 1$.

Ex. 2.   Divide $a^4 + a^2x^2 + x^4$ by $a^2 - ax + x^2$.

$$
\begin{array}{l}
a^4 + a^2x^2 + x^4 \\ \hline
a^4 - a^3x + a^2x^2 \\ \hline
 a^3x + x^4 \\
 a^3x - a^2x^2 + ax^3 \\ \hline
 a^2x^2 - ax^3 + x^4 \\
 a^2x^2 - ax^3 + x^4
\end{array}
\quad
\begin{array}{|l}
a^2 - ax + x^2 \\ \hline
a^2 + ax + x^2
\end{array}
$$

Let the pupil check the work.

Ex. 3. Divide $a^3 + 8\,x^3$ by $a - 2\,x$.

$$
\begin{array}{l|l}
a^3 + 8\,x^3 & a - 2\,x \\
\underline{a^3 - 2\,a^2x} & a^2 + 2\,ax + 4\,x^2 + \dfrac{16\,x^3}{a - 2\,x} \\
\quad 2\,a^2x + 8\,x^3 & \\
\quad \underline{2\,a^2x - 4\,ax^2} & \\
\qquad 4\,ax^2 + 8\,x^3 & \\
\qquad \underline{4\,ax^2 - 8\,x^3} & \\
\qquad\qquad 16\,x^3 &
\end{array}
$$

Let the pupil check the work by letting $a = 3$ and $x = 1$.

EXERCISE 50

Divide :

1. 8.4 by $- .02$.
2. $- 2.6$ by $- 1.3$.
3. $.8\,a^2b^2$ by $- .4\,ab^2$.
4. $- 30\,x^3y^2$ by $- 6\,x^2y$.
5. $- 7\,az^3$ by $.7\,z^3$.
6. $- 3\,x^2$ by $- 1$.

7. $- 8\,ab$ by $\frac{1}{2}\,a$.
8. $- 8\,ax$ by $- .2\,x$.
9. $- \frac{1}{4}\,\pi r^2$ by $\frac{1}{2}\,\pi r^2$.
10. $20(x+y)^3$ by $-4(x+y)$.
11. $-1.4(a-b)^5$ by $-2(a-b)^3$.

Perform the operation indicated in the following :

12. $4.5 \div (-3)$.
13. $-66 \div (.2)$.
14. $-8 \div (-5)$.
15. $(15\,a^2b - 10\,a^2b^2 - 5\,ab^2) \div (-5\,ab)$
16. $(-3\,x^2 - 2\,x + 5) \div (-1)$.
17. $(2.4\,x^2 - .12\,x + 9) \div (-.3)$.

Arrange, divide, and check :

18. $4\,x + 6\,x^5 + 3\,x^2 - 11\,x^3 - 4$ by $3\,x^2 - 4$.
19. $- x^3y - 11\,xy^3 - 2\,x^2y^2 + 6\,x^4 - 6\,y^4$ by $2\,x - 3\,y$.
20. $4\,y^3 + 6\,x^6 - 13\,x^4y$ by $3\,x^2 - 2\,y$.
21. $9\,x - 18\,x^3 + 8\,x^4 - 13\,x^2 + 2$ by $4\,x^2 + x - 2$.
22. $10 - x^3 - 27\,x^2 + 12\,x^4 - 3\,x$ by $x + 4\,x^2 - 2$.
23. $22\,x^2 - 13\,x^3 + 10\,x^5 - 18\,x^4 + 5\,x - 6$ by $x + 5\,x^2 - 2$.
24. $14\,x^2y^3 - 16\,x^3y^2 + 6\,x^5 + y^5 + 5\,x^4y - 6\,xy^4$ by $3\,x^2 + y^2 - 2\,xy$.
25. $5\,a^4b - 3\,a^3b^2 - a^2b^3 + 3\,a^5 - 4\,b^5$ by $a^2 + 3\,ab + 2\,b^2$.

Divide and check:

26. $a^3 + b^3$ by $a + b$.   **29.** $a^5 + 32\,x^5$ by $a + 2\,x$.

27. $x^3 - 8\,y^3$ by $x - 2\,y$.   **30.** $x^6 - y^6$ by $x + y$.

28. $x^4 - 16\,y^4$ by $x - 2\,y$.   **31.** $256\,x^8 - y^8$ by $4\,x^2 - y^2$.

32. $x^4 + x^2 + 1$ by $x^2 + x + 1$.

33. $x^4 + x^2 y^2 + y^4$ by $x^2 + xy + y^2$.

34. $x^2 - a^2 - 2\,ab - b^2$ by $x - a - b$.

35. $p^2 - 2\,pq + q^2 - x^2$ by $p - q + x$.

36. $a^2 + 2\,am + m^2 - k^2$ by $a + m - k$.

37. $a^2 - x^2 + 4\,xy - 4\,y^2$ by $a + x - 2\,y$.

38. $a^3 + b^3 + c^3 - 3\,abc$ by $a + b + c$.

39. Show that $x^3 + y^3$ is not exactly divisible by $x - y$.

40. Show that there is a remainder when $x^4 + y^4$ is divided by $x - y$.

41. Divide 672 by 32.   Also divide $6\,t^2 + 7\,t + 2$ by $3\,t + 2$.   Show that the two divisions are the same if t^2 stands for 100 and if t stands for 10.

42. Make up and work an example in which a trinomial is exactly divided by a binomial.

EXERCISE 51

1. If a room is l feet long and w feet wide, find a formula for the number of yards (C) of carpet, three feet wide, necessary, if 10 per cent is added for waste in cutting.

2. If a ten per cent allowance is made for waste in cutting, find the number of yards of carpet 3 ft. wide required for a room 25 ft. long and 18 ft. wide.   Use the formula obtained for C in Ex. 1.

3. Also for a room 13 ft. 6 in. wide and 24 ft. long.

4. At c cents per square yard obtain a formula for the cost in dollars (C) of painting the walls and ceiling of a room whose length, width, and height in feet are l, w, h, respectively.

5. By use of this formula, at 6 cents per sq. yd. find the cost of painting the walls and ceiling of a room 24 × 18 ft. and 8½ ft. high.

6. If one bushel of potatoes occupies 1¼ cu. ft., obtain a formula for the number of bushels of potatoes in a rectangular bin a feet long, b feet wide, and c feet deep.

7. By use of the formula obtained in Ex. 6, find how many bushels of potatoes will be held by a bin 30 × 5½ × 6 ft.

8. A certain bin is to contain 500 bushels of potatoes. If the bin is to be 25 ft. long and 6 ft. wide, how deep must it be? Use the formula found in Ex. 6.

9. The following formula is often used in computing the horse power of a gasolene engine, $H.P. = \dfrac{D^2 N}{2.5}$. Find the value of H.P. when $D = 4$ and $N = 6$.

EXERCISE 52

1. During the months of a certain year a boy's earnings and spendings were as follows:

MONTH	JAN.	FEB.	MAR.	APR.	MAY.	JUNE
Earnings . .	\$3.15	\$2.50	\$3.25	\$2.10	\$3.15	\$4.60
Spendings .	2.12	1.75	3.10	3.40	3.25	2.16

MONTH	JULY	AUG.	SEPT.	OCT.	NOV.	DEC.
Earnings . .	\$10.20	\$12.50	\$8.60	\$2.70	\$3.20	\$2.75
Spendings .	1.15	2.05	4.20	2.15	1.76	4.60

On the same diagram make two graphs, one of the earnings and the other of the spendings. The boy worked on a farm during the summer. How is the effect of this shown on the two graphs?

Shade the area between the two graphs. What does the shaded area represent?

2. Make up and work a similar example concerning a girl's earnings and spendings during a certain year.

3. During a certain year the receipts and expenses per month of a firm engaged in business were as follows:

Month	Jan.	Feb.	Mar.	Apr.	May	June
Receipts . .	$2416	$2822	$3105	$2710	$2292	$2132
Expenses. .	1572	1483	1642	1590	1480	1520

Month	July	Aug.	Sept.	Oct.	Nov.	Dec.
Receipts . .	$1560	$1430	$2172	$2872	$3200	$3610
Expenses. .	1470	1480	1590	1670	1580	1705

Graph these two sets of numbers on the same diagram. (Let each space on the vertical axis stand for $200 and let the lowest mark be $1400.)

On August 1 a new manager for the business was appointed. How is the effect shown on the graphs?

4. In a certain business the sales per month for three successive years were as follows:

	Jan.	Feb.	Mar.	Apr	May	June
1st Year . .	$8542	$9654	$8320	$7900	$6842	$5922
2d Year . .	9320	10130	8890	8200	7512	6213
3d Year . .	10412	10800	9460	8430	7600	6320

	July	Aug.	Sept.	Oct.	Nov.	Dec.
1st Year . .	$4216	$3176	$4896	$6213	$7916	$9784
2d Year . .	4340	3380	5216	7219	8824	10130
3d Year . .	4408	3520	5807	8016	9472	12842

On the same diagram make graphs of these three sets of numbers. (Let each space on the vertical axis stand for $1000 and let the lowest mark on the scale be $3000.)

From the diagram what inference can be drawn as to the state of the business during the summer months?

5. The adjoining table gives in millions of dollars the exports and imports of the United States for the years as specified. Make a graph of the exports and also of the imports on the same diagram. (Let each

Year	Exports	Imports	Year	Exports	Imports
1905	1627	1179	1914	2114	1789
1906	1798	1321	1915	3547	1779
1907	1923	1423	1916	5482	2392
1908	1753	1116	1917	6228	2953
1909	1728	1476	1918	6125	3050
1910	1866	1563	1919	7081	3096
1911	2092	1532	1920	7949	5238
1912	2399	1818	1921	6386	3654
1913	2484	1793	1922	3700	2608

space on the vertical axis stand for 500 millions of dollars; let each space on the horizontal axis stand for 1 year.)

The Great War began in the year 1914. What was its effect on each of the graphs? Can you suggest a reason for the decline of the graphs in 1921 and 1922? Shade the area between the two graphs. What does the shaded area represent?

CHAPTER VI

SHORT MULTIPLICATION; SIMPLIFICATIONS

PART I

54. Short Multiplication. Certain simple cases of multiplication are used so often that it is important to study them till the pupil can write out the products at once without the labor of actual multiplication. These cases of short multiplication should be mastered as thoroughly as the multiplication table in arithmetic.

55. I. Square of the Sum of Two Numbers. Let $a + b$ be the sum of any two algebraic numbers. By actual multiplication,

$$
\begin{array}{r}
a + b \\
a + b \\
\hline
a^2 + ab \\
+ ab + b^2 \\
\hline
a^2 + 2\,ab + b^2. \quad \textit{Product.}
\end{array}
$$

Or, in brief, $(a + b)^2 = a^2 + 2\,ab + b^2$,

which, stated in general language, is the rule:

The square of the sum of two numbers equals the square of the first, plus twice the product of the first by the second, plus the square of the second.

Ex. 1. $(2\,x + 3\,y)^2 = 4\,x^2 + 12\,xy + 9\,y^2.$ *Product.*

Ex. 2. $104^2 = (100 + 4)^2 = 100^2 + 8 \times 100 + 4^2$
$= 10{,}000 + 800 + 16 = 10{,}816.$ *Ans.*

56. II. Square of the Difference of Two Numbers. By actual multiplication, $a - b$

$$a - b$$
$$\overline{a^2 - \quad ab}$$
$$\underline{\quad - \quad ab + b^2}$$
$$a^2 - 2\,ab + b^2. \quad \textit{Product.}$$

Or, in brief, $(a - b)^2 = a^2 - 2\,ab + b^2,$

which, stated in general language, is the rule:

The square of the difference of two numbers equals the square of the first, minus twice the product of the first by the second, plus the square of the second.

Ex. 1. $(2\,x - 3\,y)^2 = 4\,x^2 - 12\,xy + 9\,y^2. \quad \textit{Product.}$

Ex. 2. $998^2 = (1000 - 2)^2$
$$= 1000^2 - 4 \times 1000 + 2^2$$
$$= 1,000,000 - 4000 + 4$$
$$= 996,004. \quad \textit{Ans.}$$

EXERCISE 53

Write at sight the value of each of the following and check:

1. $(a + x)^2.$	15. $(c + d)^2.$	29. $(10\,x^3y + 1)^2.$
2. $(a + 5)^2.$	16. $(3\,x + 2)^2.$	30. $(1 - 7\,xy^5)^2.$
3. $(p + q)^2.$	17. $(3\,x - 5)^2.$	31. $(\tfrac{1}{2}x + y)^2.$
4. $(p + 7)^2.$	18. $(5 + ab)^2.$	32. $(\tfrac{1}{8}x - 2\,y)^2.$
5. $(4 + a)^2.$	19. $(3 + 4\,x)^2.$	33. $(\tfrac{1}{4}x + \tfrac{1}{2}y)^2.$
6. $(a - x)^2.$	20. $(8\,a - 1)^2.$	34. $(\tfrac{1}{2}a - \tfrac{2}{3}b)^2.$
7. $(a - 5)^2.$	21. $(3\,x + 4\,y)^2.$	35. $(\tfrac{2}{3}x + \tfrac{3}{2}y)^2.$
8. $(p - q)^2.$	22. $(3\,x - 2\,y)^2.$	36. $(3\,x^2 - \tfrac{1}{2}y)^2.$
9. $(7 - x)^2.$	23. $(7\,a - b^3)^2.$	37. $(.2\,x + y)^2.$
10. $(9 + a)^2.$	24. $(3\,a^2 + 7\,b^2)^2.$	38. $(.2\,p + .3\,q)^2.$
11. $(9 - a)^2.$	25. $(4\,x^3 - 1)^2.$	39. $(.4\,x + .5\,y)^2.$
12. $(x + y)^2.$	26. $(5\,ab + 9\,y^3)^2.$	40. $(5\,r - .8\,s)^2.$
13. $(x - 8)^2.$	27. $(3\,x^4 + 5\,x^3)^2.$	41. $(5\,a + .4)^2.$
14. $(c - d)^2.$	28. $(6x^2y - 7\,xy^2)^2.$	42. $(.02\,x - .3\,y)^2.$

43. Write the product of $(5x^2 + 3y)^2$ by the short rule. Now multiply $(5x^2 + 3y)^2$ in full. Compare the amount of work in the two processes.

44. At sight write the value of $[(a + \dot{b}) + 5]^2$.

Sug. To get a method for the short multiplication of $[(a + b) + 5]^2$, let p (the first letter of the word " parenthesis ") stand for $(a + b)$.

Then we have $(p + 5)^2 = p^2 + 10p + 25$.

Now substitute $(a + b)$ for p.

Then $[(a + b) + 5]^2 = (a + b)^2 + 10(a + b) + 25$
$= a^2 + 2ab + b^2 + 10a + 10b + 25.$ **Ans.**

At sight write value of each of the following :

45. $[(x + y) + 5]^2$.

46. $[(a + b) - 4]^2$.

47. $[(a + b) + x]^2$.

48. $[(a - x) + 5]^2$.

49. $[(x + y) - 3]^2$.

50. $[(p + q) + y]^2$.

51. $[(2a + 3b) + 5]^2$.

52. $[(3x - y) + 5]^2$.

53. $[(2x + 3y) - 4]^2$.

54. $[(ax + 5) - y]^2$.

55. $[(x + 5a) - 2y]^2$.

56. $[(2x - 3b) + 5]^2$.

57. Write out the value of $[(2a + 3b) - 5]^2$ by inspection. Now multiply $2a + 3b - 5$ by $2a + 3b - 5$ in full. Compare the amount of the work in the two processes.

By the short method obtain the value of the following :

58. 102^2. **60.** 10.3^2. **62.** 5.3^2. **64.** 997^2. **66.** 99^2.

59. 51^2. **61.** 52^2. **63.** 999^2. **65.** 49^2. **67.** 99.6^2.

57. III. Product of the Sum and Difference of Two Numbers.

By actual multiplication, $a + b$
$$\begin{array}{r} a + b \\ a - b \\ \hline a^2 + ab \\ -ab - b^2 \\ \hline a^2 \qquad - b^2. \quad \textit{Product.} \end{array}$$

Or, in brief, $(a + b)(a - b) = a^2 - b^2$,
which, stated in general language, is the rule:

The product of the sum and difference of two numbers equals the square of the first minus the square of the second.

Ex. 1. $(2x + 3y)(2x - 3y) = 4x^2 - 9y^2$.   *Product.*

Ex. 2.   Multiply 93 by 87 by aid of the above rule.

$$93 \times 87 = (90 + 3)(90 - 3)$$
$$= 8100 - 9 = 8091. \quad Ans.$$

EXERCISE 54

Write by inspection the value of each of the following products, and check the work for each result :

1. $(a + x)(a - x)$.
2. $(b + y)(b - y)$.
3. $(x + 3)(x - 3)$.
4. $(a + 5)(a - 5)$.
5. $(x^3 + a)(x^3 - a)$.
6. $(y^2 + 7)(y^2 - 7)$.
7. $(p + x)(p - x)$.
8. $(6 - a)(6 + a)$.
9. $(a^2 - y)(a^2 + y)$.
10. $(x^3 - 1)(x^3 + 1)$.
11. $(a^4 - 2)(a^4 + 2)$.
12. $(y + 15)(y - 15)$.
13. $(x^3 - 9)(x^3 + 9)$.
14. $(8 + x^2)(8 - x^2)$.
15. $(3 - ab)(3 + ab)$.
16. $(10 - x)(10 + x)$.
17. $(8a - 7)(8a + 7)$.
18. $(3x^3 - 7y)(3x^3 + 7y)$.
19. $(4ab - 9)(4ab + 9)$.
20. $(3a^2x^3 + 5)(3a^2x^3 - 5)$.
21. $(7abc + 1)(7abc - 1)$.
22. $(4p^2q - xy)(4p^2q + xy)$.
23. $(1 - 11x^3)(1 + 11x^3)$.
24. $(\frac{1}{2}x - \frac{2}{3}y)(\frac{1}{2}x + \frac{2}{3}y)$.
25. $(.4x + .5y)(.4x - .5y)$.
26. $(.3x^2 + .5y^2)(.3x^2 - .5y^2)$.
27. $(5x - .08)(5x + .08)$.
28. $(.3ab + .8)(.3ab - .8)$.

29. $[(a + b) + 3][(a + b) - 3]$.

Sug. $(p + 3)(p - 3) = p^2 - 9$.

Hence, $[(a + b) + 3][(a + b) - 3] = (a + b)^2 - 9$
$$= a^2 + 2ab + b^2 - 9. \quad Ans.$$

30. $[(a + b) + 5][(a + b) - 5]$.
31. $[(x + y) + 4][(x + y) - 4]$.
32. $[(p + q) + a][(p + q) - a]$.

33. $[(2x+3y)+b][(2x+3y)-b]$.
34. $[(1+2x)+5a][(1+2x)-5a]$.
35. $[(3a^2+b)-4][(3a^2+b)+4]$.

Find the value of each of the following in the short way :

36. 92×88.
37. 103×97.
38. 105×95.

39. 1005×995.
40. 74×66.
41. 153×147.

Find in the shortest way :

42. The area of a rectangle 102 ft. long and 98 ft. wide.
43. The cost of 32 doz. eggs at 28 cents per dozen.
44. The cost of 67 yd. of cloth at 73 cents a yard.
45. Make up and work an example similar to Ex. 43.
46. Work Ex. 36 in full. How much of this labor is saved by using the short method of multiplication?

Write at sight the product for each of the following miscellaneous examples :

47 $(x+2a)^2$.
48. $(x+2a)(x-2a)$.
49. $(x-2a)^2$.
50. $(3x-1)(3x+1)$.
55. $[(x+2y)+5][(x+2y)-5]$.
56. $(.3x+.5y)(.3x-.5y)$.
58. 998×1002.

51. $(3x-1)^2$.
52. $(3a^2-2b^3)^2$.
53. $(3a^2-2b^3)(3a^2+2b^3)$.
54. $[(x+2y)+5]^2$.
57. 998^2.
59. 97^2.
60. 97×103.

58. Products of the Form $(x+a)(x+b)$.

By actual multiplication,

$x + 5$	$x - 5y$	$x + a$
$x + 3$	$x + 3y$	$x + b$
$x^2 + 5x$	$x^2 - 5xy$	$x^2 + ax$
$+ 3x + 15$	$+ 3xy - 15y^2$	$+ bx + ab$
$x^2 + 8x + 15$	$x^2 - 2xy - 15y^2$	$x^2 + (a+b)x + ab$

By comparing each pair of expressions with their product, we observe the following relation:

The product of two expressions of the form x + a and x + b consists of three terms:

The first term is the square of the first term of the expressions;

The last term is the product of the second terms of the expressions;

The middle term consists of the first term of the expressions with a coefficient equal to the algebraic sum of the second terms of the expressions.

Ex. 1. Multiply $x - 8$ by $x + 7$.

$$-8 + 7 = -1. \qquad -8 \times 7 = -56.$$

$$\therefore (x - 8)(x + 7) = x^2 - x - 56 \quad \textit{Product.}$$

Ex. 2. Multiply $(x - 6a)(x - 5a)$.

$$(-6a) + (-5a) = -11a. \quad (-6a) \times (-5a) = +30a^2.$$

$$\therefore (x - 6a)(x - 5a) = x^2 - 11ax + 30a^2. \quad \textit{Product.}$$

EXERCISE 55

At sight write the product of the following:

1. $(x + 2)(x + 3)$.
2. $(x + 2)(x + 7)$.
3. $(p + 4)(p + 6)$.
4. $(r + 1)(r + 5)$.
5. $(y + 9)(y + 20)$.
6. $(x - 2)(x - 5)$.
7. $(y - 3)(y - 8)$.
8. $(a - 14)(a - 11)$.
9. $(p - 7)(p - 1)$.
10. $(b^2 - 14)(b^2 - 5)$.
11. $(x + 5)(x - 2)$.
12. $(a + 7)(a - 3)$.
13. $(x^2 + 9)(x^2 - 1)$.
14. $(y^4 + 6)(y^4 - 4)$.
15. $(ab + 7)(ab - 11)$.
16. $(x - 7)(x + 3)$.
17. $(x + 15)(x - 9)$.
18. $(a^2 + 20)(a^2 - 5)$.
19. $(p^2q^2 + 3)(p^2q^2 - 1)$.
20. $(y^3 + 12)(y^3 - 1)$.
21. $(x - 9)(x - 10)$.
22. $(x + 7)(x - 3)$.
23. $(x + 5a)(x + 3a)$.
24. $(x - 3a)(x - 2a)$.

25. $(x - 6\,a)(x + 3\,a)$.
26. $(x^2 - 7\,b)(x^2 + 6\,b)$.
27. $(y + \frac{1}{2})(y + \frac{3}{2})$.
28. $(y - \frac{2}{3})(y - \frac{1}{3})$.
29. $(b + \frac{1}{4})(b - \frac{3}{4})$.
30. $(a - 2)(a - \frac{3}{2})$.
31. $(a - 2)(a + \frac{3}{2})$.

32. $(x + .5)(x + .2)$.
33. $(x + .4)(x - .5)$.
34. $(x + .3)(x + .2)$.
35. $(y - .2)(y - .05)$.
36. $(x + a)(x + b)$.
37. $(y + c)(y - d)$.
38. $(x - p)(x + q)$.

39. $[(a + b) + 5][(a + b) + 3]$.

Sug. Hence $(p + 5)(p + 3) = p^2 + 8p + 15$

or, $[(a + b) + 5][(a + b) + 3] = (a + b)^2 + 8(a + b) + 15$
$$= a^2 + 2\,ab + b^2 + 8\,a + 8\,b + 15.$$
$$Ans.$$

40. $[(a + b) + 7][(a + b) + 2]$.
41. $[(x + y) + 4][(x + y) - 3]$.
42. $[(p + q) - 5][(p + q) - 4]$.
43. $[(a + x) - 7][(a + x) + 3]$.
44. Write the product of $x + 7$ by $x - 9$ by the short rule. Now multiply $x + 7$ by $x - 9$ in full. Compare the amount of work in the two processes.

EXERCISE 56

Write at sight the product for each of the following miscellaneous examples :

1. $(x + 5)(x - 5)$.
2. $(x + 5)^2$.
3. $(x - 5)^2$.
4. $(x + 5)(x - 3)$.
5. $(x - 3\,a)(x + 3\,a)$.
6. $(x - 3\,a)(x + 5\,a)$.
7. $(x - 5\,a)^2$.
8. $(3\,x + 5)(3\,x - 5)$.
9. $(3\,x - 5)^2$.

10. $(x + 7)(x + 5)$.
11. $(x - 8)(x + 3)$.
12. $(3\,a + 4)^2$.
13. $(3\,a + 4)(3\,a - 4)$.
14. $(3\,a - 4)^2$.
15. $(7\,x - 5)(7\,x + 5)$.
16. $(x - 6)(x - 3)$.
17. $(x - 4\,a)(x + 3\,a)$.
18. $(x^3 - 1)^2$.

19. $(x^3 - 1)(x^3 - 2)$.
20. $(ab + 3)(ab - 3)$.
21. $(x + .8)^2$.
22. $(x - .5)(x - .3)$.
23. $(4 + 9a)(4 - 9a)$.
24. $(5x - .6)^2$.
25. $(x + .5)(x - .3)$.
26. $(ab + .4)(ab - .4)$.
27. $[(x + y) + 7][(x + y) - 3]$.
28. $[(a + b) - 4][(a + b) + 4]$.
29. $[(a + b) - 5]^2$.
30. $[7 + (x + y)]^2$.
31. 54^2.
32. 46^2.
33. 105×95.
34. 203×197.
35. 107^2.
36. 93^2.

SIMPLIFICATION

59. The **simplification** of an algebraic expression in which multiplications are indicated by means of the parenthesis means performing the operations indicated and collecting terms.

Ex. Simplify $3(x - 2y)(x + 2y) - 2(x - 2y)^2$.

$$3(x - 2y)(x + 2y) - 2(x - 2y)^2$$
$$= 3(x^2 - 4y^2) - 2(x^2 - 4xy + 4y^2)$$
$$= 3x^2 - 12y^2 - (2x^2 - 8xy + 8y^2)$$
$$= 3x^2 - 12y^2 - 2x^2 + 8xy - 8y^2$$
$$= x^2 + 8xy - 20y^2. \quad Ans.$$

To check this result let $x = 1$ and $y = 2$.

Then $3(x - 2y)(x + 2y) - 2(x-2y)^2 = 3(1-4)(1+4) - 2(1-4)^2$
$$= -45 - 18 = -63.$$
$$x^2 + 8xy - 20y^2 = 1 + 16 - 80 = -63.$$

EXERCISE 57

Simplify and check:

1. $22 + 2(4 + 3)$.
2. $(22 + 2)(4 + 3)$.
3. $84 - 3(5 + 2)$.
4. $(84 - 3)(5 + 2)$.
5. $7 - 5(a - 3)$.
6. $(7 - 5)(a - 3)$.
7. $8 - 3(x + 4)$.
8. $x - 2(x + 1)$.

9. $(x-2)(x+1)$.

10. $(2x+3)(5x-4)$.

11. $2x+3(5x-4)$.

12. $7a-3(4a+8)$.

13. $9a+5(3a+4)$.

14. $(x+2)^2+(x-2)^2$.

15. $(x+2)^2-(x-2)^2$.

16. $(a+2b)^2+(a-2b)^2$.

17. $(a+2b)^2-(a-2b)^2$.

18. $(3x-2)^2+(3x+2)^2$.

19. $(3x-2)^2-(3x+2)^2$.

20. $2(a+b)^2-(a-b)^2$.

21. $3(x+2y)^2+2(x-2y)^2$.

22. $4(a+b)^2-3(a-b)^2$.

23. $7(2a-3b)^2-4(2a+3b)^2$.

24. $(x+5)^2-(x+2)(x-2)$.

25. $(x+8)(x-3)-(x-4)^2$.

26. $(3a-4)^2-5(a+3)(a-3)$.

27. $3x(x^2-2)-2x(x^2+3)$.

28. $(2a-3b+5)^2-(2a+3b-5)^2$.

29. $3x-2(3x^2-5x+2)$.   30. $(x-2)(x-1)(x+3)$.

31. $(x-y-z)^2-x(x-2y+2z)$.

32. $2x^2-3(x-1)^2+(x-2)^2$.

33. $3x^2-x(1-x)(2+x)+x^3$.

34. $2-3(x-2)^2-2(3-2x)(1+x)$.

35. $a^2-[x(a-x)-a(x-a)]-x^2$.

36. $(x-1)(x-2)-(x-2)(x-3)+(x-3)(x-4)$.

Find in the shortest way you can the value of :

37. $(3x+2)^2(3x+2)$.   39. $(a+b)^4$.

38. $(3a-4b)^3$.   40. $(3x-2y)^4$.

If $a=3$, $b=2$, $c=1$, $x=-4$, find the value of the following in the shortest way you can :

41. $3a+2(4b+x)(2a-c)$.

Sug. $9+2(8-4)(6-1)$
$=9+2(4)5$
$=9+8\times5=49$.   *Ans.*

42. $5b+(6c+x)(2a-b)$.

43. $(2a+x)(3a+x)-(2b+c)(3a+x)$.

44. $(a+b)^2-(a-b)^2.$ **45.** $(2a+c)^2+(a-2c)^2.$

46. $3(ab+c)^2-4(ab-c)^2.$

47. $(2+b)^2-(a+c)(3b+x).$

48. $2(4c+b)-(2a+c)(2a-c).$

EXERCISE 58

VERBAL PROBLEMS

1. By how much does 20 exceed x?

2. A woman bought s yards of silk at $4 a yard and w yards of woolen cloth at $3 a yard. How much did she pay in all?

3. A man is y years old now. How old will he be in 3 times c years?

4. A train travels m miles an hour for h hours. Express the distance in miles. In rods.

5. How many cents in 5 dimes? In h dimes? In h dimes and p nickels together?

6. A farmer buys m cows for x dollars. What is the average price per cow?

7. How long must a boy work to earn $10 if he receives c cents per day?

8. How many hours in 140 minutes? In a minutes?

9. Separate the number n into two parts of which one part is 4.

10. If d dollars are divided among b boys, how many dollars will each boy receive? How many cents?

EXERCISE 59

1. 36 multiplied by $5=180$. In this multiplication which number is the multiplicand? Which the multiplier? Which the product?

Hence, we have the rule, *the product equals the multiplicand multiplied by the multiplier.*

Hence, if p stands for the product, M for the multiplicand, and m for the multiplier, $p = Mm$.

2. By use of the formula $p = Mm$, find the multiplier when the product is 504 and the multiplicand is 14.

3. By use of $p = Mm$, find the multiplicand when the multiplier is $\frac{2}{5}$ and the product is $\frac{32}{5}$.

4. By the use of $p = Mm$, find the number which when multiplied by 1.5 produces 12.3.

5. If d stands for divisor, D for dividend, q for quotient, and r for remainder, show by a numerical example that $D = dq + r$.

6. Find the divisor when the dividend, quotient, and remainder are 87, 5, and 7, respectively. Use $D = dq + r$.

7. Find the quotient when the dividend, divisor, and remainder are 68, 13, and 3, respectively. Use $D = dq + r$.

8. If the entire area of the walls and ceiling of a room is denoted by a, the length of the room by l, the width by w, and the height by h, find a formula for a in terms of the other letters.

9. By use of this formula find the area of the walls and ceiling of a room 20 ft. long, 16 ft. wide, and 9 ft. high.

10. Allowing 32 sq. ft. for a roll, find a formula for the number of rolls (R) of paper required for the walls and ceiling of a room whose dimensions are l, w, and h feet.

11. By use of this formula, compute the number of rolls of paper required for the walls and ceiling of a room 22 ft. long, 16 ft. 6 in. wide, and 8 ft. 4 in. high.

12. If 200 cu. ft. of air space are allowed in a schoolroom for each pupil, obtain a formula for the number of pupils (P) that can be accommodated in a room whose dimensions are l, w, h, feet.

13. By use of the formula for P obtained in **Ex. 12**, find how many pupils are allowable in a room $30 \times 20 \times 15$ ft.

14. Also in a room 30 ft. 6 in. long, 24 ft. 3 in. wide, and 14 ft. high.

15. Using the formula obtained in **Ex. 12**, find the number of pupils allowable in your own schoolroom.

16. A schoolroom is to be 30 ft. long and 25 ft. wide. How high should the ceiling be in order that the room may accommodate 45 pupils? Use formula found in **Ex. 12**.

PART II

60. Examples in Short Multiplication.

Ex. 1. Multiply $x+a+b$ by $x-a-b$ by a short method.

$$(x + a + b)(x - a - b) = [x + (a + b)][x - (a + b)]$$
$$= x^2 - (a + b)^2$$
$$= x^2 - a^2 - 2\,ab - b^2. \quad Ans.$$

Let the pupil check the work by letting $x = 1$, $a = 2$, $b = 2$.

Ex. 2. Multiply $x+y-z$ by $x-y+z$.

$$(x + y - z)(x - y + z) = [x + (y - z)][x - (y - z)]$$
$$= x^2 - (y - z)^2.$$
$$= y^2 - (y^2 - 2\,yz + z^2).$$
$$= x^2 - y^2 + 2\,yz - z^2. \quad Ans.$$

Check by letting $x = 3$, $y = 2$, $z = 1$.

EXERCISE 60

At sight write the product of the following:

1. $[(a+b)+4]^2.$

2. $[(a+b)-3]^2.$

3. $[(a+b)+c]^2.$

4. $[(2\,a-x)+3\,y]^2.$

5. $[3+(a+b)]^2.$

6. $[5\,a-(x+y)]^2.$

7. $[2\,a^2-(b-2\,c)]^2.$

8. $[(x+y)-(a+b)]^2.$

9. $[(x+y)+3][(x+y)+5].$

10. $[(x+y)-3][(x+y)+5]$.

11. $(x+y-3)(x+y+5)$.

12. $(a+2b+5)(a+2b+3)$.

13. $(2x+3y+3a)(2x+3y-5a)$.

14. $(2x+a-3b)(2x-a+3b)$.

15. $[(a+b)+3][(a+b)-3]$.

16. $[(x+y)+a][(x+y)-a]$.

17. $[(2x-1)+y][(2x-1)-y]$.

18. $[4+(x+1)][4-(x+1)]$.

19. $[2x+(3y-5)][2x-(3y-5)]$.

20. $(a+b+3)(a+b-3)$.

21. $(x+y+a)(x+y-a)$.

22. $(4+x+a)(4-x-a)$.

23. $(2x+3y-5)(2x-3y+5)$.

24. $(4+x+y)(4-x-y)$.

25. $(x^2+3x+2)(x^2+3x-2)$.

26. $(a+b+3x)(a+b-3x)$.

27. $(a+b-3x)(a-b+3x)$.

28. $(x+y+5a)(x+y-5a)$.

29. $(x+y+5a)^2$.   30. $(x+y+5a)(x+y+3a)$.

31. $(x+y+5a)(x+y-3a)$.

32. Find the product of $x+y+6$ and $x+y-3$ by multiplying in full. Then find the same product by the method of § 60. About how much of the work of multiplication is saved by use of the latter method?

33. Show that $a^2=(a+b)(a-b)+b^2$.

34. By use of the relation proved in Ex. 33, obtain the value of $(7\frac{1}{2})^2$ in a short way.

Sug. We have $\quad (7\frac{1}{2})^2=(7\frac{1}{2}+\frac{1}{2})(7\frac{1}{2}-\frac{1}{2})+(\frac{1}{2})^2$
$\qquad\qquad\qquad = 8\times 7 + \frac{1}{4} = 56\frac{1}{4}.$ *Ans.*

Using the method just shown find the value of:

35. $(8\frac{1}{2})^2$.   **38.** $(15\frac{1}{2})^2$.   **41.** $(7.5)^2$.   **44.** $(75)^2$.

36. $(19\frac{1}{2})^2$.   **39.** $(49\frac{1}{2})^2$.   **42.** $(19.5)^2$.   **45.** $(195)^2$.

37. $(199\frac{1}{2})^2$.   **40.** $(99\frac{1}{2})^2$.   **43.** $(99.5)^2$.   **46.** $(995)^2$.

     47. $(9.7)^2$.   Sug. $(9.7)^2 = 10 \times 9.4 + .3^2$.

48. $(9.8)^2$.   **49.** $(9.6)^2$.   **50.** $(4.8)^2$.   **51.** 98^2.

EXERCISE 61

By short multiplication find the value of:

1. $(5a + .04)^2$.   **4.** $(.3a + .04b)^2$.

2. $(.4a - .05)^2$.   **5.** $(.3x + .05y)(.3x - .05y)$.

3. $(.02x - .03y)^2$.   **6.** $(5x - .08)(5x + .08)$.

7. $(x + .3)(x + .02)$.   **9.** $(x - .05)(x - .02)$.

8. $(y - .2)(y - .05)$.   **10.** $(a - .08)(a - .5)$.

11. 1003^2.   **14.** 9.5^2.   **17.** 1005×995.

12. 10.02^2.   **15.** 9997^2.   **18.** $103^2 - 97^2$.

13. 9998^2.   **16.** 17.5^2.   **19.** $(17.31)^2 - (2.69)^2$.

   20. $(3 + x + y)(3 - x - y)$.

   21. $(4a + x - y)(4a - x + y)$.

   22. $(a + b - 7)(a + b - 3)$.

   23. $(a + b + x)(a + b - x)$.

   24. $(x + y + 5a)^2$.

   25. $(5 - 2a - 3y)(5 + 2a + 3y)$.

   26. $(a - 2b + 5c)(a + 2b - 5c)$.

Compute in the shortest way:

27. The area of a field 103 rd. long and 97 rd. wide.

28. The area of a square field each side of which is 98 rd.

29. The cost of 62 yd. of cloth at 58 cents per yard.

30. The cost of 85 A. of land at $95 per acre.

Simplify:

1. $7 - 5(.3 - .04)$.

2. $(7 - 5)(.3 - .04)$.

3. $x - 3(x - .5)$.

4. $(x - 3)(x - .5)$.

5. $4.2 - .3(x - .3)(x - .5)$.

6. $3(x - .2)^2 - 5(x + .2)^2$.

7. $5x(x - .2) - 4x(x - .5)$.

8. $8(x + .5)^2 - 4(x - .5)^2$.

9. $3x - 5(x - .1)(x + .2)$.

10. $3y^2 - (.1y - .2z)(.3y + .4z)$.

11. $(x - y - z)^2 - x(x - 2y + 2z)$.

12. $.2x^2 - .3(x - 1)^2 + (x - 2)^2$.

13. $3(x - y)^2 - 2\{(x + y)^2 - (x - y)(x + y)\} + 2y^2$.

14. $(2x - 7y)(2x + 7y) - 4(x - 2y)^2 + 13y(5y - x)$.

15. $(3x^2 + 5)^2 + x^2(10 - 3x)(10 + 3x) - (5 + 13x^2)^2$.

16. $(a - c + 1)(a + c - 1) - (a - 1)^2 + 2(c - 1)^2$.

17. $(x + y - xy)(x - y - xy) + x^2y - (x - y^2)(x + y^2)$.

18. $3[(a + 2b)x + 2my] - 5[(m - c)y + bx] - 4[(x - a)a + cy]$.

19. $26ab - (9a - 8b)(5a + 2b) - (4b - 3a)(15a + 4b)$.

20. Multiply the sum of $(a - 2x)^2$ and $(2a - x)^2$ by $3a - 2(a - x)$.

21. Subtract $(x - 2y)^3$ from $x^3 - 8y^3$ and divide the remainder by $x - 2y$.

22. Find the value of $3 \times 0 + 4$. Of $8 - 7 \times 0$.

If $a = 3$, $b = 0$, $x = -2$, and $y = -5$, find the value of:

23. $2ax$.

24. bx^3y.

25. $3x^2 + aby$.

26. $bxy - ax^2$.

27. $\frac{1}{3}abx + \frac{1}{2}xy$.

28. $by^2 + 3x(x - y)$.

Sug. Do not multiply $3x$ into $x - y$, but proceed thus:
$$0(25) + 3(-2)(-2 + 5) = 0 + 3(-2)(3), \text{ etc.}$$

29. $4x^2 - abx(4x - y)$.

30. $3x - 5(2x + 3)$.

31. $2(x^2 + y) - aby + ax^3$.

32. $a(b + x) - 5(x + by)$.

33. $2(1 - 2x)^2 + (x + y)(a^2 + x)$.

34. $(x - 1)^2 - 3(x + 1)(x + 2) - x(x^2 - 2)(y - 2x)$.

35. $3a(a - 2x) - \{a - (a - 1)(x + 1) - (a + x)^2\} + 5ax$.

Find the value of the following:

36. $(x + a)^2 - (x - a)^2$ when $x = 2a$.

37. $5(x + p)^2 - (x + p)(x - 2p)$ when $x = 3p$.

38. $3x^2 + 4x - 5(x - 1)^2$ when $x = ab$.

39. If $x = 2$ and $y = 1$, find the value of $(x + y)^3$. Also of $x^3 + y^3$.

EXERCISE 63

1. If the cost price of an article is denoted by c, the selling price by s, and the per cent of profit expressed decimally by p, obtain a formula for s in terms of c and p.

2. By use of the formula for s obtained in Ex. 1, find the selling price of an article which is bought for 35 cents and is to be sold at an advance of 20 per cent.

3. An article is to be sold for $225 at a gain of 25%. By use of the formula for s obtained in Ex. 1, find what the dealer can afford to pay for it.

4. A dealer wishes to sell shoes for $3 a pair, and at the same time to gain 20 per cent. What can he afford to pay for them? Use the formula for s found in Ex. 1.

5. At a 10 cent store it is proposed to sell a certain article for 10 cents, making a profit of 20 per cent. What is the limit of the cost? Use the formula for s found in Ex. 1.

6. Find a formula for the entire surface of a box-shaped solid whose dimensions are a, b, and c units of length.

7. By use of the formula obtained in Ex. 6, find the area of the entire surface of a brick $8 \times 4 \times 2$ in.

8. Also of a block of ice $2\frac{1}{2}$ ft. long, $1\frac{1}{2}$ ft. wide, and 7 in. thick.

9. If the weight of a cubic foot of iron is 600 lb., obtain a formula for the weight of an iron bar whose length is l feet, whose width is w inches, and thickness is t inches.

10. By use of the formula obtained in Ex. 9, find the weight of an iron bar 10 ft. long, 8 in. wide, and $1\frac{1}{2}$ in. thick.

11. Also find the weight of an iron bar 12 ft. long, 8.4 in. wide, and 1.6 in. thick.

12. If an iron bar is to be 10 in. wide, how thick must it be in order that a part of it 1 ft. long shall weigh 50 lb.? Use the formula obtained in Ex. 9.

13. If a body is moving at the rate of 100 ft. per second, how many miles will it go in one hour?

14. If a body is moving at the rate of v feet per second, obtain a formula for d, the number of miles it will go in 1 hour.

15. A certain boy can run 100 yd. in 10 seconds, that is, at the rate of 30 ft. in 1 second. If he could continue to run at the same rate, how many miles would he go in 1 hour? In $\frac{1}{2}$ hour? In 2 hours? Solve by use of the formula found in Ex. 14.

16. A certain athlete can run the quarter mile in 48 seconds. At this rate how many miles would he go in 1 hour? In 4 hours? Solve by use of the formula found in Ex. 14.

CHAPTER VII

EQUATIONS (*continued*)

PART I

61. Members of an Equation. The algebraic expression to the left of the sign of equality is called the *first member* of the equation; that to the right of the sign of equality is called the *second member*.

Thus, in the equation $3x - 1 = 2x + 3$, the first member is $3x - 1$; the second member is $2x + 3$.

The members of an equation are sometimes called *sides* of the equation.

The members of an equation are similar to the pans of a set of weighing scales which must be kept balanced.

62. The Transposition of a Term is moving the term from one member of an equation to the other member. We shall see that when a term is transposed, the sign of the term must be changed.

Ex. 1. Find the value of x in $x - 5 = 7$.

PROCESS WITHOUT TRANSPOSITION

We have given $\qquad x - 5 = 7.$

$$5 = 5.$$

Adding 5 to each of the equals, $\quad x = 7 + 5,$

or $x = 12.$ *Ans.*

PROCESS WITH TRANSPOSITION

We have given $\qquad\qquad x - 5 = 7.$

Transferring 5 to the right-hand member of the equation and changing its sign, $\qquad\quad x = 7 + 5,$

$\qquad\qquad\qquad\qquad\qquad\qquad x = 12.$ *Ans.*

Hence transposition is a short way of adding equal numbers to the two members of an equation. The labor saved by means of transposition is more evident when several terms are to be transposed at the same time.

Ex. 2. Solve $5x + (x - 3) = 8 - (x + 4)$.

Removing parentheses, $5x + x - 3 = 8 - x - 4.$

Transposing terms, $\quad 5x + x + x = 8 - 4 + 3.$

Collecting terms, $\quad\quad\quad 7x = 7.$

$$x = 1. \textit{ Ans.}$$

Let the pupil check the work.

63. The Method of Solving a Simple Equation may now be stated as follows:

Clear the equation of parentheses by performing the operations indicated by them;

Transpose the unknown terms to the left-hand side of the equation, the known terms to the right-hand side;

Collect terms;

Divide both members by the coefficient of the unknown number.

EXERCISE 64

ORAL

Solve at sight by use of transposition:

1. $x - 3 = 2.$	**9.** $x - a = b.$	**17.** $x + a = 3a.$
2. $x - 7 = 4.$	**10.** $2x - 4 = 6.$	**18.** $3x = 6 - x.$
3. $x - 11 = -5.$	**11.** $3x - 2 = 7.$	**19.** $4x = 12 - 2x.$
4. $x + 2 = 5.$	**12.** $2x + 4 = 12.$	**20.** $4x = 12 + 2x.$
5. $x + 7 = 3.$	**13.** $5x - 4 = 16.$	**21.** $5x = 18 - x.$
6. $x + 8 = -5.$	**14.** $2x - 3 = -11.$	**22.** $6x = 15 + x.$
7. $x + a = 5.$	**15.** $3x + 2 = -10.$	**23.** $2x + 3b = 9b.$
8. $x + a = b.$	**16.** $x - 3 = -5.$	**24.** $2x - 3b = 9b.$

25. When $x = 2$, does $x + 4 = 7$?

26. When $x = 5$, does $2x - 3 = 7$?

27. When $x = -2$, does $3x + 4 = 12$?

28. When $x = \frac{1}{2}$, does $4x = 2x + 1$?

29. When $x = 1$ does $5x - 2 = 4x - 1$?

30. Is 3 the root of $2x + 6 = 11$?

EXERCISE 65

Solve by use of transposition and check:

1. $2x = 15 - 3x$.

2. $15 + 3x = 27$.

3. $4x - 11 = 29$.

4. $16x + 3 = 15x + 7$.

5. $14x - 10 = 12x - 3x$.

6. $3x - 7 = 14 - 4x$.

7. $2x - 7 = 8 + 5x$.

8. $2x - (x - 1) = 5$.

9. 2 ft. $+ x = 12$ ft.

10. 7 in. $+ x = 2$ ft.

11. $3x - 1 + 2x = 3x + 7$.

12. $12 - 2x + 19 = 15 - 7x + 6$.

13. $7 + 2x + 2 = 4x - 6 + 3x$.

14. $3x + 5 + 2 + 2x = 10x - 8 - 2$.

15. $3(x + 2) = 2(x - 3) + 17$.

16. $x^2 - x(x + 5) = x + 12$.

17. $2x - 3(x - 3) + 2 = 0$.

18. $3 - 2(3x + 2) = 7$.

19. $(x - 8)(x + 12) - (x + 1)(x - 6) = 0$.

20. $5(x - 3) - 7(6 - x) + 3 = 24 - 3(8 - x)$.

21. $3(x - 1)(x + 1) = x(3x + 4)$.

22. $4(x - 3)^2 = (2x + 1)^2$.

Find the value of the letter in each of the following:

23. $5 - 2b = 1$.

24. $5(c - 1) = 12 - c$.

25. $3(y - 4) = 5(2 - y)$

26. $r - 3(r - 1) = 5$.

27. $10p + 5 - 7p = 9p - 7 - 4p$.

28. $11y + 5 - 8y = 9y - 7 - 5y$.

29. $5a - (2a + 3) = 12$.

30. $17w = 18 - 5(2 - 3w)$.

31. $9\,d = 3 + 2(1 + 4\,d)$.

32. $2(4 - r) - 3(r - 6) = 0$.

33. Solve the equation $3\,x - 7 = 18 - 2\,x$ without the use of transposition. Now solve by the use of transposition. By counting the number of symbols in each process compare the amount of work in the two methods.

34. Find whether $x + (2\,x - 1) = 14$ when $x = 1$. When $x = 5$. When $x = 7$.

EXERCISE 66
VERBAL PROBLEMS

1. In how many hours can a boy walk x miles at a miles an hour?

2. How many dimes in x dollars and y half-dollars?

3. What number is 40 less than x? What number is x less than 40?

4. By how much does $a + b$ exceed x?

5. How many cents did a girl have left if she had \$5 and spent 15 cents? If she had a dollars and spent b cents?

6. A boy had a dollars, received b cents, and then spent c cents. How many cents did he have left?

7. What is the interest on a dollars at b per cent for c years?

8. Express algebraically the following statement: a divided by b gives c as a quotient and d as a remainder.

9. Express the following as an equation: twice x exceeds 24 by 6.

10. Draw a rectangle and mark its width as 5. If its length exceeds its width by 5, what is its length? What expression would you put inside the rectangle to denote its area? What would the perimeter of the rectangle be?

11. Draw a square and mark one side as 3. What is the perimeter of the square? Its area?

12. Draw a square and mark one of its sides as *x*. What is its perimeter? Its area?

13. If the side of a square is *x*+3. what is its perimeter? Its area?

14. If a rod is 9 ft. long, and 3 ft. are cut off, how many feet are left?

15. If a rod is *x* feet long and 3 feet are cut off, how many feet are left?

16. If a line is 8 inches long and one part is denoted by *x*, what denotes the other part?

17. If a school contains *x* pupils and 46 of them are boys, how many of them are girls?

18. If a school contains *x* pupils and *b* of them are boys, how many of them are girls?

EXERCISE 67

1. Separate $84 into two parts such that one part is three times as large as the other.

2. Separate $84 into two parts such that one part exceeds the other by $12.

3. Separate $84 into three parts such that the first part is twice as large as the second, and the second part is twice as large as the third.

4. A and B pay together $100 in taxes; if A pays $22 more than B, what does each pay?

5. Two boys made $67.50 one summer by taking passengers on a launch. The boy who owned the launch received twice as large a share of the profits as the other boy. How much did each receive?

6. How many grains of gold are there in a gold dollar, if the gold dollar weighs 25.8 grains and 9 parts of the dollar are gold and 1 part copper?

7. A ball nine has played 64 games and won 12 more than it has lost. How many games has it won?

8. A man left $21,000 to his wife and four daughters. If the wife received three times as much as each daughter, how much did each receive?

9. If he had left $21,000 so that the wife received $10,000 more than each daughter, how much would each have received?

10. Find three consecutive numbers whose sum is 63.

11. The difference of the squares of two consecutive numbers is 43. Find the numbers.

12. John solved a certain number of examples, and William did 12 less than twice as many. Together they solved 96. How many did each solve?

13. Three boys earned together $98. If the second earned $11 more than the first, and the third $28 less than the other two together, how many dollars did each earn?

14. The sum of three numbers is 50. The first is twice the second, and the third is 16 less than three times the second. Find the numbers.

15. A farmer bought a cow and a horse. For the horse he paid $13 more than twice as much as he paid for the cow. For both together he paid $94. How much did he pay for each?

16. In reducing iron ore in a furnace, 7 times as many carloads of coke as of limestone are used, and 8 times as many carloads of iron ore as of limestone. If 800 carloads in all are used on a certain day, how many carloads of each are used?

17. A man spent $3.24 for coffee and sugar, buying the same number of pounds of each. If the sugar cost 5 cents a pound and the coffee 22 cents, how many pounds of each did he buy?

18. If a certain number is diminished by 24 and the result multiplied by 3, the final result will be 78. Find the number.

19. If a certain sum of money is increased by $150 and the result multiplied by 4, the final result will be $1000. What is the original sum of money?

20. Find the number whose double exceeds 24 by 6.

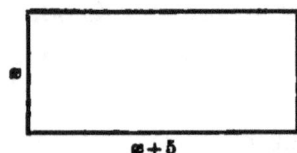

21. The perimeter of a given rectangle is 26 feet, and the length of the rectangle exceeds the width by 5 feet. Find the dimensions of the rectangle.

22. The perimeter of a given rectangle is 18 yd., and the length exceeds the width by 3 ft. Find the dimensions.

23. The length of a rectangle exceeds a side of a given square by 3 inches and the width of the rectangle is 2 inches less than a side of the square. If the area of the rectangle equals the area of the square, find a side of the square.

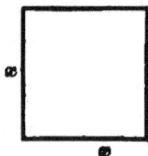

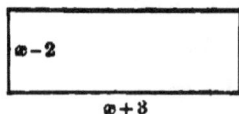

SUG. Denote the sides of the square and rectangle as in the diagram.

Since the areas of the two figures are equal,

$$x^2 = (x+3)(x-2), \text{ etc.}$$

24. The length of a rectangle exceeds a side of a given square by 8 ft. and the width of the rectangle is 4 ft. less than a side of the square. If the area of the rectangle equals the area of the square, find a side of the square.

SUG. First, draw a diagram of both the square and rectangle.

25. If one side of a square is increased by 4 yd., and an adjacent side by 3 yd., a rectangle is formed whose area exceeds that of the square by 47 sq. yd. Find a side of the square. (First draw a diagram of both the square and rectangle.)

26. The perimeter of a rectangle is 120 ft., and the rectangle is twice as long as it is wide. Find its dimensions. (Draw diagrams.)

27. A rectangle is 8 ft. longer than it is wide and the perimeter is 120 ft. Find the dimensions of the rectangle.

28. What number subtracted from 100 gives a result equal to the sum of 14 and the number?

29. Find two consecutive integers such that the first plus 5 times the second equals 53.

30. Two men, A and B, start from places 35 miles apart and walk toward each other at the rate of 4 miles and 3 miles an hour respectively. How many hours will it be before they meet?

Sug.	NUMBERS DEALT WITH	SYMBOLS
	Number of hours A travels	$= x.$
	Number of hours B travels	$= x.$
	Number of miles A travels	$= 4\,x.$
	Number of miles B travels	$= 3\,x.$
	Total no. of mi. traveled by A and B	$= 35.$

The formation of an equation is greatly facilitated by drawing a diagram; thus:

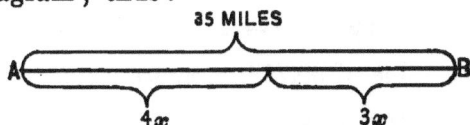

35 MILES

A   B

$4x$ $3x$

31. Two men, A and B, start from places 42 miles apart and walk toward each other, at the rate of 4 and 3 miles per hour respectively. How many hours will it be before they meet?

32. Make up and work an example similar to Ex. 31.

33. Two bicyclists, A and B, start respectively from New York and Philadelphia, 90 miles apart, and ride toward each other. A rides 8 miles per hour, and B rides 12. How long and how far will A ride before meeting B?

34. Boston is 234 miles from New York. If two auto trucks start from the two cities at the same time and travel toward each other at the rate of 12 and 14 miles per hour respectively, how far will each go before they meet?

EXERCISE 68

REVIEW

1. Express the following in as few terms as possible: $6x^3 - 3ab + x^2 - 4y^3 - 7ab - c + 7b - 5x^3 + 10ab - 6y^3 - 5b + 4c + 8x^2$.

2. Subtract $3a^2 - 5b^2 + 5 + c$ from $5a^2 - 2b^2 - 7 + 3b$.

3. Simplify $7 - [5 - (a - b - 3) + 4 - 3b]$.

4. Simplify $3(2x + 5y)^2 - 2(2x - 5y)^2$.

5. Solve $x^2 - 2x(x + 3) = 9 - x^2 - 3x$.

6. Two partners made $8500 in one year. The older partner received $1000, more than twice as much as the other partner. How much did each partner receive?

7. Multiply $3x^2 - 2x + 5$ by $2x - 3$.

8. Divide $12a^3 - 14a^2 - 19a + 15$ by $3a - 5$.

9. Solve for y: $(2y - 3)^2 - (2y + 3)^2 = 96$.

10. Find the value of 99.7^2 in a short way.

11. Find in a short way the cost of 127 acres at $133 per acre.

12. Find the numerical value of $p + prt$ when $p = 350$, $r = .06$, and $t = 2\frac{1}{2}$.

13. Also of $\pi r(r + l)$ when $\pi = \frac{22}{7}$, $r = 8$, and $l = 13$.

14. Also of $vt + \frac{1}{2} gt^2$ when $v = 200$, $t = 8$, and $g = 32$.

15. Reduce $(2\,aaa - 5\,bb)\,(2\,aaa + 5\,bb)$ to its simplest form.

16. By use of the formula $c = \frac{4}{7}r$, find the circumference of a wheel 1 ft. 2 in. in diameter.

17. The circumference of a hot air pipe is measured and found to be 27 in. Using the formula $c = \frac{4}{7}r$, find the diameter of the pipe to two decimal places.

18. An army camp was to be rectangular in shape and was to contain 6 sq. mi. If its front was to be $2\frac{1}{4}$ mi., how deep must the tract be? Solve by use of the formula $a = lw$.

19. If 12 is added to a certain number, the result equals three times the given number. Find the number.

20. A cord of wood is a pile 8 ft. long, 4 ft. wide, and 4 ft. high. Find a formula for the number of cords (C) in a rectangular pile, l feet long, w feet wide, and h feet high.

21. By use of the formula obtained in Ex. 20, find the number of cords in a pile 20 ft. long, 12 ft. wide, and $6\frac{1}{2}$ ft. high.

22. The wheat crops of the United States at 5 year intervals from 1875 to 1920 are given in the adjoining table. Make a graph of these facts. (Let each space on the vertical axis represent 50 million bushels.)

Year	Million of Bushels	Year	Million of Bushels
1875	292	1900	522
1880	499	1905	693
1885	357	1910	635
1890	399	1915	1026
1895	467	1920	833

23. $12 \times 14 = (13 - 1)(13 + 1) = 169 - 1 = 168$.

Hence commit to memory the following:

$13^2 = 169$	$15^2 = 225$	$17^2 = 289$	$19^2 = 361$
$14^2 = 196$	$16^2 = 256$	$18^2 = 324$	$20^2 = 400$

and then give at sight the value of the following products:

(1) 13×15 (3) 15×19 (5) 13×19 (7) $13\frac{1}{2} \times 14\frac{1}{2}$
(2) 17×19 (4) 13×17 (6) $14\frac{1}{2} \times 15\frac{1}{2}$ (8) $14\frac{1}{4} \times 15\frac{3}{4}$

24. Fill in each blank space in the following table:

SUBJECT	RULE	FORMULA
Area of rectangle	Area equals length multiplied by width	
Volume of solid		$v = lwh$
Interest	Interest equals principal multiplied by rate, by time	
Moving body		$d = vt$
Percentage	Percentage equals base multiplied by rate	
Moving body	Distance equals rate multiplied by time	
Cubic yards in cellar		$Y = \dfrac{lwh}{27}$
Multiplication		$P = Mm$
Division		$D = dq + r$

25. Express the following as an equation and then solve the equation: twice x exceeds 24 by 6.

26. A wholesale fruit dealer found during a certain season that when the price of strawberries was 20 cents per quart he sold 100 qt. per day; when the price was 15 cents he sold 500 qt. daily; when the price was 10 cents he sold 1000 qt.; when price was 5 cents he sold 3000 qt. Construct a graph for these facts. (Let each space on the horizontal axis represent 5 cents and each space on the vertical axis represent 500 qt.)

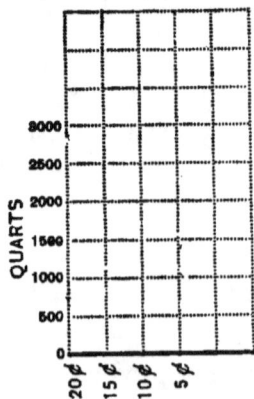

A graph of facts like these is called a *demand curve.*

27. A certain house cost d dollars and rents for m dollars per month. The taxes are t dollars per year and the other expenses are e dollars per year. Find a formula

for i, the annual income from the house; also another formula
for r, the rate per cent of income which the investment pays.

28. A certain house cost $\$4000$ and rents for $\$40$ a month.
The taxes are $\$60$ and the other expenses $\$140$ per year.
By use of the formula found in Ex. 27, compute the per cent
of income from the investment.

29. A farmer wishes to make a cattle yard twice as long as
it is wide and he has 96 yd. of fencing. What will be the
length and width of the yard? Draw a diagram of the yard,
using a convenient scale.

PART II

64. Aids in Solving Equations. It is well to note at
this point that in solving problems and equations we have
been making use of the following principles :

1. *The same number or equal numbers may be added to
both members of an equation without changing the value of
the root of the equation.*

2. *The same number or equal numbers may be subtracted
from both members of an equation without changing the root
of the equation.*

3. *Both members of an equation may be multiplied by any
known number (except zero) without changing the value of
the root of the equation.*

4. *Both members of an equation may be divided by any known
number (except zero) without changing the value of the root.*

Ex. Solve $3x - 5 = x + 7$ by aid of these principles
and point out each principle as it is used.

$$3x - 5 = x + 7$$
$$\underline{+5 \qquad +5}$$
$$3x = x + 12 \qquad \text{(Principle 1)}$$
$$\underline{-x \quad -x}$$
$$2x = 12 \qquad \text{(Principle 2)}$$
$$x = 6. \quad \textit{Ans.} \quad \text{(Principle 4)}$$

Solve and check :

1. $.3x + .2 = .5x - 2.$ 3. $.5(x-2) = .2(2x+3)-1.$
2. $3x + 7 = .2x + 35.$ 4. $.25x - 2 = .2x + 3.$
5. $(.2x + .2)(.4x - .3) = (.4x - .4)(.2x + .3).$
6. $4\{5x - 4(x-1)\} = 3\{4(x+1) - 7x\}.$
7. $(2x+1)(2x+3) - (x+1)(x+3) = 3x^2 + 20.$
8. $12 = (x+6)(6x+1) - (2x+3)(3x-2).$
9. $(2x+7)(3x-5) - (6x+7)(x-5) = 10x - 12.$

Solve the following equations, using short methods of multiplication wherever possible :

10. $(2x+1)^2 - (2x-1)^2 = 16.$
11. $4(x-2)^2 - 5[3x + (x-4) + 72] = 4x^2.$
12. $(5x+2)(3x-4) = 15(x-1)^2 + 9.$
13. $(x+1)(x+2)(x-3) - (x-2)(x+1)^2 + 3x + 8 = 0.$
14. $(3x+1)^2 - (3x-5)(3x+7) = 2x - 7.$
15. $3(2x+3)^2 = 12(x+3)^2 - (x+11).$

In each of the following, find the value of the letter to two decimals (that is, to the nearest hundredth):

16. $.7(a + .12) = .9(a - .12).$
17. $.38(2y + 5) - .16(y + 20) = 0.$
18. $.01(2p-5) = .012(p+5).$
19. $2.3(3.5r - 1.5) - 1.5(2.5r + 1.5) = 0.$

If $x=4$, fill in the second member of the following:

20. $5(x+5) - 4(x-6) = (?).$
21. $x^2 - (x+2)^2 - (x+3)(x-1) = (?).$
22. If $x = 3$, does $3x - (x-2) = 7$?
23. Is .2 the root of the equation $5x + 2 = 4$?
24. Solve $5x - 7 = x + 16$ by use of the principles stated in § 64, and point to each principle as it is used.

65. Identities and Equations. If we take the expression $(x-2)(x+2) = x^2 - 4$, and substitute $x = 1$, we obtain $-3 = -3$. The two members of the expression are found to be equal.

Similarly, they are found to be equal if we let $x = 2, 3, 4$, etc.; $0, -1, -2$, etc.; or any number. An expression having this characteristic is termed an *identity*.

An **identity** is an equality whose two members are equal for all values of the unknown quantity (or quantities) contained in it.

An **equation** in one unknown is an equality which is true for only one value of x.

Hence, the sign $=$ is used in two senses in elementary algebra, viz. to indicate sometimes an equation, and sometimes an identity. The context enables us to decide readily which of these two meanings the sign $=$ has in any given case.

Later it will be found useful to use the mark $\equiv$ to indicate an identity, and $=$ to indicate a conditional equation, or equation proper.

EXERCISE 70

1. Does $(x-3)(x+3) = x^2 - 9$ when $x = 2$? When $x = 1$? When $x = 0$? When $x = 4$? When $x = 9$? Is it therefore an identity or an equation?

2. Does $(x-2)(x+4) = x^2 + 2x - 8$ when $x = 1$? When $x = 2$? When $x = 3$? For what values of x does $(x-2)(x+4) = x^2 + 2x - 8$?

3. For what values of x does $(x-3)^2 = x^2 - 6x + 9$?

4. For how many values of x does $(x-3)^2 = x^2$?

5. State which of the two following is an identity:
(1) $(x-2)(x+4) = x^2 + 2x - 8.$
(2) $(x-2)(x+4) = x^2.$

State which of the following are identities and which equations:

6. $(x-3)(x+4) = x^2 + x - 12.$ **11.** $(x+5)^2 = 17 + x^2.$

7. $(x-3)(x+4) = x^2.$

8. $(x+5)^2 = x^2 + 10x + 25.$ **12.** $\dfrac{x^2 - 25}{x - 5} = x + 5.$

9. $(x-2)(x+2) = x^2 - 4.$

10. $3x - 7 = x + 9.$ **13.** $\dfrac{x^2 + 1}{x} = x + \dfrac{1}{x}.$

14. Show that the difference between the squares of any two consecutive numbers equals the sum of the numbers.

Sug. Denote the numbers by x and $x + 1$.

15. Write an identity and an equation of which the first members are the same.

16. Prove that the sum of any three consecutive numbers equals three times the middle number.

Sug. Let the three numbers be indicated by $n - 1$, n, and $n + 1$.

17. Find a similar result for the sum of five consecutive numbers. Of seven consecutive numbers.

EXERCISE 71

VERBAL PROBLEMS

1. How many times is m contained in a?

2. What is the shortest way of expressing the sum of twelve x's?

3. If the perimeter of an equilateral triangle is $2x$, what is the length of one side of the triangle?

4. What number must be added to m to obtain x?

5. A rectangle is 12 by 16 ft. Its length is increased by f feet, and its width by i inches. What is its new area in square feet?

6. If John is x years old and Mary is two years more than one half as old, express Mary's age. What was her age 3 years ago? What will her age be 2 years hence?

7. By how many inches does f feet, 3 inches exceed t feet, i inches?

8. A man earns d dollars a month and spends s dollars a month. How many dollars will he save in 3 years?

9. If John has m marbles and James has n marbles and they divide equally, how many will each boy have?

10. Is $2n$ an even or an odd number? Which is $2x+1$?

11. A boy had d dimes and n nickels and spent 15 cents. How many cents had he left?

12. If a boy has $\frac{2}{3}(x+4)$ dollars and spends one half of what he has, how much will he have left?

13. A man can do a piece of work in d days. He works n days. What part of the work does he do in the n days?

14. If 25 men each give x dollars toward a certain fund, how much do they give all together? If 20 men each give $x+1$ dollars, how much do they give all together?

15. If a boy walks at the rate of 3 miles an hour, how far will he walk in 5 hours? In a hours? In x hours? In $x+2$ hours?

16. A boy starts at a given time and walks 5 hours. Another boy then starts and rides a bicycle x hours until he overtakes the first boy. How many hours does the first boy walk?

17. One boy walks at the rate of a miles per hour, and another boy at the rate of b miles an hour. If they start at the same place and walk in opposite directions, how far apart will they be in x hours?

EXERCISE 72

1. Find five consecutive numbers whose sum shall be 3 less than six times the least.

2. Find three consecutive odd numbers whose sum is 63.

3. Find the number which is exceeded by 12 by as much as three times the number is exceeded by 24.

4. Find five consecutive numbers such that the last is twice the first.

5. A telegram at a 25–2 rate cost 47 cents. How many words were in the telegram?

Sug. A 25–2 rate means a cost of 25 cents for the first 10 words and 2 cents for each additional word.

6. Make up and work an example, similar to Ex. 5, concerning a telegram sent at a 40–3 rate.

7. A talk over a long distance telephone at a 50–7 rate cost 85¢. How many minutes did the talk last?

Sug. A 50–7 rate over a long distance telephone means a cost of 50 cents for the first 3 minutes and 7 cents for each additional minute.

8. For every nickel which a girl put in her savings bank, her father put in a dime. If her bank contained $18.75 at the end of one year, how many nickels did the girl save in that time?

9. For every dime that a boy spent for books, his father gave him a quarter to spend in the same way. If he spent $52.50 in all, how much did his father give him?

10. A purse contains $10.50 in dollar bills and quarters, but there are twice as many quarters as bills. How many are there of each?

11. How can $2.25 be paid in 5 and 10 cent pieces so that the same number of each is used?

12. How can $5.95 be paid in dimes and quarters, using the same number of each?

13. Separate $110 into four parts such that twice the sum of one part and $2 equals each of the other parts.

14. A man walked 15 mi., rode a certain distance, and then took a boat for twice as far as he had previously traveled. All together he went 120 mi. How far did he go by boat?

15. If 5 is subtracted from a certain number and the difference is subtracted from 115, the result equals three times the given number. Find the number.

16. A certain rectangle is three times as long as it is wide. If 20 ft. is added to its length and 10 ft. is deducted from its width, the area is diminished by 400 sq. ft. Find the dimensions of the rectangle.

17. A rectangle is 4 in. longer than it is wide. If its length is increased by 4 in., and its width diminished by 2 in., its area remains unchanged. Find the dimensions of the rectangle.

18. A tennis court is 42 ft. longer than it is wide   If a margin of 15 ft. on each end and of 10 ft. on each side is added, the area of the court is increased by 3240 sq. ft. Find the dimensions of the court.

19. A football field is 140 ft. longer than it is wide. Adding 20 ft. on each side and end of the field, increases the area by 20,000 sq. ft. Find the dimensions of the field.

20. A boy is three times as old as his brother. Five years hence he will be only twice as old. Find the present age of each.

21. A man is twice as old as his brother. 5 years ago he was three times as old. Find the present age of each.

22. How many pounds of coffee at 30 cents a pound must be mixed with 12 lb. of coffee at 20 cents a pound to make a mixture worth 24 cents a pound?

23. How many pounds of tea at 60 cents a pound must be mixed with 25 lb. of tea at 40 cents a pound, to make a mixture worth 45 cents a pound?

24. A man is 48 years old and his son is 18. How many years ago was the father four times as old as the son? Also how many years hence will the father be twice as old as the son?

25. A boy sold a certain number of newspapers on Monday, twice as many on Tuesday, on Wednesday 5 more than on Monday, and on Thursday 7 less than on Tuesday. If he sold 310 newspapers on the four days, how many did he sell on each of the days?

26. Twenty-five men agreed to pay equal amounts in raising a certain sum of money. Five of them failed to pay their subscriptions, and as a result each of the other twenty had to pay one dollar more. How much did each man subscribe originally?

27. A boy starts from a certain place and walks at the rate of 3 miles an hour. Three hours later another boy starts after the first boy, riding on a bicycle at the rate of 6 miles an hour. In how many hours will the second boy overtake the first? (Draw a diagram.)

28. If the boys had traveled in opposite directions, how many hours after the second boy started would it have been before they were 81 miles apart?

29. One boy starts from New York on a bicycle and travels toward Philadelphia, 90 mi. distant, at the rate of 8 miles an hour. One hour later another boy starts from Philadelphia and goes toward New York at the rate of 6 miles an hour. How long before they will meet?

30. New York and Washington are 228 miles apart. A train starts from New York at a given time and goes at the rate of 26 miles an hour, and two hours later a train starts from Washington and proceeds at the rate of 34 miles an hour? How long before they will meet?

31. Two boys start from New York and Philadelphia at the same time and travel toward each other until they meet. If one goes twice as fast as the other and they meet in $7\frac{1}{2}$ hours, what is the rate per hour of each boy?

32. A set out from a town, P, to walk to Q, 45 miles distant, an hour before B started from Q toward P. A walked at the rate of 4 miles an hour, but rested 2 hours on the way; B walked at the rate of 3 miles an hour. How many miles did each travel before they met?

EXERCISE 73

REVIEW

1. Express in as few terms as possible: $3.2\,x^2 - 2.5\,xy + .16\,y^2 + 1.5\,x^2 - .8\,y^2 - .32\,xy + .4\,y^2 - 1.5\,x^2 + .4\,xy$.

2. Subtract $.15\,a^2 + .3\,b^2 - 2.5\,ab$ from $-7\,a^2 - 4\,ab - 1.5\,b^2$.

3. Add $2\frac{1}{5}p^2 - 1.5\,p + \frac{7}{8}$, $.75\,p^2 + \frac{2}{3}p - .4$, $\frac{3}{4} - 6\frac{3}{4}p^2 - .5\,p$.

4. Simplify $3.2\,x^2 - [.8\,x^2 + (3.5\,x - \frac{1}{2} - .2\,x^2) - 1.5 - 3\,x]$.

5. Solve $.3\,x - 4 = .2\,x + .5$.

6. What is the root of an equation? How do you check your solution of an equation? Check Ex. 5.

7. Simplify $5\,x - 3\,(x - 2)(x + 7) + 3\,(x - 2)^2$.

If $a = 0$, $b = 1$, $c = 4$, $x = -2$, find the value of

8. $a\,(b + c) - 3\,x$.　　　　**9.** $\dfrac{3\,a + 5\,(2 + x)}{b + c}$.

10. $(c + 2\,x)\,(b - a) - 3\,(x + 4)\,(x + 5)$.

11. If it costs c cents per cubic yard to dig a cellar, find the number of dollars (D) it will cost to dig a cellar l feet long, w feet wide, and d feet deep.

12. By use of this formula find the cost of digging a cellar 30.5 ft. long, 19.5 ft. wide, and 6 ft. deep at 20 cents per cubic yard.

13. Multiply $x - 2 + 4\,x^2$ by $2\,x^2 - 1 - 5\,x$.

14. Find three consecutive numbers whose sum is 33.

15. Divide $x^4 + 3 - 6\,x^2 + x^5 + 8\,x - 11\,x^3$ by $x^2 - 3 - 2\,x$.

16. The record time for the 100-yd. swim at a certain date was $55\frac{2}{5}$ sec. This was $7\frac{2}{5}$ sec. more than 5 times that for the 100-yd. dash. What was the record time for the latter?

17. Show that there is a remainder when $x^4 + y^4$ is divided by $x + y$. Also when it is divided by $x - y$.

18. Insert the last three terms of $3x^2 - 5x^2 + 2ab - b^2$ in a parenthesis preceded by a minus sign.

19. For the eight half years of his high school course a boy's grades in science were as follows : 65, 68, 74, 78, 82, 81, 88, 94. Make a graph of these numbers. On the same diagram make a graph of his general averages which were as follows : 76, 74, 77, 78, 80, 81, 80, 82. What can be inferred from these two graphs as to the kind of life work for which the boy is best fitted ?

20. For the eight half years of her high school course a girl's grades in art and English, and her general average, were as fóllows :

Art	54	58	65	70	71	78	82	90
English . . .	68	66	64	62	70	66	72	70
Gen. average .	72	74	77	76	78	75	80	78

On the same diagram graph these three sets of numbers. What inference can be drawn from these graphs ?

21. Add (1) $(a + b)^2x$ (2) $(a + b)^2x$ (3) $(2a - 3b)^2x$

$(a - b)^2x$ $(a^2 - b^2)x$ $(2a + 3b)^2x$

22. Find the value in the shortest way of

(1) $2\frac{5}{6} + 3\frac{4}{7} + 4\frac{1}{6}$.

SUG. First add $2\frac{5}{6}$ and $4\frac{1}{6}$.

(2) $3\frac{3}{8} + 7\frac{1}{3} + 2\frac{5}{8}$.

(3) $6\frac{7}{12} + 4\frac{1}{2} - 2\frac{1}{12}$.

(4) $\frac{2}{3}x^2 + \frac{1}{2}x^2 + \frac{1}{3}x^2$

(5) $\frac{5}{6}a^2b + \frac{5}{8}a^2b + \frac{1}{6}a^2b$.

The above examples illustrate the important principle that, in adding terms, the order of the terms may be changed and the terms may be grouped together in any way that is convenient and efficient (called the commutative and associative laws of algebra).

23. In the shortest way find the value of

(1) $4 \times 97 \times 25$.

SUG. First multiply 25 by 4.

(2) $25 \times 62 \times 4$.

(3) $8 \times 92 \times 12\frac{1}{2}$.

(4) $16 \times 193 \times 37\frac{1}{2}$.

(5) $(a + b)(3a^2 - 2b^2)(a - b)$

24. Find the remainder when $x^5 + 1$ is divided by $x - 1$.

25. The length of a given rectangle exceeds the width by 5 ft. and the perimeter of the rectangle is 38 ft. Find the dimensions of the rectangle. Draw a diagram of the rectangle on squared paper, letting one linear space on the squared paper stand for 2 ft.

26. In the formula $L = \dfrac{.0045(P - t)(T - t)C}{P}$, find L when $P = 200$, $t = 4$, $T = 6$, and $C = 600$.

27. During a given year the cash and credit sales per month of a certain business were as follows:

MONTH	JAN.	FEB.	MAR.	APR.	MAY	JUNE
Cash Sales .	\$ 2500	\$ 2805	\$ 2720	\$ 2425	\$ 1916	\$ 1820
Credit Sales .	3010	3215	3618	3822	3592	2733

MONTH	JULY	AUG.	SEPT.	OCT.	NOV.	DEC.
Cash Sales .	\$ 1542	\$ 1432	\$ 2091	\$ 2172	\$ 2440	\$ 2856
Credit Sales .	2260	2142	3130	3792	3617	3592

On the same diagram make a graph of these two sets of figures. (Let one space on the vertical axis represent \$ 200, and let the lowest mark be \$1400.)

28. If the length of a barn is l feet, the width w feet, the height at the eaves e feet, and the height at the peak p feet, obtain the formula for the number of cows (C) which the barn will accommodate, if 500 cu. ft. of space be allowed for one cow.

29. By use of this formula determine the number of cows allowable in a barn 40 ft. long, 20 ft. wide, $12\frac{1}{2}$ ft. high at the eaves and $22\frac{1}{2}$ ft. high at the peak.

30. Form the power whose base is 5 and exponent 2. Also form the power whose base is 2 and exponent 5. Find the difference in value between these two powers.

CHAPTER VIII

FACTORING

PART I

66. Factors. In arithmetic, 3 and 2, for instance, are called the factors of 6, since $3 \times 2 = 6$. Similarly, in algebra, the factors of an expression are the quantities whose product is the given expression.

Thus, $x + 2y$ and $x - 2y$ are the factors of $x^2 - 4y^2$, since $(x + 2y)(x - 2y) = x^2 - 4y^2$.

The factors composing a single term are readily obtained by inspection   Thus the factors of $3\,a^2b$ are 3, a, a, b.

In what follows in this chapter we study the factors of expressions containing more than one term.

CASE I

67. A Factor Common to all the Terms of an Expression.

Ex. 1.   Factor $3x^2 + 6x$.

$$3\,x)\underline{3\,x^2 + 6x} = 3\,x(x + 2). \quad \textit{Factors.}$$
$$x + 2$$

After some practice the pupil may perform the short divi sion mentally and write the process in the following form:

$$3\,x^2 + 6\,x = 3\,x(x + 2). \quad \textit{Factors.}$$

Ex. 2.   Factor $6\,x^3y^2 - 9\,x^2y^3 + 3\,y^2$.

$$3\,y^2)\underline{6\,x^3y^2 - 9\,x^2y^3 + 3\,y^2} = 3\,y^2(2\,x^3 - 3\,x^2y + 1). \quad \textit{Factors.}$$
$$2\,x^3 - 3\,x^2y + 1$$

Let the pupil check the work in two ways.

Hence, in general,

Divide all the terms of the given expression by their largest common factor;

The factors will be the divisor and quotient.

EXERCISE 74

Factor the following and check the work for each example either by substitution or by multiplication or by both as the teacher may direct:

1. $5x + 5y$.
2. $5a + 15b$.
3. $ax + ay$.
4. $6a - 12b$.
5. $18x - 15y$.
6. $5 - 10x$.
7. $x^2 + xy$.
8. $bx - xz$.
9. $2a + 20b$.
10. $18x^2 - 54x$.
11. $3x + 3x^2$.
12. $p + prt$.
13. $12a^2b - 18ab^2$.
14. $abc + bcx$.
15. $3xyz + 6ayz$.

16. $abc - abcd$.
17. $25a^2bc + 20ab^2c$.
18. $a^3b^2c^3 - a^2b^3c^4$.
19. $15 - 75a^2b^2c^2$.
20. $2x^3 + 5x^2yz$.
21. $7a + 14a^3$.
22. $3a^3x^2 - 15a^2x^3$.
23. $25a^3b^3 - 15a^2b^2$.
24. $6x - 18y + 12z$.
25. $ax + ay + az$.
26. $ab^2 + ac^2 - ad^2$.
27. $3a^2 - 6ax + 9x^2$.
28. $2x + 4x^2 - 6x^3$.
29. $x^2 - x^3 - x^4 + x^5$.
30. $33a^3b^2c^3 - 22a^2b^3c^4 + 44a^3b^3c^5$.
31. $36ax^2y^3 + 60a^2x^2y^3 - 72ax^2y$.
32. $127 \times 75 + 136 \times 75 - 119 \times 75$.
33. $\frac{22}{7} \times 139 + \frac{22}{7} \times 124 + \frac{22}{7} \times 150$.

In the shortest way find the numerical value of:

34. $847 \times 915 - 847 \times 913$.
35. $312.75 \times 87 - 312.75 \times 84$.
36. $8 \times 11 \times 23^2 + 7 \times 11 \times 23^2 - 5 \times 11 \times 23^2$.
37. $\pi R^2 + \pi r^2$ when $\pi = \frac{22}{7}$, $R = 8$, and $r = 6$.
38. $\pi R^2 - \pi r^2$ when $\pi = \frac{22}{7}$, $R = 410$, and $r = 60$

CASE II

68. A Three-termed Expression that is a Square. By § 55 a three-termed expression is a square when it can be arranged so that its first and last terms are squares and positive, and the middle term is twice the product of the square roots of the end terms.

Thus $x^2 + 8x + 9$ is not a square since the middle term, $8x$, is not twice the product of x and 3. But if we change $8x$ to $6x$ the expression becomes a square.

Ex. 1. Factor $16x^2 + 24xy + 9y^2$.

$$16x^2 + 24xy + 9y^2 = (4x + 3y)(4x + 3y). \quad Ans.$$

Ex. 2. Factor $4x^4 - 12x^3y + 9x^2y^2$.

$$4x^4 - 12x^3y + 9x^2y^2 = x^2(4x^2 - 12xy + 9y^2)$$
$$= x^2(2x - 3y)^2. \quad Ans.$$

Hence in general, to factor a three-termed expression that is a square, after arranging terms as above,

Take the square roots of the first and last terms, and connect these by the sign of the middle term;
Take the result as a factor twice.

EXERCISE 75

Supply a missing term which will make each of the following a square :

1. $x^2 + (\ ) + 4$.

2. $x^2 + (\ ) + 9$.

3. $16a^2 + (\ ) + 1$.

4. $4x^2 + (\ ) + 9y^2$.

5. $16a^2 + (\ ) + 9$.

6. $4x^2 - (\ ) + 25y^2$.

7. $1 - (\ ) + 36x^2$.

8. $25a^2 - (\ ) + 9b^2$.

9. $a^2 + 6ab + (\ )$.

10. $x^2 - 12x + (\ )$.

11. $9p^2 + 12pq + (\ )$.

12. $(\ ) - 20ax + 4x^2$.

13. $81a^2 - (\ ) + 16b^2$.

14. $100x^2 - 180x + (\ )$.

Factor and check :

15. $x^2 + 2\,ax + a^2$.

16. $y^2 + 2\,y + 1$.

17. $a^2 + 6\,a + 9$.

18. $p^2 + 4\,pq + 4\,q^2$.

19. $25\,x^2 + 10\,xy + y^2$.

20. $9\,x^2 + 6\,x + 1$.

21. $1 + 8\,a + 16\,a^2$.

22. $25\,a^2 - 30\,ax + 9\,x^2$.

23. $x^2 - 12\,x + 36$.

24. $9\,a^2 - 30\,ab + 25\,b^2$.

25. $256\,x^2 - 32\,x + 1$.

26. $9\,m^2n^2 - 42\,mn + 49$.

27. $16\,a^2 - 24\,ay + 9\,y^2$.

28. $25\,x^2 - 10\,x + 1$.

29. $x^2 - 20\,xy + 100\,y^2$.

30. $49\,c + 28\,bc^2 + 4\,b^2c^3$.

31. $a^3b^2 + 4\,a^3b + 4\,a^3$.

32. $xy^2 + 2\,xy + x$.

33. $2\,m^2n - 4\,mn + 2\,n$.

34. $4\,x^3 + 44\,x^2y^2 + 121\,xy^4$.

35. $81\,a^3b + 126\,a^2b^2 + 49\,ab^3$.

36. $8\,a^2y - 40\,axy + 50\,x^2y$.

37. $30\,x^2y + 3\,x^4 + 75\,y^2$.

38. $a^3x + ax^3 - 2\,a^2x^2$.

39. $x^{10} + 2\,x^5y + y^2$.

40. $\frac{4}{9}\,x^2 + 4\,xy + 9\,y^2$.

41. $\frac{1}{4}\,x^2 + \frac{1}{3}\,xy + \frac{1}{9}\,y^2$.

42. $.04\,a^2 - .12\,ab + .09\,b^2$.

43. $25\,a^2 - 30\,ax + 9\,x^2$.

44. $p^2q^3 - 2\,pq^2x + qx^2$.

Factor the following examples in Case I and II, and check each result:

45. $x^3 - 4\,x^2 + 4\,x$.

46. $x^3 - 3\,x^2 + 4\,x$.

47. $m^5 - 2\,m^4n + m^3n^2$.

48. $a^2x^2 - 6\,a^2x + 4\,a^2$.

49. $a^3x^2 - 8\,a^3x + 16\,a^3$.

50. $10\,a^2 - 20\,a + 10$.

51. $20\,a^2 - 10\,a + 10$.

52. $16\,a^2p^2 - 24\,a^2p + 9\,a^2$.

53. $8\,x^3 + 16\,x^2y + xy^4$.

54. $16\,m^2n^3 - 9\,mn^3 + n^3$.

CASE III

69. The Difference of Two Squares.

From § 57, p. 103, $(a + b)(a - b) = a^2 - b^2$.

Hence, $a^2 - b^2 = (a + b)(a - b)$.

But any algebraic quantities may be used instead of a and b. Hence, in general, to factor the difference of two squares,

Take the square root of each square ;

The factors will be the sum of these roots and their difference.

Ex. 1.   Factor $x^2 - 16\,y^2$.

$$x^2 - 16\,y^2 = (x + 4\,y)(x - 4\,y). \quad \textit{Factors.}$$

Ex. 2.   Factor $x^4 - y^4$.

$$x^4 - y^4 = (x^2 + y^2)(x^2 - y^2)$$
$$= (x^2 + y^2)(x + y)(x - y). \quad \textit{Factors.}$$

EXERCISE 76

Factor and check :

1. $x^2 - 4$.   3. $9\,a^2 - b^2$.   5. $25 - a^2$.   7. $121 - x^2$.

2. $a^2 - x^2$.   4. $16 - y^2$.   6. $y^2 - 49$.   8. $1 - x^2$.

9. $1 - 9\,x^2$.   19. $100 - 81\,m^4$.   29. $242 - 2\,x^2$.

10. $64 - x^2$.   20. $16\,x^4 - y^4$.   30. $3\,x^3 - 75\,xy^6$.

11. $9\,a^2b^2 - 49\,x^2$.   21. $4\,x^6 - a^8$.   31. $a^4 - 81\,b^4$.

12. $81\,x^2 - y^2$.   22. $16\,q^4 - 49\,t^2$.   32. $ax^6 - ax^2$.

13. $16\,x^2 - 25\,y^2$.   23. $a^4 - x^4$.   33. $a^3x - x$.

14. $1 - 144\,a^2$.   24. $x^4 - 1$.   34. $225\,x^2 - y^2$.

15. $9\,p^2 - 64\,q^2r^2$.   25. $16\,a^4 - y^4$.   35. $2\frac{1}{4}\,x^2 - \frac{4}{9}\,y^2$.

16. $4\,a^2m^2 - 9\,n^2$.   26. $a^8 - x^8$.   36. $\frac{4}{25}\,a^2 - 9\,b^2$.

17. $25 - 16\,a^2$.   27. $x^8 - 1$.   37. $.09\,x^2 - .16\,y^2$.

18. $4\,a^2 - 49\,b^2x^2$.   28. $x^4 - 9\,a^2x^2$.   38. $.25\,y^2 - \frac{1}{25}\,b^2$.

39. $.01\,a^2 - .04\,b^2$.   40. $.81\,x^2 - .0025\,b^2$.

41. Compute the value of $123^2 - 77^2$ in a short way.

Sug. $123^2 - 77^2 = (123 + 77)(123 - 77)$
$$= 200 \times 46$$
$$= 9200. \quad \textit{Ans.}$$

42. $138^2 - 62^2$.   44. $214^2 - 186^2$.   46. $8.06^2 - 7.94^2$.

43. $512^2 - 488^2$.   45. $1.56^2 - 1.44^2$.   47. $27.3^2 - 12.7^2$.

Factor the following miscellaneous examples:

48. $a^2 - 4a$.

49. $a^2 - 4$.

50. $a^2 - 4a + 4$.

51. $a^3 - 4a$.

52. $a^3 - 4a^2 + 4a$.

53. $a^4 - 4a^3 + 4a^2$.

54. $x^3 - 6x$.

55. $x^3 - 9x$.

56. $a^3 + 9a^2x + 6ax^2$.

57. $a^3 + 6a^2x + 9ax^2$.

58. $a^5 - ax^4$.

59. $a^5 + ax^4$.

60. $9x^3 - x$.

61. $9x^2 - 6x + 1$.

62. $16m^2x^2 - x^2$.

63. $25x^2y^2 - 49z^2$.

CASE IV

70. An Expression of the Form $x^2 + bx + c$.

It was found in § 58, p. 105, that on multiplying two binomials like $x + 3$ and $x - 5$, the product, $x^2 - 2x - 15$, was formed by taking the algebraic sum of $+3$ and -5 for the coefficient of x (viz. -2), and their product (-15) for the last term of the result. Hence, in undoing this work to find the factors of $x^2 - 2x - 15$, the essential part of the process is to find two numbers which, added together, will give -2 and, multiplied together, will give -15.

Ex. 1. Factor $x^2 + 11x + 30$.

The pairs of numbers whose product is 30 are 30 and 1, 15 and 2, 10 and 3, 6 and 5. Of these, that pair whose sum is 11 is 6 and 5.

Hence, $x^2 + 11x + 30 = (x + 6)(x + 5)$. *Factors.*

Ex. 2. Factor $x^2 - x - 30$.

It is necessary to find two numbers whose product is -30, and whose sum is -1.

Since the sign of the last term is minus, the two numbers must be one positive and the other negative; and since their sum is -1, the greater number must be negative.

$$x^2 - x - 30 = (x - 6)(x + 5). \quad Factors.$$

Ex. 3.   Factor $x^2 + 3\,xy - 10\,y^2$.

Since $5\,y$ and $-2\,y$ added give $3\,y$ and multiplied give $-10\,y^2$,

$$x^2 + 3\,xy - 10\,y^2 = (x + 5\,y)(x - 2\,y). \quad Factors.$$

Hence, in general, to factor an expression of the form

$$x^2 + bx + c,$$

Find two numbers which, multiplied together, produce the third term of the expression and, added together, give the coefficient of the second term;

x(or whatever takes the place of x) plus the one number, and x plus the other number, are the factors required.

EXERCISE 77

Factor and check :

1. $x^2 + 3\,x + 2$.
2. $a^2 + 4\,a + 3$.
3. $y^2 + 5\,y + 6$.
4. $y^2 + 6\,y + 5$.
5. $a^2 + 7\,a + 12$.
6. $p^2 + 10\,p + 16$.
7. $p^2 + 10\,p + 21$.
8. $p^2 + 10\,p + 24$.
9. $x^2 - 9\,x + 20$.
10. $r^2 - 11\,r + 28$.
11. $y^2 - 16\,y + 63$.
12. $y^2 - 24\,y + 63$.
13. $x^2 + 2\,x - 15$.
14. $x^2 + 2\,x - 3$.
15. $x^2 - 2\,x - 8$.
16. $y^2 + 2\,y - 80$.

17. $p^2 - 3\,p - 70$.
18. $x^2 + 5\,xy + 6\,y^2$.
19. $x^2 - x - 6$.
20. $x^5 + 7\,x^4 - 44\,x^3$.
21. $x^2 - 11\,x + 30$.
22. $x^4 + x^3 - 30\,x^2$.
23. $x^2 + 6\,xy - 16\,y^2$.
24. $p^2 + 5\,p - 36$.
25. $x^6 - 5\,x^3 - 36$.
26. $x^4 - 5\,x^2y^2 - 36\,y^4$.
27. $x^4 + 3\,x^2 - 28$.
28. $x^4 - 2\,x^3 - 48\,x^2$.
29. $x^2 - 8\,xy - 48\,y^2$.
30. $x^2 + 13\,xy - 48\,y^2$.
31. $x^2 - 22\,x - 48$.
32. $x^2 - 4\,x - 165$

EXERCISE 78

Factor and check the following miscellaneous examples:

1. $3\,a + 3\,b.$
2. $a^2 - 6\,ax + 9\,x^2.$
3. $4 - 9\,x^2.$
4. $x^2 - 5\,x + 6.$
5. $4\,a^2 + 4\,ax + x^2.$
6. $3\,x^2y - 6\,xy^2.$
7. $7\,a - 14\,b - 21\,c.$
8. $a^2 - 25.$
9. $x^2 + 10\,x + 25.$

10. $a^2 - a - 20.$
11. $x^2 - 16\,x + 64.$
12. $a^2 - ax - 12\,x^2.$
13. $y^2 + 4\,y - 5.$
14. $16 - x^4.$
15. $x^2 - 4\,x + 3.$
16. $x^3 - x^2 - 6\,x.$
17. $x^3 - 4\,x.$
18. $x^3 + 6\,x^2 + 9\,x.$

19. $a^4 - 4\,x^2y^2.$
20. $a^4 - 4\,a^2y + a^2y^4.$
21. $a^8 - 1.$
22. $x^2 + 5\,ax + 6\,a^2.$
23. $x - x^5.$
24. $a^4 - 7\,a^2 + 12.$
25. $4\,x^3 - 4\,x.$
26. $xy^2 + 2\,xy + x^2.$
27. $x^3 - 9\,a^2x.$

28. $a^3 - 9\,a^2x + 20\,ax^2.$
29. $2\,m^2n - 4\,mn + 2\,n.$
30. $3\,y^2 - 3\,y - 18.$
31. $m - 64\,a^2m.$
32. $a^3b^2 + 4\,a^3b + 4\,a^3.$
33. $7\,a^2 + 42\,a + 63.$
34. $a^5 + 4\,a^4 + a^3.$
35. $242 - 2\,x^2.$
36. $ax^2 + 6\,axy - 16\,ay^2.$
37. $2\,x^4 - 8\,x^3 + 8\,x^2.$
38. $4 - 64\,a^4.$
39. $20\,x^2 - 45\,p^2q^2.$
40. $6\,x^2 - 24\,xy + 54\,y^2.$

41. $49\,c + 28\,bc^2 + 4\,b^2c^3.$
42. $3\,x^3 - 30\,xy^6.$
43. $6\,x^2 - 36\,x - 96.$
44. $4\,x^3 + 44\,x^2y^2 + 121\,xy^4$
45. $16\,a^2 - 16\,a - 96.$
46. $81\,a^3b + 126\,a^2b^2 + 49\,ab^3.$
47. $a^8x - x.$
48. $7\,a^2 + 21\,ab + 28\,b^2.$
49. $8\,a^2y - 40\,axy + 50\,x^2y.$
50. $\pi r^2 + \pi r^3 + \pi r^4.$
51. $30\,x^2y + 3\,x^4 + 75\,y^2.$
52. $x^{12}y^9 - yz^{16}.$
53. $x^5 - 18\,x^3 + 81\,x.$

54. Make up and work an example in each of the cases of factoring.

Make up and work an example in factoring where the following cases are combined:

55. Cases I and II.
56. Cases I and III.
57. Cases I and IV.
58. Cases II and III.

148   A FIRST BOOK IN ALGEBRA

EXERCISE 79

In the following examples, some of the graphs are
curved lines like those in Exercise 15; others are broken
lines like those in Exercise 49.   In working each example,
the pupil should note carefully, before drawing the graph
line, whether it is to be a broken or curved line.

1. On a certain winter day the temperatures at Minne-
apolis at 2-hour intervals were as follows:

Midnight $-12°$	8 A.M. $-2°$	4 P.M. $8°$
2 A.M. $-13°$	10 A.M. $4°$	6 P.M. $2°$
4 A.M. $-14°$	Noon $6°$	8 P.M. $-4°$
6 A.M. $-8°$	2 P.M. $10°$	10 P.M. $-8°$

Make a graph of these temperatures.

2. The following were the market prices of a certain
$50 bond on Jan. 1 for a number of years as indicated:
1910, $54; 1911, $58; 1912, $62; 1913, $64; 1914, $70;
1915, $62; 1916, $48; 1917, $42; 1918, $32; 1919, $35:
1920, $42; 1921, $46; 1922, $44; 1923, $42; 1924, $45.
Graph these numbers.

How is the effect of the Great War (1914–1918) shown
on the graph?   During which two years was the price of
the bond lowest?

3. At a certain school during ten weeks the average
attendance of pupils per week was as follows: 212, 224,
216, 182, 172, 156, 162, 180, 210, 204.   Make a graph of
these numbers.   At one time during the period graphed,
there was an epidemic of influenza in the town.   How is
its effect shown on the graph?   During which two weeks
was the epidemic at its worst?

4. The corn crop in the United States in millions of
bushels was as follows in the years specified:

Year	Millions of Bushels	Year	Millions of Bushels
1850	592	1890	1689
1860	839	1900	2151
1870	1094	1910	2886
1880	1717	1920	3209

Graph the above facts.

5. At a certain telephone exchange in a business district of a city the number of telephone calls for the different hours of the day were as indicated at the right:

Make a graph of these numbers.

For Hour Ending at	Number of Calls	For Hour Ending at	Number of Calls
7 A.M.	40	1 P.M.	316
8 A.M.	120	2 P.M.	416
9 A.M.	280	3 P.M.	580
10 A.M.	460	4 P.M.	520
11 A.M.	625	5 P.M.	420
NOON	610	6 P.M.	180

How do you account for the drop in the curve between noon and 2 P.M.?

By examining the two peaks of the graph, find at what hours the greatest amount of telephone business occurred.

6. During a given year the office expenses per month of a certain business were as follows:

Month . . .	Jan.	Feb.	Mar.	Apr.	May	June
Expenses . .	$327	$380	$370	$320	$340	$315

Month . . .	July	Aug.	Sept.	Oct.	Nov.	Dec.
Expenses . .	$300	$322	$205	$190	$182	$180

On September 1, a machine for addressing letters and other labor-saving devices were installed in the office. How is the result shown on the graph?

7. Graph the growth of the population of the United States using the following table:

Year	1790	1800	1810	1820	1830	1840	1850
Millions . . .	4	5	7	10	13	17	23

Year	1860	1870	1880	1890	1900	1910	1920
Millions . . .	31	39	50	63	76	92	106

From your graph determine as accurately as you can the population during the War of 1812. During the Mexican War (1846–1847).

By extending your graph determine as accurately as you can what the population is likely to be in 1930. Also in 1940.

It is to be noted that, in connection with graphs, the business man uses the term curve somewhat loosely to cover broken lines as well as curves. Also in constructing graphs, he often draws a graph as a broken line, where more properly it would be a curve. The broken line is easier to make and answers his needs.

PART II

71. A Prime Quantity in algebra is one which cannot be divided by any quantity except itself and unity; as a, b, $a^2 + b^2$, 17.

In all work in factoring, prime factors are sought, unless the contrary is stated.

Thus, $x^2 - y^2$ is a factor of $x^4 - y^4$, but it is not a prime factor. The prime factors of $x^4 - y^4$ are $x^2 + y^2$, $x + y$, $x - y$.

CASE I

72. A Polynomial All of Whose Terms Contain a Common Factor.

EXERCISE 80

Factor and check:

1. $\frac{1}{2} bx + \frac{1}{2} by.$ 4. $\frac{1}{2} hb_1 + \frac{1}{2} hb_2.$ 7. $vt + \frac{1}{2} gt^2.$

2. $\frac{1}{3} ap - \frac{1}{3} aq.$ 5. $2 \pi r^2 + 2 \pi rh.$ 8. $.4 x - .8 y.$

3. $\frac{3}{4} hx - \frac{3}{4} hy.$ 6. $108^3 - 108^2.$ 9. $\dfrac{an}{2} + \dfrac{bn}{2} - \dfrac{pn}{2}.$

10. $\dfrac{5x}{9} - \dfrac{160}{9}.$ 12. $\frac{2}{3} ax - \frac{2}{3} ay + \frac{2}{3} az.$

11. $.3 a - 1.2 b + .9 c.$ 13. $7(a + b)x + 5(a + b)y.$

14. $7(a + b)x^2 y + 5(a + b)xy^2.$

15. $9(2 x - a)^3 - 12(2 x - a)^5.$

16. $2 a(x - y) + b(x - y).$

17. $\frac{1}{3} a(x + y) + \frac{1}{2} b(x + y).$

18. $\frac{1}{2} h(a + b) - \frac{1}{2} h(x + y).$

19. $a(x + y) - b(x + y) - c(x + y).$

20. $a^2 - b^2 + 2(a - b).$ 22. $ax - ay + bx - by.$

21. $x^2 - y^2 + x - y.$ 23. $xy + 2 x + 3 y + 6.$

Compute in a short way by aid of Case I of Factoring:

24. $3.1416 \times 289 - 3.1416 \times 225.$

25. $8 \times 11 \times 23^2 + 7 \times 11 \times 23^2 - 5 \times 11 \times 23^2.$

26. $218 \times \frac{1}{365} \times .05 + 312 \times \frac{1}{365} \times .05 + 200 \times \frac{1}{365} \times .05.$

27. Collect the coefficients of x, y, and z in the expression, $3 x - 4 y + 5 z - ax - by - cz - bx + ay + az.$

Sug. The complete coefficient of x is $(3 - a - b)$; of y, $(- 4 - b + a)$ or $- (4 + b - a)$; of z, $(5 - c + a).$

Hence, the expression may be written,

$$(3 - a - b)x - (4 + b - a)y + (5 - c + a)z. \quad Ans.$$

In like manner collect the coefficients of x, y, and z in:

28. $mx - ny + 3 z + 2 x + nz - 4 y.$

29. $x - y - 2 z - ax + by - az - bx - ay + cz.$

30. $-7x + 12y - 10z - 2ax + 3bz - cy + 2bx - 6dy.$

31. $5y - 3acx - 5cdz - 4abx - 3cdy + 2cx - 4z - 5ax.$

Collect the coefficients of x^3, x^2, and x in :

32. $3x^3 + x - 2x^2 - ax^3 - 5 + ax^2 - 2ax - cx^3 - cx^2 - cx.$

33. $-x^2 - x - ax^3 + x^3 - ax + bx^2 - ax^2 - 3bx - 2bx^3 + 3a.$

34. $a^2x^2 - ax - a - b^3x^3 - 2b^2x^2 + 3bx - a^3x^3 - cx^2 + cx - c.$

CASE II

73. A Trinomial that is a Square.

Ex. 1. $.09x^2 - 1.2xy + 4y^2 = (.3x + 2y)^2.$ *Ans.*

Ex. 2.   Factor $(a + b)^2 + 4(a + b)x + 4x^2.$

$$(a + b)^2 + 4(a + b)x + 4x^2 = [(a + b) + 2x]^2$$
$$= (a + b + 2x)^2. \quad Ans.$$

EXERCISE 81

Factor and check :

1. $a^2 + 1.2a + .36.$ **5.** $.49 - 1.4y + y^2.$

2. $x^2 - 1.8x + .81.$ **6.** $(x + y)^2 + 6(x + y) + 9.$

3. $a^2 + .4a + .04.$ **7.** $(a - b)^2 - 4(a - b) + 4.$

4. $.25 + a + a^2.$ **8.** $(x - y)^2 + 2(x - y) + 1.$

9. $(3p - q)^2 + 10(3p - q) + 25.$

10. $(x + y)^2 - 12(x + y) + 36.$

11. $(5x - y)^2 - 12(5x - y) + 36.$

12. $(a - b)^2 - 2c(a - b) + c^2.$

13. $9(x + y)^2 + 12z(x + y) + 4z^2.$

14. $16(2a - 3)^2 - 16ab + 24b + b^2.$

15. $25(x - y)^2 - 120xy(x - y) + 144x^2y^2.$

16. $a^2 + b^2 + c^2 + 2ab + 2ac + 2bc.$

CASE III

74. The Difference of Two Squares.

Ex. 1.   Factor $(a + b)^2 - 9.$

$$(a + b)^2 - 9 = [(a + b) + 3][(a + b) - 3]$$
$$= (a + b + 3)(a + b - 3). \quad Ans.$$

Ex. 2. Factor $(3x + 4y)^2 - (2x + 3y)^2$.

$(3x + 4y)^2 - (2x + 3y)^2 = [(3x + 4y) + (2x + 3y)][(3x + 4y)$
$$- (2x + 3y)]$$
$$= (3x + 4y + 2x + 3y)(3x + 4y - 2x - 3y)$$
$$= (5x + 7y)(x + y). \quad Factors.$$

Let the pupil check the above examples.

Ex. 3. Compute the value of $72.9^2 - 47.1^2$ in a brief way by the aid of factoring.

$$72.9^2 - 47.1^2 = (72.9 + 47.1)(72.9 - 47.1)$$
$$= 120 \times 25.8$$
$$= 3096. \quad Ans.$$

EXERCISE 82

Factor the following and check :

1. $9 - .25 a^2$.
2. $1.44 x^2 - .49 y^2$.
3. $.16 a^2 - .25 b^2$.
4. $x^4 - .36 y^2$.
5. $.01 a^2 - 16 y^2$.
6. $.81 x^2 - .0025 s^2$.
7. $12.25 y^2 - \frac{1}{25} b^2$.

8. $x^8 - y^6$.
9. $(x + y)^2 - 1$.
10. $x^2 - (y + 1)^2$.
11. $(x - y)^2 - 9$.
12. $4(x - y)^2 - 25$.
13. $1 - 36(x + 2y)^2$.
14. $(x + 2y)^2 - (x + 1)^2$.

15. $25(2a - b)^2 - (a - 3b)^2$.

16. $x^{12}y^9 - yz^{16}$.
17. $81 x^{12} - 16 y^4$.
18. $x^5 - 144 xy^2z^6$.
19. $(a - b)^2 - 4(c + 1)^2$.
20. $1 - 100(x^2 - x - 1)^2$.
21. $a^2 + 2ab + b^2 - x^2$.
22. $a^2 - 2ab + b^2 - 4x^2$.
23. $a^2 - x^2 - 2xy - y^2$.
24. $9a^2 - x^2 - 4xy - 4y^2$.
25. $16a^2 - x^2 + 2xy - y^2$.
26. $a^2 + b^2 + 2ab - 4x^2$.

27. $a^2 + b^2 - 4x^2 + 2ab$.
28. $a^2 - 4x^2 + b^2 + 2ab$.
29. $x^2 - 2ab - a^2 - b^2$.
30. $4ab + x^2 - a^2 - 4b^2$.
31. $x^2 + a^2 - y^2 - 2ax$.
32. $a^4 - x^4 - 2x^2y - y^2$.
33. $9x^2 - y^2 - 1 - 2y$.
34. $1 + 2xy - x^2 - y^2$.
35. $a^2 + b^2 - c^2 - 2ab$.
36. $2ab + a^2b^2 + 1 - x^2$.
37. $2z^2 - 4z - 2z^4 + 2$.

38. $a^2 + 2\,ab + b^2 - c^2 - 2\,cd - d^2$.

39. $x^2 + 4\,y^2 - 9\,z^2 - 1 - 4\,xy - 6\,z$.

40. $9\,a^2 - 25\,x^2 + 4\,b^2 - 1 - 10\,x - 12\,ab$.

41. $a^2 - 9\,b^2x^2 - 1 + 6\,bx - 10\,ab + 25\,b^2$.

By the aid of factoring, compute the following in a short way :

42. $7.5^2 - 2.5^2$. **44.** $5.8^2 - 4.2^2$. **46.** $67.52^2 - 22.48^2$.

43. $12.5^2 - 7.5^2$. **45.** $19.7^2 - 5.3^2$. **47.** $87.23^2 - 52.77^2$.

48. If $\pi = \frac{22}{7}$, $R = 25$, and $r = 15$, find the value of $\pi R^2 - \pi r^2$ in a short way by the aid of factoring.

49. If $\pi = 3.1416$, $R = 64$, $r = 36$, find the value of $\pi R^2 - \pi r^2$ in the snortest way you can.

Factor the following miscellaneous examples :

50. $4\,x^2 - .09$. **54.** $x^8 - 1$.

51. $4\,x^2 + .4\,x + .01$. **55.** $x^2 + 2\,xy + y^2 - 9$.

52. $x^4 - 2\,x^3y + x^2y^2$. **56.** $ax^3 - axy^2$.

53. $9(x + y)^2 - 4$. **57.** $.4\,x^3 - .25\,xy^2$.

58. $a^2 + 2\,a\,(x + y) + (x + y)^2$.

59. $(a + b)^2 - (2\,a - 3\,b)^2$.

60. $\frac{1}{3}\,ah_1 + \frac{1}{3}\,ah_2$. **61.** $\frac{2}{3}\,br^2 - \frac{2}{3}\,hr^2$.

62. $238 \times \frac{1}{365} \times .02 - 128 \times \frac{1}{365} \times .02$.

Case IV

75. A Trinomial of the Form $x^2 + bx + c$.

Ex. 1. $x^2 - .1\,x - .72 = (x - .9)(x + .8)$.

Ex. 2. $(a + b)^2 + 7(a + b) + 12$.

$$= [(a + b) + 4]\,[(a + b) + 3]$$
$$= (a + b + 4)(a + b + 3). \quad Ans.$$

Let the pupil check the preceding examples.

Factor the following and check:

1. $x^2 - .5x + .06.$
2. $a^2 + .8a + .12.$
3. $b^2 - .1b - .12.$
4. $x^2 + x - 56.$
5. $a^2x^2 - 9ax + 18.$
6. $x^2 - 2x - 63.$
7. $a^2 - 23a + 132.$
8. $p^2 - p - 132.$
9. $x^4 - 5x^2 + 6.$
10. $a^4 + 7a^2 + 12.$
11. $x^2y^2 - 23xy + 132.$
12. $x^2 - 5ax - 24a^2.$
13. $x^4 - 9x^2 + 8.$
14. $2a - 14ax - 60ax^2.$
15. $a^2b^2 - 11abc^2 - 26c^4.$
16. $(a+b)^2 - 5(a+b) + 6.$
17. $(a+b)^2 - 2(a+b) - 15.$
18. $(a+x)^2 - (a+x) - 2.$
19. $x^2 + (a+b)x + ab.$
20. $x^2 + (2a - 3b)x - 6ab.$
21. $(2a+b)^2 + 10(2a+b) + 21.$
22. $(x - 2y)^2 - 7(x - 2y) + 12.$
23. $(x - y)^2 - 3(x - y) - 18.$
24. $25 - 10(a - b) + (a - b)^2.$

Factor and check the following miscellaneous examples:

1. $6a - 12b - 6c.$
2. $x^3 - .09x.$
3. $4x^3 + 4x^2 + x.$
4. $a^4 + 4a^3 + 3a^2.$
5. $a^4 + 8a^3x + 16a^2x^2.$
6. $x^3 - x^2y - 6xy^2.$
7. $a^2 - (x + 2y)^2.$
8. $x^2 - .7x + .12.$
9. $x^4 - 9x^2 + 18.$
10. $4(x + y)^2 - a^2.$
11. $(a+b)^2 + 2(a+b) + 1.$
12. $2x^3 - 8x^2y + 8xy^2.$
13. $x^3 - 25x^2 + 144x.$
14. $x^4 - 1 + 2y - y^2.$
15. $1 - 5a^2 + 4a^4.$
16. $x^5 - 4x^4 - 45x^3.$
17. $7a - 7a^3b^4.$
18. $x^4 - (x - 2)^2.$
19. $x^4 - 2x^2 + 1.$
20. $a^{16} - 1.$
21. $4x^4 - 12x^2y^2 + 9y^4.$
22. $\frac{1}{2}hx + \frac{1}{2}hy - \frac{1}{2}hz.$
23. $x^2 - .7x + .1.$
24. $1 - 4(x + y)^2.$
25. $x^4 - 256.$
26. $16a^4 - 8a^2 + 1.$

27. $(2a-b)^2-4(2a-b)x+4x^2$.

28. $5(a+b)x+10(a+b)y$.

29. $14(a-b)^2-7(a-b)^3$.

30. $9(x+y)^2-4(a+b)^2$.

31. $.14x^3y-.28x^2y^2+.42xy^3$.

32. $\frac{1}{2}b(x+y)+\frac{1}{2}a(x+y)$.

33. $249 \times \frac{1}{365} \times .03 - 139 \times \frac{1}{365} \times .03$.

Write and factor the following :

34. A binomial expression the prime factors of which are two binomials.

35. A binomial whose prime factors are three binomials.

36. A trinomial whose prime factors are two equal binomials.

37. A binomial which has two factors one of which is a monomial and the other a binomial.

38. A trinomial which has three factors one of which is a monomial and the other two of which are unlike binomials.

39. A trinomial whose prime factors are four binomials.

Supply the missing factor in each of the following expressions :

40. $\frac{3x}{2}+\frac{3y}{2}=\frac{3}{2}(?).$ **41.** $\frac{5a^2}{9}-\frac{5b^2}{9}+\frac{5ac}{9}=\frac{5}{9}(?).$

42. $100h+10t+u=h+t+u+99h+9t$
$$=h+t+u+9\ (?).$$

43. If h denotes the number of hundreds in a number of three figures, t the tens, and u the units, what principle concerning the divisibility of numbers is proved by the final expression obtained in Ex. 42 ?

44. Make up and work an example in factoring where Cases III and IV are combined.

45. Also where Cases I, III, and IV are combined.

76. Graphs: Two Scales on One Axis. Often by using two different scales on the vertical axis (that is, using a special scale for each graph) it is possible to put two graphs on the same diagram, and thus make it much easier to compare the graphs.

Ex. A newsboy sold papers during the following numbers of hours on the successive days of a week : Monday, $1\frac{1}{2}$ hr.; Tuesday, 2 hr.; Wednesday, 1 hr.; Thursday, $1\frac{3}{4}$ hr.; Friday, $1\frac{1}{4}$ hr.; Saturday, 2 hr. The number of papers sold by him on these days were as follows : 47, 42, 54, 68, 72, 56. Graph these two sets of numbers on the same diagram.

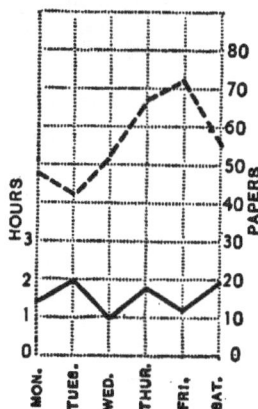

In graphing the first set of numbers, we let one space on the vertical axis represent 1 hour; while in graphing the second set of numbers, we let one space on the vertical axis represent 10 papers.

From an inspection of this chart how can we form an idea of the relative efficiency of the newsboy on the different days ?

EXERCISE 85

1. One week on successive workdays, a newsboy sold papers during the following number of hours : 1, $1\frac{1}{2}$, $2\frac{1}{2}$, $1\frac{1}{4}$, 1, $1\frac{1}{2}$, $2\frac{1}{2}$. The papers sold per day numbered as follows : 47, 72, 64, 76, 84, 92, 67. On the same diagram construct graphs of these two sets of facts.

2. The cost per month in dollars of running an auto truck for the first six months of a year and the number of miles run per month were as follows :

Month . . .	Jan.	Feb.	Mar.	Apr.	May	June
Cost . . .	$31	$34	$43	$32	$38	$28
Miles run . .	632	476	462	610	705	800

Graph these two sets of facts on the same diagram.
(For the first graph let one space on the vertical axis
represent $10; and for the second graph, let one space
represent 100 miles.)

From the chart what inference can be drawn as to the
increasing or decreasing efficiency of the auto truck?

3. For the first nine months of a given year the sales
and the advertising expenses of a certain business were as
follows:

Month. . .	Jan.	Feb.	Mar.	Apr.	May
Sales . . .	$7620	$8200	$7420	$7530	$8940
Ad. Expenses	225	240	316	320	460

Month. . . .	June	July	Aug.	Sept.
Sales	$10,725	$12,010	$15,240	$17,903
Ad. Expenses .	530	515	468	420

Graph these two sets of facts on the same diagram.   (For
the first graph let one space on the vertical axis represent
$2000 ; and for the second let one space represent $100.)

On May 1, a special advertising agent was employed.
How does the chart show the result?

4. For a period of years as specified, the population of
a given city and the sales of a business establishment in
the city increased as follows:

Year . . .	1918	1919	1920	1921
Population .	1,260,000	1,280,000	1,301,000	1,346,000
Sales . . .	$84,000	$98,000	$120,000	$170,000

Year	1922	1923	1924
Population	1,390,000	1,410,000	1,412,000
Sales	$192,000	$236,000	$392,000

Graph the facts on one diagram. (For the population curve let one space represent 200,000; for the sales curve let one space represent $50,000.) What inference can you draw from the chart?

5. Example 7, page 150, gives the population of the United States at 10 year intervals. Graph these facts. On the same diagram make a graph of the growth of the wealth of the United States, using the following facts as to the nation's wealth in billions of dollars at 10 year intervals:

Year	1850	1860	1870	1880	1890	1900	1910
Wealth in Billions .	7	16	30	44	65	89	165

What inference can you draw from this chart?

6. The following table gives the amount of $1 at simple interest, and also at compound interest at 4% for 5, 10, 15, 20, etc. years. On the same diagram draw (1) a graph of the amounts at simple interest; (2) a graph of the amounts at compound interest.

Years . .	0	5	10	15	20	25	30	35
Amounts at Simple Int.	$1	$1.20	$1.40	$1.60	$1.80	$2.00	$2.20	$2.40
Amounts at Com. Int.	1	1.22	1.48	1.80	2.19	2.67	3.24	3.95

The amounts of $1 at 5% for the same periods of time at compound interest are $1, $1.28, $1.63, $2.08, $2.65, $3.39, $4.32, $5.52. On the chart just formed make a graph of these amounts.

7. At a certain place on a given day the readings of a barometer in inches were as follows:

8 A.M. 28.32 | 11 A.M. 29.36 | 2 P.M. 28.24 | 5 P.M. 29.72
9 A.M. 28.47 | Noon 29.12 | 3 P.M. 28.9 | 6 P.M. 30.05
10 A.M. 29.05 | 1 P.M. 28.88 | 4 P.M. 29.4 | 7 P.M. 30.23

Make a graph of these facts. (See Ex. 8, p. 38.)

CHAPTER IX

FRACTIONS

PART I

77. Fractions in Algebra have in general the same properties as fractions in arithmetic.

For example, in arithmetic $\frac{8}{12} = \frac{2}{3}$.

Similarly, in algebra $\quad \dfrac{5\,ax}{5\,ay} = \dfrac{x}{y}$.

TRANSFORMATIONS OF FRACTIONS

I. TO REDUCE A FRACTION TO ITS LOWEST TERMS

78. A Fraction in its Lowest Terms is a fraction whose numerator and denominator have no common factor.

To reduce a fraction to its lowest terms, as in arithmetic,

Resolve the numerator and denominator into their prime factors, and cancel the factors common to both.

Ex. 1. Reduce $\dfrac{36\,a^3x^2}{48\,a^2x^3y^2}$ to its lowest terms.

Dividing both numerator and denominator by $12\,a^2x^2$,

$$\frac{36\,a^3x^2}{48\,a^2x^3y^2} = \frac{3\,a}{4\,xy^2}. \quad Ans.$$

Ex. 2. $\quad \dfrac{9\,ab - 12\,b^2}{12\,a^2 - 16\,ab} = \dfrac{3\,b(3\,a - 4\,b)}{4\,a(3\,a - 4\,b)} = \dfrac{3\,b}{4\,a}. \quad Ans.$

Check by letting $\quad a = 2$ and $b = 1$.

Thus, $\quad \dfrac{9\,ab - 12\,b^2}{12\,a^2 - 16\,ab} = \dfrac{18 - 12}{48 - 32} = \dfrac{6}{16} = \dfrac{3}{8}.$

Also $\quad\quad\quad \dfrac{3\,b}{4\,a} = \dfrac{3}{8}.$

Notice particularly that in reducing a fraction to its lowest terms it is allowable to cancel a *factor* which is common to both denominator and numerator, but it is not allowable to cancel a *term* which is common unless this term is a factor.

Thus, $\dfrac{ab}{ac}$ reduces to $\dfrac{b}{c}$;

but in $\dfrac{a+x}{a+y}$, a of the numerator will not cancel a of the denominator.

This is a principle very frequently violated by beginners.

EXERCISE 86

ORAL

At sight give the simplest form of the following:

1. $\frac{9}{12}$.
2. $\frac{8}{12}$.
3. $\frac{14}{21}$.
4. $\frac{15}{20}$.
5. $\frac{8}{24}$.
6. $\frac{3a}{3b}$.
7. $\frac{4\,ay}{5\,az}$.
8. $\frac{7\,ab}{5\,ab}$.
9. $\frac{9\,x}{12\,x}$.
10. $\frac{4\,x^2\cdot}{5\,x^3}$.
11. $\frac{3\,y^2}{3\,y^6}$.
12. $\frac{3\,xy}{9\,x^2y^2}$.

13. $\frac{7\,ab^2}{7\,a^2b}$.
14. $\frac{5\,x^3y}{15\,xy^3}$.
15. $\frac{6\,xy}{9\,yz}$.
16. $\frac{8\,a^2}{16\,a^3}$.
17. $\frac{12\,x^2y}{18\,x^2y^2}$.
18. $\frac{9\,abx^2}{12\,abx^3}$.
19. $\frac{7\,abx}{21\,acx}$.
20. $\frac{3\,lmn}{6\,lm^2n}$.

21. $\frac{4(a+b)}{5(a+b)}$.
22. $\frac{4(x+y)}{8(x+y)}$.
23. $\frac{5(x-y)}{15(x-y)^2}$.
24. $\frac{6(a+b)x}{8(a+b)y}$.
25. $\frac{3(a+b)^2}{5(a+b)}$.
26. $\frac{8(x+y)}{12(x+y)^2}$.

EXERCISE 87

Reduce each of the following to its simplest form:

1. $\dfrac{27}{36}$.
2. $\dfrac{108}{144}$.
3. $\dfrac{72}{150}$.
4. $\dfrac{8\,a^3x^4}{12\,a^2x^5}$.
5. $\dfrac{12\,x^4yz^5}{15\,x^3y^2z^5}$.
6. $\dfrac{3\,a^2x}{6\,a^2-9\,a^3x}$.

7. $\dfrac{72\,x^2y^3z^4}{96\,xy^5z^3}.$

12. $\dfrac{x^2-y^2}{(x+y)^2}.$

17. $\dfrac{6\,xy}{9\,x^2y-12\,xy^2}.$

8. $\dfrac{3\,a^3-6\,a^2b}{4\,a^2b^2-8\,ab^3}.$

13. $\dfrac{12\,a^2x^2-8\,a^2xy}{18ax^3-12ax^2y}.$

18. $\dfrac{6\,a^2b^2+12\,ab^3}{9\,a^3b+18\,a^2b^2}.$

9. $\dfrac{2\,a}{4\,a^2-2\,a}.$

14. $\dfrac{8(x^2-1)}{12\,x-12}.$

19. $\dfrac{2\,x^2-3\,xy}{4\,x^3-9\,xy^2}.$

10. $\dfrac{3\,x-6\,y}{6\,ax-12\,ay}.$

15. $\dfrac{45(x-y)^2}{18(x-y)^3}.$

20. $\dfrac{49\,x^2-64\,y^2}{14\,x^3-16\,x^2y}.$

11. $\dfrac{4\,x+4\,y}{4\,ax+4\,ay}.$

16. $\dfrac{a^2b+ab^2}{2\,a^2b-2\,ab^2}.$

21. $\dfrac{9\,x^2+9\,x}{6\,x^2-6}.$

Supply the missing numerator in the fraction to the right in each of the following:

22. $\dfrac{7}{12}=\dfrac{?}{36}.$

23. $\dfrac{2\,a}{3\,b}=\dfrac{?}{12\,a^2b^2}.$

24. $\dfrac{1}{x+y}=\dfrac{?}{x^2-y^2}.$

25. What is the correct value of the fraction $\dfrac{1+4}{6+4}$? If the 4's are struck out, what does the value of the above fraction become? Is it allowable, therefore, to strike out the 4's in the above fraction?

26. Which of the following fractions can be simplified by striking out the 4's?

$\dfrac{1+4}{11+4},\quad \dfrac{x+4}{y+4},\quad \dfrac{4\,x}{4(1+y)},\quad \dfrac{3\times4}{4+11},\quad \dfrac{4\,a}{x+4},\quad \dfrac{4\,ab}{4\,xd}.$

27. Make up and work an example similar to Ex. 26, involving 7's.

28. Which of the following can be simplified by striking out the b^2's?

$\dfrac{b^2+x}{b^2+y},\quad \dfrac{b^2x}{b^2y},\quad \dfrac{b^2-4}{3\,b^2+4},\quad \dfrac{a^2b^2}{a^2+b^2},\quad \dfrac{a^2b^2}{b^2x^4}.$

II. To Reduce an Improper Fraction to an Integral or Mixed Quantity

79. An **Improper Fraction** in algebra is one in which the numerator contains a power equal to or higher than the highest power of the same letter in the denominator.

Exs. $\dfrac{3x^2+5}{2x^2}$, $\dfrac{4x^3-3x^2+7}{x+2}$.

Since a fraction is an indicated division, to reduce an improper fraction to an integral or mixed expression,

Divide the numerator by the denominator;

If there is a remainder, write it over the denominator, and annex the result to the quotient with the proper sign.

Ex. 1. $\dfrac{ax+c}{x} = a + \dfrac{c}{x}.$ *Ans.*

Ex. 2. Reduce $\dfrac{5a^2b^2+10ab-a-b}{5ab}$ to a mixed number.

$$5ab)\overline{5a^2b^2+10ab-a-b}$$
$$ab+2-\frac{a+b}{5ab}. \quad Ans.$$

When the remainder is made the numerator of a fraction with the minus sign before it, as in this example, the signs of terms of the remainder must be changed, since the line between the numerator and denominator is in effect a parenthesis.

EXERCISE 88

Reduce each of the following to a mixed number:

1. $\frac{32}{5}$.

2. $1\frac{21}{9}$.

3. $1\frac{81}{17}$.

4. $\dfrac{ay+b}{y}$.

5. $\dfrac{ay-b}{y}$.

6. $\dfrac{x^2+ab}{x}$.

7. $\dfrac{6x^3-7y}{3x^2}$.

8. $\dfrac{a^3+5}{a^2}$.

9. $\dfrac{p^4+2}{p^2}$.

10. $\dfrac{10x^2+7}{5x}$.

11. $\dfrac{20x+5}{4x}$.

12. $\dfrac{3abc-5}{3ac}$.

13. $\dfrac{x^2 - 2x + 3}{x}.$

16. $\dfrac{b^2 + b - 1}{b}.$

19. $\dfrac{27\,x^3 - 6\,x^2 - 9}{3\,x^2}.$

14. $\dfrac{4\,x^3 + 6\,x - 5}{2\,x}.$

17. $\dfrac{x^3 - 2x^2 + 1}{x^2}.$

20. $\dfrac{bx - c - d}{x}.$

15. $\dfrac{16x^2 + 8x + 3}{4\,x}.$

18. $\dfrac{abx + d^2x}{ax}.$

21. $\dfrac{6\,a - x + y}{3\,a}.$

22. $\dfrac{10\,x^2 + 5\,x - 3\,a - 1}{5\,x}.$

23. $\dfrac{8\,a^2b - 4\,ab - p - q}{2\,ab}.$

III. To Reduce a Mixed Expression to a Fraction

80. To Reduce a Mixed Expression to a fraction, it is necessary simply to reverse the process of § 79.   Hence,

Multiply the integral expression by the denominator of the fraction, and add the numerator to the result, changing the signs of the terms of the numerator if the fraction is preceded by the minus sign;

Write the denominator under the result.

Ex. $x + y - \dfrac{x^2 + y^2}{x - y}$

$= \dfrac{(x + y)(x - y) - (x^2 + y^2)}{x - y}$

$= \dfrac{x^2 - y^2 - x^2 - y^2}{x - y} = \dfrac{-2\,y^2}{x - y}$

$= \dfrac{2\,y^2}{y - x}.$ *Ans.*

EXERCISE 89

Reduce to a fraction :

1. $3\frac{5}{7}.$

2. $12\frac{3}{8}.$

3. $13\frac{5}{12}.$

4. $2 + \dfrac{x}{3}.$

5. $8 + \dfrac{5}{x}.$

6. $x - \dfrac{y^2}{x}.$

7. $\dfrac{5\,a}{x} + 6.$

8. $\dfrac{7\,b}{y} - 3.$ **12.** $\dfrac{8\,x}{9} - 1.$ **16.** $x + y + \dfrac{y^2}{x}.$

9. $\dfrac{8}{3\,a} + 5.$ **13.** $\dfrac{3\,x}{5\,y} - 2\,x.$ **17.** $a + \dfrac{a^2 + b}{2\,a}.$

10. $\dfrac{7\,x^2}{3\,y} - 2\,y.$ **14.** $5\,x - 2 + \dfrac{7}{x}.$ **18.** $3\,a - \dfrac{a^2 + b}{2\,a}.$

11. $1 - \dfrac{5\,x^2}{6\,y^2}.$ **15.** $\dfrac{5}{2\,x} + 3 - x.$ **19.** $x - 3 + \dfrac{2}{x + 4}.$

20. $6\,y - 5\,x - \dfrac{5\,x - 2\,y}{4}.$ **22.** $\dfrac{(a + b)^2}{4} - ab.$

21. $x + 1 + \dfrac{1}{x - 1}.$ **23.** $a - b - \dfrac{a^2 - b^2}{a + b}.$

IV. To Reduce Fractions to Equivalent Fractions of the Lowest Common Denominator

81. **To Reduce Fractions** to their lowest common denominator, as in arithmetic, is to find the smallest number that will contain all the given denominators as factors, and to transform all the fractions till they shall have this lowest common denominator. Hence,

Factor each of the denominators;

Take each factor the highest number of times it occurs in any one denominator;

Take the product of these results.

Ex. Reduce $\dfrac{2}{3\,ax}$, $\dfrac{3}{4\,a^2x}$, and $\dfrac{5}{6\,ax^2}$ to equivalent fractions having the lowest common denominator.

The L. C. D. is $12\,a^2x^2$.

Dividing this by each of the denominators, we get the quotients $4\,ax$, $3\,x$, and $2\,a$.

Multiplying each of these quotients by the corresponding numerator and setting the results over the common denominator, we obtain

$$\frac{8\,ax}{12\,a^2x^2}, \quad \frac{9\,x}{12\,a^2x^2}, \quad \frac{10\,a}{12\,a^2x^2}. \quad Ans.$$

EXERCISE 90

Write the smallest number that will contain the following as factors:

1. $6, 9.$
2. $6, 9, 12.$
3. $5, 8, 2.$
4. $3\,a, 4\,a.$

5. $6\,a, 4\,a.$
6. $10\,x^2, 6\,x.$
7. $6\,ab, 9\,bc.$
8. $5\,b, 10, b.$

9. $2\,ab^2, a^2b, 3\,a^3.$
10. $3\,a^4, 4\,ax, 6\,a^2x^2.$
11. $m, p, q, r.$
12. $a+b, a-b.$

13. $a^2 - b^2, a - b.$
14. $(a-b)^2, a-b.$
15. $x(1+x), x(1-x^2).$
16. $x^2 - 9, x^2 - 6\,x + 9.$
17. $x^2, x - y, x^2 - xy.$

18. $ab, a(a-b), b(a^2 - b^2).$
19. $2 - 2\,x, 3 + 3\,x.$
20. $a^2b, ab^2, bc^2.$
21. $x, (1-x), x^2(1+x).$
22. $a^3b, a^2b^2, a^3 - ab^2.$

Reduce to their lowest common denominator:

23. $\frac{5}{8}, \frac{7}{12}, \frac{5}{6}.$
24. $\frac{3}{5}, \frac{4}{15}, \frac{9}{20}.$
25. $\frac{2\,x}{9}, \frac{5\,x}{6}, \frac{x}{12}.$
26. $\frac{12\,a}{5\,b}, \frac{7}{10}, \frac{a}{b}.$
27. $\frac{1}{2\,ab^2}, \frac{2}{a^2b}, \frac{1}{ab}.$
28. $\frac{2}{3\,a^2}, \frac{3}{4\,ax}, \frac{1}{x}.$
29. $\frac{ac}{bd}, \frac{ab}{cd}, \frac{bc}{ad}, \frac{ad}{bc}.$

30. $\frac{1}{a^2 - a}, \frac{3}{a-1}.$
31. $\frac{x}{1+x}, \frac{1}{x}, \frac{1}{x+x^2}.$
32. $\frac{1}{4\,x^2 - 9}, \frac{1}{2\,x+3}, \frac{1}{x}.$
33. $\frac{1+x}{2 - 2\,x}, \frac{1-x}{3+3\,x}.$
34. $\frac{3}{a^2b + ab^2}, \frac{4}{a^2b - ab^2}.$
35. $\frac{1}{3\,x-6}, \frac{5}{2\,x+4}, \frac{3}{x^2-4}.$

36. $\frac{2}{x-x^3}, \frac{x}{3+3\,x}, \frac{x}{2-2\,x}.$

PROCESSES WITH FRACTIONS

I. ADDITION AND SUBTRACTION OF FRACTIONS

82. The Method of Adding or Subtracting Fractions, as in arithmetic, is as follows :

Reduce the fractions to their lowest common denominator ;

Add their numerators, changing the signs of the numerator of any fraction preceded by the minus sign ;

Set the sum over the common denominator ;

Reduce the result to its lowest terms.

Ex. 1. $\dfrac{3\,x}{5\,a^2} - \dfrac{7}{10\,ab} - \dfrac{4\,y}{15\,b^2}$

$$= \frac{18\,b^2x - 21\,ab - 8\,a^2y}{30\,a^2b^2}. \quad \textit{Ans.}$$

Let the pupil check the work.

Ex. 2. $\dfrac{x^2}{x^2 - 1} + \dfrac{x}{x + 1} + \dfrac{x}{x - 1}$

$$= \frac{x^2 + x(x - 1) + x(x + 1)}{x^2 - 1}$$

$$= \frac{x^2 + x^2 - x + x^2 + x}{x^2 - 1} = \frac{3\,x^2}{x^2 - 1}. \quad \textit{Ans.}$$

Let the pupil check the work by letting $x = 2$.

EXERCISE 91

ORAL

At sight simplify :

1. $\frac{1}{2} + \frac{3}{4}$.

2. $\frac{1}{2} - \frac{1}{8}$

3. $\frac{3}{4} - \frac{1}{2}$.

4. $\frac{1}{2} - \frac{1}{6}$.

5. $\frac{3}{2} - \frac{5}{6}$.

6. $\dfrac{a}{2} + \dfrac{a}{4}$.

7. $\dfrac{b}{2} - \dfrac{b}{6}$.

8. $\dfrac{3\,a}{4} - \dfrac{a}{2}$.

9. $\dfrac{1}{2\,a} - \dfrac{1}{4\,a}$.

10. $\dfrac{1}{3\,a} - \dfrac{1}{6\,a}$.

11. $\dfrac{a}{3} + \dfrac{a}{6}$.

12. $\dfrac{a}{3} - \dfrac{a}{6}$.

13. $\dfrac{x}{2} - \dfrac{x}{3}$.

14. $\dfrac{2x}{3} + \dfrac{x}{6}$.

15. $\dfrac{x^2}{2} - \dfrac{3x^2}{8}$.

16. $\dfrac{3p}{4} - \dfrac{p}{8}$.

17. $\dfrac{1}{4a} - \dfrac{1}{8a}$.

18. $\dfrac{1}{a} + \dfrac{1}{b}$.

19. $\dfrac{1}{a} - \dfrac{1}{b}$.

20. $\dfrac{a+b}{2} + \dfrac{a-b}{2}$.

21. $\dfrac{a+b}{2} - \dfrac{a-b}{2}$.

22. $\dfrac{3a}{2} - \dfrac{a+b}{2}$.

23. $\dfrac{x+y}{4} - \dfrac{y}{4}$.

24. $\dfrac{2x+y}{4} + \dfrac{x-y}{4}$.

25. $\dfrac{2a+b}{2} - \dfrac{2a-b}{2}$.

EXERCISE 92

Simplify and check the following:

1. $\tfrac{5}{12} - \tfrac{3}{8}$.

2. $\tfrac{5}{16} + \tfrac{7}{24}$.

3. $-\tfrac{3}{8} + \tfrac{7}{10}$.

4. $\dfrac{2a}{5} - \dfrac{5b}{8}$.

5. $\dfrac{3}{4x} - \dfrac{5}{8x}$.

6. $\dfrac{3a}{2b} - \dfrac{5x}{6b}$.

7. $\dfrac{x}{ab} + \dfrac{y}{bc}$.

8. $\dfrac{7}{3y^2} - \dfrac{5}{2y}$.

9. $\dfrac{3a}{5b} - \dfrac{6c}{7d}$.

10. $\dfrac{a}{2} + \dfrac{a}{3} + \dfrac{b}{4}$.

11. $\dfrac{3}{2x} + \dfrac{2}{x} - \dfrac{1}{3x}$.

12. $\dfrac{2}{3a} - \dfrac{3}{4ax} + \dfrac{1}{x}$.

13. $\dfrac{5}{2ac} - \dfrac{1}{3ab} - \dfrac{1}{bc}$.

14. $\dfrac{1}{x} + \dfrac{1}{y} + \dfrac{1}{z}$.

15. $\dfrac{3a}{5} + \dfrac{7a}{10} + \dfrac{4a}{15}$.

16. $\dfrac{5}{a} + \dfrac{7}{ab} + \dfrac{3}{abc}$.

17. $\dfrac{1}{a^2bc} + \dfrac{1}{ab^2c} + \dfrac{1}{abc^2}$.

18. $\dfrac{1}{x+3} + \dfrac{1}{x-4}$.

19. $\dfrac{2}{a-3} + \dfrac{3}{a+4}$.

20. $\dfrac{1}{x-2} - \dfrac{1}{x-3}$.

21. $\dfrac{3}{a-b} - \dfrac{4}{a+b}$.

22. $\dfrac{x+y}{5} - \dfrac{x-y}{3}$.

23. $\dfrac{a+x}{3} - \dfrac{5}{a-x}$.

24. $\dfrac{1}{a-b} - \dfrac{a}{a^2-b^2}.$

25. $\dfrac{a-x}{a+x} - \dfrac{a+x}{a-x}.$

26. $\dfrac{a+2\,b}{2\,ab} - \dfrac{6\,a-b}{6\,a^2}.$

27. $\dfrac{4}{3\,x-3} + \dfrac{1}{2\,x-2}.$

28. $\dfrac{3\,a}{a^2-16} - \dfrac{2}{a-4}.$

29. $\dfrac{2\,x-1}{2\,x+1} + \dfrac{2\,x+1}{2\,x-1}.$

30. $\dfrac{2\,a^2x+3}{4\,ax^2} - \dfrac{3\,a+x}{6\,x} + \dfrac{7-3\,x}{3\,x^2}.$

31. $\dfrac{a}{3+a} + \dfrac{a}{3-a} + \dfrac{2\,a^2}{9-a^2}.$

32. $\dfrac{1}{2\,x+1} - \dfrac{1}{2\,x-1} - \dfrac{4\,x}{4\,x^2-1}.$

33. $\dfrac{4}{x^2-x} + \dfrac{3\,x}{x-1} - \dfrac{2}{x}.$

34. $\dfrac{3\,x}{x+2} - \dfrac{2\,x}{x-2} + \dfrac{10\,x^2}{x^2-4}.$

35. $\dfrac{3}{x+1} - \dfrac{4}{x+2} + \dfrac{2}{x+3}.$

II. MULTIPLICATION OF FRACTIONS

83. The Method of Finding the Product of two or more fractions, as in arithmetic, is as follows:

Multiply the numerators together for a new numerator, and multiply the denominators together for a new denominator, canceling factors that are common to the two products.

Ex. $\quad \dfrac{x+y}{x} \times \dfrac{x^2-y^2}{x^3+xy^2} \times \dfrac{4\,x^2}{(x+y)^2}$

$= \dfrac{x+y}{x} \times \dfrac{(x+y)(x-y)}{x(x^2+y^2)} \times \dfrac{4\,x^2}{(x+y)(x+y)}$

$= \dfrac{4(x-y)}{x^2+y^2}. \quad \textit{Ans.}$

EXERCISE 93

Oral

At sight multiply and simplify when possible:

1. $\frac{2}{5} \times 3$.

2. $\frac{3}{7} \times 2$.

3. $2 \times \frac{5}{3}$.

4. $4 \times \frac{5}{6}$.

5. $2 \times \frac{a}{b}$.

6. $\frac{x}{y} \times 3$.

7. $\frac{3x}{5y} \times 4$.

8. $-\frac{2}{3} \times 4$.

9. $(-\frac{5}{6})(-7)$.

10. $\frac{2a}{5x} \times 4$.

11. $\frac{3p}{7q} \times 2r$.

12. $\frac{3}{4} \times \frac{5}{6}$.

13. $\frac{a}{4} \times \frac{2}{3}$.

14. $\frac{a}{b} \times \frac{c}{d}$.

15. $\frac{6}{x^2} \times \frac{x}{3}$.

16. $\frac{ab}{cd} \times \frac{c}{b}$.

17. $\frac{5a}{7x^3} \cdot \frac{x}{a}$.

18. $\frac{a+b}{10} \times 5$.

19. $\frac{a-b}{7} \times \frac{3}{a-b}$.

20. $\frac{x^2-y^2}{5} \times \frac{5}{x-y}$.

21. $\frac{3(x+y)}{4} \cdot \frac{4}{5(x+y)}$.

EXERCISE 94

Simplify:

1. $\frac{5}{6} \times \frac{12}{25}$.

2. $\frac{25}{27} \times \frac{36}{125}$.

3. $\frac{3}{5} \times \frac{4}{9} \times \frac{15}{16}$.

4. $\frac{5x^2y}{14a^3c} \cdot \frac{28a^2b^3}{15xy^2}$.

5. $\frac{16a^4b^2}{21x^3y^2} \times \frac{3x^2y}{4a^2b}$.

6. $\left(\frac{9a^2b}{8c^2x}\right)\left(\frac{28ax^2}{15b^2c}\right)\left(\frac{10bc^3}{21a^3x}\right)$.

7. $\frac{5ac^3}{21b^2} \cdot \frac{3bc^2}{8a^2} \cdot \frac{7a^2b^3}{30c^4}$.

8. $\frac{x^2-y^2}{8a^3} \cdot \frac{16a^2}{(x-y)^2}$.

9. $\frac{(x-y)^3}{x+y} \cdot \frac{(x+y)^2}{(x-y)^4}$.

10. $\frac{a^2b^2+3ab}{4a^2-1} \cdot \frac{2a+1}{ab+3}$.

11. $\frac{x^2-9}{x^2+x} \cdot \frac{x^2-1}{x-3}$.

12. $\frac{x^2-121}{x^2-4} \cdot \frac{x+2}{x+11}$.

13. $\frac{3a^2+a}{2a-2} \cdot \frac{a-1}{2a+6a^2}$.

14. $\frac{m^2-1}{m^2+1} \cdot \frac{m^4+2m^2+1}{m-1}$.

15. $\frac{4x^2-9}{9x^2-1} \times \frac{6x+2}{12x-18}$.

16. $\frac{(a-1)^3}{a(x+1)^2} \cdot \frac{x+1}{(a-1)^2}$.

17. $\dfrac{a^2 - 4\,b^2}{9\,x^2} \cdot \dfrac{a^2 + 2\,ab}{(a - 2\,b)^2} \cdot \dfrac{27\,x^3(a - 2\,b)}{a^2}.$

18. $\dfrac{x^2 - 3\,x + 2}{p + q} \cdot \dfrac{p^2 x - q^2 x}{x^2 - 4} \cdot \dfrac{x + 2}{(p - q)x^3}.$

III. Division of Fractions

84. The Method of Dividing one fraction by another is the same as in arithmetic.

Hence, to divide one fraction by another,

Invert the divisor and proceed as in multiplication.

Ex. 1. $\dfrac{3\,a^2}{5\,x^3} \div \dfrac{6\,a^3}{10\,x} = \dfrac{3\,a^2}{5\,x^3} \cdot \dfrac{10\,x}{6\,a^3} = \dfrac{1}{ax^2}.$ *Ans.*

Ex. 2. $\dfrac{a + 1}{a^2 b^2} \times \dfrac{a - 1}{6\,ab^3} \div \dfrac{a^2 - 1}{3\,a^3 b^3}$

$\qquad = \dfrac{a + 1}{a^2 b^2} \times \dfrac{a - 1}{6\,ab^3} \times \dfrac{3\,a^3 b^3}{(a + 1)(a - 1)}$

$\qquad = \dfrac{1}{2\,b^2}.$ *Ans.*

EXERCISE 95

Oral

At sight state the quotient for each of the following:

1. $\frac{4}{5} \div 2.$

2. $\frac{4}{7} \div 2.$

3. $\frac{3}{7} \div 2.$

4. $\frac{5}{9} \div 4.$

5. $\dfrac{a}{5} \div 2.$

6. $\dfrac{a}{b} \div 2.$

7. $5 \div \frac{1}{2}.$

8. $a \div \frac{2}{3}.$

9. $3\,b \div \frac{1}{5}.$

10. $2 \div \dfrac{a}{b}.$

11. $\frac{2}{3} \div \frac{3}{4}.$

12. $\frac{1}{2} \div \frac{1}{4}.$

13. $\frac{2}{3} \div \frac{1}{2}.$

14. $\dfrac{a}{b} \div \dfrac{1}{2}.$

15. $\dfrac{a}{b} \div \dfrac{2}{3}.$

16. $\dfrac{3}{4} \div \dfrac{p}{q}.$

17. $\dfrac{x^2}{y} \div x.$

18. $x^2 \div \dfrac{x}{b}.$

19. $\dfrac{a + b}{2} \div 2.$

20. $\dfrac{3}{2} \div \dfrac{x - y}{2}.$

EXERCISE 96

Simplify:

1. $\dfrac{6\,x^2y}{5\,ab^2} \div \dfrac{18\,xy}{25\,a^2b^2}$.

2. $\dfrac{5\,p^2q}{7\,xy^2} \div \dfrac{15\,bp^3}{14\,xy^3}$.

3. $\dfrac{12\,a^3b^2c^5}{7\,x^2y^3z^5} \div \dfrac{18\,a^5b^2c^3}{35\,x^3y^2z^2}$.

4. $\dfrac{x+y}{2} \div \dfrac{x^2-y^2}{6}$.

5. $\dfrac{x+1}{a^2-1} \div \dfrac{x+1}{a-1}$.

6. $\dfrac{y^2+2\,y+1}{3\,ab} \div \dfrac{y^2-1}{6\,ab^2}$.

7. $\dfrac{a+3\,y}{2\,b+1} \div \dfrac{a^2-9\,y^2}{4\,b^2-1}$.

8. $\dfrac{x^2+x}{(a-b)^2} \div \dfrac{ax+a}{a^2-b^2}$.

9. $\dfrac{p^2-16\,q^2}{p^2-25\,q^2} \div \dfrac{p+4\,q}{p-5\,q}$.

10. $\dfrac{(x+1)^2}{4} \div \dfrac{x^2-1}{8\,a}$.

11. $\dfrac{a^2-1}{4} \div \dfrac{a-1}{a+2}$.

12. $\dfrac{a^3-ab^2}{b^2} \div \dfrac{a^2-b^2}{ab}$.

13. $\dfrac{x^2-y^2}{10\,ab} \div \dfrac{x^2-xy}{5\,b^2}$.

14. $\dfrac{x^2-x-2}{a^2-b^2} \div \dfrac{x^2-1}{a(a+b)}$.

15. $\dfrac{9\,a^3x}{4\,bcd} \times \dfrac{2\,cd}{3\,abx} \div \dfrac{3\,acx}{2\,b^2}$.

16. $\dfrac{2\,x}{y} \cdot \dfrac{3(x^2-y^2)}{5} \div \dfrac{x^2-xy}{10}$.

17. $\dfrac{7\,xy}{3\,ab} \cdot \dfrac{a^2+2\,ab-3\,b^2}{a+b} \div \dfrac{x(a^2-b^2)}{a^2+ab}$.

18. $\dfrac{22\,a^2x^2}{5\,yz} \cdot \dfrac{15\,xyz^2}{11\,x^2-55\,xy+66\,y^2} \cdot \dfrac{x^2-4\,y^2}{6\,a^3x^3} \div \dfrac{x^2+2\,xy}{10\,a^2yz^2}$.

EXERCISE 97

Solve the following miscellaneous examples in fractions:

1. Reduce $\dfrac{3\,a^2b+6\,ab^2}{9\,a^2+36\,ab+36\,b^2}$ to its lowest terms.

2. Reduce $\dfrac{x^2+3\,xy-a}{x}$ to a mixed number.

3. Reduce $x^2+x-1+\dfrac{5}{x}$ to an improper fraction.

4. Divide $\tfrac{8}{15}$ by 2.   5. Divide $\dfrac{4\,a}{7\,b}$ by 2.

Simplify:

6. $\dfrac{5}{3\,a-3} - \dfrac{3}{2\,a-2}$.

7. $\dfrac{a^3-9\,a}{a^2b+4\,ab} \div \dfrac{a^2b^2-3\,ab^2}{a^2-16}$.

8. $\dfrac{3\,a^4}{10\,b^4} \times \dfrac{15\,b^6}{7\,c} \div \dfrac{9\,a^2}{28\,c^3}$.

9. $\dfrac{1}{a(a+1)} + \dfrac{1}{a(a-1)}$.

10. $\dfrac{2\,x+14}{3\,y-9} \div \dfrac{9\,x^2+63\,x}{y^2-9}$.

11. $\dfrac{a-2\,b}{b} - \dfrac{3\,a-2\,b}{2\,b} - \dfrac{5\,a}{4\,b}$.

12. $\dfrac{3\,ab^2}{x+y} \times \dfrac{x^2-y^2}{6\,a^2b}$.

13. $\dfrac{x^2-1}{14\,a} \div \dfrac{(x-1)^2}{7\,a^2}$.

14. $\dfrac{3}{a+b} - \dfrac{2\,b+a}{a^2-b^2}$.

15. $\dfrac{x^2-y^2}{4} \div \dfrac{x+y}{2}$.

16. $\dfrac{a+b}{a^2-3\,ab} + \dfrac{a-b}{ab-3\,b^2}$.

17. $\dfrac{a^2-x^2}{a^2} \cdot \dfrac{a^3}{(a-x)^2}$.

18. $\dfrac{3\,x}{x+2\,y} - \dfrac{2\,x}{x-2\,y} + \dfrac{8\,xy}{x^2-4\,y^2}$.

19. Does $\dfrac{x+2}{3} + \dfrac{x-5}{x}$ equal $\dfrac{2\,x-3}{3\,x}$ when $x=2$?

20. Does $\dfrac{x+2}{3} + \dfrac{x-5}{x}$ equal $\dfrac{x^2+5\,x-15}{3\,x}$ when $x=2$?

21. Does $\dfrac{3\,x+2}{5\,x-2} = 7$ when $x=\frac{1}{2}$?

22. Find the value of $\dfrac{a+b}{1-ab}$ when $a=\frac{1}{2}$ and $b=\frac{1}{3}$.

EXERCISE 98

VERBAL PROBLEMS

1. State the fraction formed by dividing 5 by 7. x by 7. x by y. $x+2$ by $3\,x-5$.

2. If 1 orange costs 3 cents, how many oranges can be bought for 12 cents? For x cents? For $x+y$ cents?

3. If 1 orange costs a cents, how many oranges can be bought for 25 cents? For x cents? For $x+y$ cents?

4. If 1 acre of land costs x dollars, what will $\frac{1}{2}$ an acre cost? $\frac{2}{3}$ of an acre? $\frac{3}{4}$ of an acre?

5. What is $\frac{3}{50}$ of x? Express as a fraction 6 per cent of x. Also 35 per cent of x.

6. Express as a fraction a per cent of b. x is what per cent of $x - 5$?

7. If n represents a number, write the sum of a third and fourth part of it.

8. A man earned d dollars in x days. Find the average number of dollars which he earned per day.

9. A mixture of concrete contains a parts cement, b parts sand, and c parts gravel. What fractional part of the whole mixture is the cement? The sand? The gravel?

10. If the expense for tires on an automobile averages c cents per mile, what is the tire expense on a 100 mile trip?

11. A bag of feed will last c cows d days. How long will the bag last x cows?

12. An alloy consists of a parts of copper and t parts of tin. What per cent of the whole is the copper? The tin?

EXERCISE 99

1. If three boys weigh a, b, c pounds respectively, what is their average weight?

2. Using your result as a formula, find the average weight of three boys whose respective weights are 67, 72, and 74 lb.

3. If four boys can run the quarter mile in p, q, r, s seconds respectively, what is their average time?

4. Using your result as a formula, find the average time for the quarter mile for four boys whose records are 56, 54, 58 seconds, and 1 minute, 2 seconds.

5. How many acres are there in a field a feet long and b feet wide?

6. How many acres are there in a field c rods × d rods? In one f yards × e feet? p feet × q rods?

7. Using one of these results as a formula, find the number of acres in a field 100 yd. long and 250 ft. wide.

8. If sugar is worth a cents a pound, find a formula for P, the number of pounds of sugar which can be obtained in exchange for b pounds of butter worth c cents a pound.

9. Using the formula obtained in Ex. 8, find the number of pounds of sugar which can be obtained in exchange for 6 pounds of butter worth 42 cents a pound, if sugar is selling at 8 cents a pound.

10. If coal is worth d dollars a ton, find a formula for C, the number of tons of coal which can be obtained in exchange for h tons of hay worth m dollars a ton.

11. By means of this formula, find the number of tons of coal which can be obtained in exchange for 19 tons of hay worth $15.50 a ton, when coal costs $9.50 a ton.

12. Make up and work examples similar to Exs. 10 and 11, concerning c calves worth a dollars each, exchanged for chairs worth d dollars each.

13. In the subject of interest in arithmetic you learned that the amount equals the principal plus the interest. Hence if a represents the amount, p the principal, r the rate expressed decimally, and t the time in years, $a = p(1 + rt)$.

By use of this formula, find the amount of $250 after it has been at interest for $4\frac{1}{2}$ years at 6 per cent.

14. Using $a = p(1 + rt)$, find the number of years it will take $300 at interest at 5 per cent to amount to $360.

15. By use of $a = p(1 + rt)$ find the rate per cent at which $650 will amount to $741 in $3\frac{1}{2}$ years.

16. By addition of fractions $\dfrac{1}{a} + \dfrac{1}{b} = \dfrac{a + b}{ab}$. Regarding this as a formula, by use of it find the sum of $\frac{1}{12}$ and $\frac{1}{15}$.

17. Express $\frac{1}{a}+\frac{1}{b}=\frac{a+b}{ab}$ as a rule for finding the sum of two fractions whose numerators are 1. By use of this rule find the sum of $\frac{1}{5}$ and $\frac{1}{7}$. Also of $\frac{1}{x+y}$ and $\frac{1}{x-y}$.

18. Express $\frac{1}{a}-\frac{1}{b}=\frac{b-a}{ab}$ as a rule for finding the difference of two fractions whose numerators are 1.

By use of this rule find the value of $\frac{1}{10}-\frac{1}{13}$. Also of $\frac{1}{x+y}-\frac{1}{x-y}$.

19. In the formula $R=\frac{gs}{g+s}$, find the value of R to the nearest thousandth, when $g=16$ and $s=2.5$.

20. In the formula $F=\frac{mv^2}{r}$, find F when $m=20$, $v=24$, and $r=2\frac{1}{2}$.

21. In $C=\frac{5}{9}(F-32)$, find C when $F=1580$.

PART II

85. A Fraction is the indicated quotient of two algebraic expressions. This quotient is usually indicated in the form $\frac{a}{b}$.

Note that the dividing line of a fraction takes the place of a parenthesis and is in effect a *vinculum*.

Thus, $\frac{a+b}{3}$ has the same meaning as $(a+b)\div 3$.

Another method of writing the first fraction given above is a/b. This is called the *solidus notation*. It is convenient in printing mathematical expressions, and is much used in European mathematical literature. $\frac{x+1}{3x-5}$ written in the solidus notation would be $(x+1)/(3x-5)$.

The *numerator* of a fraction is the dividend and the *denominator* is the divisor of the indicated quotient.

86. Terms of a fraction is a general name for both numerator and denominator.

87. An integral expression is one which does not contain a fraction; as $3\,x^2 - 2\,y$.

An expression like $5\,x^2 + \frac{3}{2}\,x + \frac{1}{4}$ in which fractions occur only in the numerical coefficients is sometimes regarded as an integral expression.

88. A mixed expression is one which is part integral, part fractional.

$$\text{Thus, } 3\,x^2 + x - 5 + \frac{x+1}{3\,x^2 - 2}.$$

89. Sign of a Fraction. A fraction has its own sign, which is distinct from the sign of both numerator and denominator. It is written to the left of the dividing line of the fraction.

The sign of $-\dfrac{a}{b}$ is $-$, and the sign of $\dfrac{-a}{b}$ is $+$ understood.

90. Law of Signs. By the laws of signs for multiplication and division (see § 41, p. 77, and § 45, p. 83),

$$\frac{a}{b} = \frac{-a}{-b}, \quad -\frac{a}{b} = \frac{-a}{b} = \frac{a}{-b}, \quad \frac{a}{bc} = -\frac{a}{-b \times c} = \frac{a}{-b \times -c}$$

$$\frac{x+y}{y-x} = \frac{x+y}{-(x-y)} = -\frac{x+y}{x-y}.$$

Or, in general,

*The signs of any **even** number of factors of the numerator and denominator of a fraction may be changed without changing the sign of the fraction.*

*But if the signs of an **odd** number of factors are changed, the sign of the fraction must be changed.*

1. Reduce $\dfrac{(x-2)^2}{4-x^2}$ to its lowest terms.

$$\frac{(x-2)^2}{4-x^2} = \frac{(x-2)(x-2)}{(2+x)(2-x)} = \frac{(2-x)(2-x)}{(2+x)(2-x)} = \frac{2-x}{2+x}.$$

CHECK.   Let $x = 1$, then, $\dfrac{(x-2)^2}{4-x^2} = \dfrac{(-1)^2}{4-1} = \dfrac{1}{3}$.

Also, $\dfrac{2-x}{2+x} = \dfrac{2-1}{2+1} = \dfrac{1}{3}$.

EXERCISE 100

1. Simplify : $-\dfrac{8}{24}$; $\dfrac{-.8}{2.4}$; $\dfrac{-8}{-12}$; $\dfrac{36}{-10.8}$; $\dfrac{-2.4}{-7.2}$.

Reduce to the simplest form and check :

2. $\dfrac{x^2-9}{x^2-6x+9}$.

3. $\dfrac{(a+b)^2-c^2}{a^2-(b-c)^2}$.

4. $\dfrac{4x^2-12xy+9y^2}{2x^3-3x^2y}$.

5. $\dfrac{x^2-a^2-2ab-b^2}{(x+a+b)^2}$.

6. $\dfrac{x^4-9x^2}{x^4-x^3-6x^2}$.

7. $\dfrac{(a-b)x-(a-b)y}{a^2-b^2}$.

8. $\dfrac{(x+y)^2-(p+q)^2}{(x-p)^2-(q-y)^2}$.

9. $\dfrac{(2x-y)^2}{y^2-4x^2}$.

10. $\dfrac{9-m^2}{m^2-7m+12}$.

11. $\dfrac{9-x^2}{(x-3)^2}$.

12. $\dfrac{a-b}{b-a}$.

13. $\dfrac{a+b-c}{c^2-(a+b)^2}$.

14. $\dfrac{6y-3x}{12ay-6ax}$.

15. $\dfrac{(a-b)x-(a-b)y}{b^2-a^2}$.

16. State in which of the following fractions the fraction may be simplified by the cancellation of the 3's :

$$\frac{a+3}{b+3}, \quad \frac{3a}{3b}, \quad \frac{3a}{3b+5}, \quad \frac{3(a+b)}{a+3}, \quad \frac{3a}{5+3b}, \quad \frac{3x}{3(x+y)}.$$

Supply the missing numerator in each of the following:

17. $\dfrac{x+y}{x-y} = \dfrac{?}{4(x-y)^2}.$ **18.** $\dfrac{3}{x-2} = \dfrac{?}{x^4-16}.$

19. $\dfrac{2a+b}{3a-b} = \dfrac{?}{9a^2-6ab+b^2}.$

20. $\dfrac{5x-1}{x-2} = \dfrac{?}{x^2-5x+6}.$

21. $\dfrac{17a^2+3}{a^2-9} = \dfrac{?}{a^3+3a^2-9a-27}.$

22. Reduce $\dfrac{5}{4a+4}$ to an equivalent fraction having $8a^2-8$ for its denominator.

23. Reduce $\dfrac{2}{x^2-5x+6}$ to an equivalent fraction having $(x-2)(x-3)(x-4)$ for its denominator.

91. Examples in the Reduction of Fractions.

Ex. 1. Reduce $\dfrac{x^3+4x^2-5}{x^2+x+2}$ to a mixed expression.

$$
\begin{array}{rl}
x^3+4x^2-5 & |\,x^2+x+2 \\
\underline{x^3+x^2+2x} & x+3 \\
3x^2-2x-5 & \\
\underline{3x^2+3x+6} & \\
-5x-11 &
\end{array}
$$

$\therefore\ \dfrac{x^3+4x^2-5}{x^2+x+2} = x+3 - \dfrac{5x+11}{x^2+x+2}.$ *Ans.*

Ex. 2. Reduce $1-\left(x-\dfrac{1}{1+x}\right)$ to an improper fraction.

$$1-\left(x-\dfrac{1}{1+x}\right) = 1-x+\dfrac{1}{1+x}$$

$$= \dfrac{1-x^2+1}{1+x} = \dfrac{2-x^2}{1+x}.\ \ Ans.$$

Reduce to a mixed expression :

1. $\dfrac{x^3 + 2}{x + 1}$.

2. $\dfrac{a^4 + 8}{a + 1}$.

3. $\dfrac{r^3 - 27}{r + 3}$.

4. $\dfrac{9\,a^3}{3\,a^2 - 2\,b}$.

5. $\dfrac{4\,p^2 + 2\,q^2}{2\,p - q}$.

6. $\dfrac{x^4 + 1}{x^2 + 1}$.

7. $\dfrac{x^5}{x^2 - x - 1}$.

8. $\dfrac{2\,x^4 + 7}{x^2 + x + 1}$.

9. $\dfrac{x^4 + 1}{x^2 + x - 1}$.

10. $\dfrac{x^2 + 3\,xy - 2\,y^2 - 1}{x + y}$.

11. $\dfrac{3\,x^4 - 13\,x - 28}{x^2 - 3}$.

12. $\dfrac{x^3 - x^2 - x + 2 - a}{x - 1}$.

13. $\dfrac{x^4 + x^2 - x - 1}{x^2 + 2}$.

14. Make up an improper fraction with a monomial denominator and reduce it to a mixed number.

15. Make up an improper fraction with a binomial denominator and reduce it to a mixed number.

Reduce to an improper fraction :

16. $x + 1 + \dfrac{1}{x - 1}$.

17. $x^2 + x - 1 - \dfrac{1}{x - 1}$.

18. $4\,x - 2 - \dfrac{y - 2}{2\,x + 1}$.

19. $\dfrac{3\,a - 1}{a - 2} + a - 1$.

20. $x - a - \dfrac{ay - a^2}{x + a} + y$.

21. $1 - \dfrac{b^2 - c^2 + a^2}{2\,bc}$.

22. $6\,a + \dfrac{(2 - 3\,a)^2}{4}$.

23. $\dfrac{(2\,ab - 1)^2}{4} + 2\,ab$.

24. $5\,x^2 - 3\,x + 2 - \dfrac{4\,x + 3}{3\,x - 1}$.

25. $1 - \left(x - x^2 + \dfrac{1}{1 + x}\right)$.

26. $x^2 - \left[x - \left(1 - \dfrac{2}{x + 1}\right)\right]$.

By substituting a numerical value for x, find

27. Whether $3\,x - 1 + \dfrac{2\,x}{x + 1}$ equals $\dfrac{5\,x - 7}{x + 1}$.

28. Whether $\dfrac{x^2 + 1}{x + 1}$ equals $x - 1 + \dfrac{2}{x + 1}$.

92. Examples in Adding Fractions.

Ex. 1. $\dfrac{a}{a-1} - a + \dfrac{1}{a^2 - a} + \dfrac{1}{a}$

$= \dfrac{a}{a-1} - \dfrac{a}{1} + \dfrac{1}{a^2 - a} + \dfrac{1}{a}$

$= \dfrac{a^2 - a^3 + a^2 + 1 + a - 1}{a(a-1)}$

$= \dfrac{-a^3 + 2a^2 + a}{a(a-1)} = \dfrac{-a^2 + 2a + 1}{a-1}.$ **Ans.**

Ex. 2. Simplify $\dfrac{x^2}{x^2 - 1} + \dfrac{x}{x+1} - \dfrac{x}{1-x}.$

The factors of $x^2 - 1$ are $x + 1$ and $x - 1$. Hence, if the sign of the denominator, $1 - x$, is changed, it will become $x - 1$, and be a factor of $x^2 - 1$. But by § 90 (p. 177), if the sign of $1 - x$ is changed, the sign of the fraction in which it occurs must also be changed. Hence, we have

$\dfrac{x^2}{x^2 - 1} + \dfrac{x}{x+1} + \dfrac{x}{x-1} = \dfrac{x^2 + x^2 - x + x^2 + x}{x^2 - 1} = \dfrac{3x^2}{x^2 - 1}.$ **Ans.**

EXERCISE 102

Simplify and check :

1. $\dfrac{a-b}{a^2 b} + \dfrac{a-b}{ab^2}.$

3. $\dfrac{(a+b)^2}{4(a-b)^2} - \dfrac{2b}{a-b}.$

2. $\dfrac{a-b}{ab} - \dfrac{c-a}{ac} + \dfrac{b-c}{bc}.$

4. $\dfrac{l-1}{l-2} - \dfrac{l+1}{l+2} + \dfrac{l-6}{l^2 - 4}.$

5. $\dfrac{2x^2 + 3}{6x^2} - \dfrac{3x^3 + 1}{12x^3} - \dfrac{3x^4 - 2}{36x^4}.$

6. $\dfrac{2x^2 y - 3z}{3x^2 y} - \dfrac{xz^2 - y^2 z}{2xy^2} + \dfrac{y - 3xz^2}{6x^2 z} - \dfrac{2}{3}.$

7. $\dfrac{1}{3x-3} - \dfrac{1}{2x+2} + \dfrac{x-5}{6x^2 - 6}.$

8. $\dfrac{x}{x^2 - 1} + 2 - \dfrac{x-1}{x+1} - \dfrac{x-2}{x-1}.$

9. $\dfrac{a^2-3}{4\,a^2-9}+\dfrac{a+2}{2\,a+3}-\dfrac{2\,a+3}{2\,a-3}.$

10. $\dfrac{4\,b^2}{a^2-b^2}-\dfrac{(a-b)^2}{a^2-b^2}+2-\dfrac{3\,a+3\,b}{3\,a-3\,b}.$

11. $\dfrac{3\,x}{x^2-1}+\dfrac{4}{1-x}+\dfrac{1}{1+x}.$ **13.** $\dfrac{1}{x-1}+\dfrac{1}{1+x}+\dfrac{2\,x}{1-x^2}.$

12. $\dfrac{2\,a}{a^2-b^2}+\dfrac{1}{a+b}+\dfrac{2}{b-a}.$ **14.** $\dfrac{x^2+y^2}{x^2-y^2}-\dfrac{x}{x+y}+\dfrac{y}{y-x}.$

15. $\dfrac{3\,xy}{x^2-4\,y^2}-\dfrac{y-x}{2\,y+x}+\dfrac{y+x}{2\,y-x}.$

16. $\dfrac{3}{8-8\,a}+\dfrac{5}{4\,a+4}-\dfrac{7\,a}{8\,a^2-8}.$

17. $\dfrac{3}{x}+\dfrac{2}{x-1}+\dfrac{5\,x}{1-x^2}-\dfrac{1}{x+1}-\dfrac{3}{x+x^2}.$

18. $\dfrac{2\,b+a}{x+a}-\dfrac{2\,b-a}{a-x}-\dfrac{4\,bx-2\,a^2}{x^2-a^2}.$

19. $\dfrac{x+1}{6\,x-6}-\dfrac{2\,x-1}{12\,x+12}+\dfrac{2}{3-3\,x^2}-\dfrac{7}{12\,x}.$

20. $\dfrac{b}{a+b}-\dfrac{ab}{(a+b)^2}-\dfrac{ab^2}{(a+b)^3}.$

21. $\dfrac{2\,xy}{x^2-y^2}+\dfrac{3\,y}{2\,x}+\dfrac{3\,x}{2\,y}-\dfrac{3\,x^2-3\,y^2}{2\,xy}.$

22. $1-\dfrac{2}{x-1}+x-\dfrac{3\,x-1}{x+1}-\dfrac{2\,x-5}{2}.$

23. $\dfrac{5\,x}{2(x-3)^2}-\dfrac{7}{3\,x+9}-\dfrac{26}{4\,x^2-36}.$

Reduce each of the following fractions to its lowest terms and collect:

24. $\dfrac{x+3}{x^2-9}-\dfrac{1}{x-4}.$ **25.** $\dfrac{x^2+x}{x^2-1}+1-\dfrac{x^2+x}{(x+1)^2}.$

26. $\dfrac{1-x^2}{9-x^2}+\dfrac{x^2-9}{3(x+3)^2}-\dfrac{x^2-4\,x+3}{5(x-3)^2}-\dfrac{2\,x}{5\,x^2-45}.$

By substituting a numerical value for x, find

27. Whether $\dfrac{x}{2} - \dfrac{3\,x}{5}$ equals $-\dfrac{x}{10}$.

28. Also whether $\dfrac{3\,x}{x^2-1} - \dfrac{4}{x-1} + \dfrac{1}{x+1}$ equals $\dfrac{5}{1-x^2}$.

29. Make up and solve an example in which three fractions with monomial denominators are added together.

30. Also one in which two fractions with binomial denominators are subtracted.

31. Also one in which one fraction with a monomial denominator and two with binomial denominators are added.

93. The **reciprocal** of a number is the result obtained by dividing unity by the given number.

Thus, the reciprocal of 2 is $1 \div 2$ or $\dfrac{1}{2}$; of x is $\dfrac{1}{x}$.

Hence, the reciprocal of a fraction is the fraction inverted.

Thus, the reciprocal of $\frac{2}{3}$ is $1 \div \frac{2}{3}$; that is, $1 \times \frac{3}{2}$, or $\frac{3}{2}$.

Similarly, the reciprocal of $\dfrac{a}{b}$ is $\dfrac{b}{a}$; of $\dfrac{a-b}{x+y}$ is $\dfrac{x+y}{a-b}$.

94. Examples in Multiplication and Division of Fractions.

Ex. 1. $\left(\dfrac{x^2-4\,x+4}{x^2+x}\right) \cdot \dfrac{(x^2-1)^2}{x-2} \div \dfrac{(x-1)^2(x^2-3\,x+2)}{(x+1)^2}$

$= \dfrac{(x-2)^2}{x(x+1)} \cdot \dfrac{(x^2-1)\,(x^2-1)}{x-2} \cdot \dfrac{(x+1)\,(x+1)}{(x-1)\,(x-1)\,(x-1)\,(x-2)}$

$= \dfrac{(x+1)^3}{x(x-1)}.$ *Ans.*

Let the pupil check the work.

Ex. 2. Simplify $\left(\dfrac{x^2}{5} - 5\right) \div \left(1 + \dfrac{x}{5}\right)$.

$\left(\dfrac{x^2}{5} - 5\right) \div \left(1 + \dfrac{x}{5}\right) = \dfrac{x^2-25}{5} \div \dfrac{5+x}{5}$

$= \dfrac{x^2-25}{5} \cdot \dfrac{5}{5+x} = x - 5.$ *Ans.*

Simplify the following :

1. $\dfrac{xy^2z^3(a-b)}{2\,a^3} \div \dfrac{5\,xy^3(a^2-b^2)}{3\,a^3}.$

2. $\dfrac{a^2-b^2}{3\,x} \cdot \dfrac{10\,xy}{a^3-ab^2} \div \dfrac{5\,x(a-b)}{3\,ax(a+b)}.$

3. $\left(1+\dfrac{a}{b}\right) \div \left(1-\dfrac{b^2}{a^2}\right).$

6. $\left(\dfrac{x}{x^2-4\,y^2}\right)\left(1+\dfrac{2\,y}{x}\right).$

4. $\left(x-\dfrac{1}{x}\right) \div \left(1+\dfrac{1}{x}\right).$

7. $\left(4-\dfrac{3}{x+1}\right) \div \left(6+\dfrac{5}{x^2-1}\right).$

5. $\left(\dfrac{a^2-b^2}{a}\right) \div \left(1-\dfrac{b}{a}\right).$

8. $\left(1+\dfrac{1}{x-1}\right) \div \left(\dfrac{1}{x-1}\right).$

9. $\left(\dfrac{a^2-x^2}{a+y}\right)\left(a+\dfrac{ax}{a-x}\right) \div \dfrac{ax+x^2}{a^2-y^2}.$

10. $\dfrac{l+2}{x-3} \cdot \dfrac{3\,x^2-27}{2\,l^2-8} \div \dfrac{lx+3\,l}{4}.$

11. $\dfrac{x^2y+xy^2}{x^2y+y^3} \div \dfrac{5\,xy(x+y)^2}{x^4-y^4}.$

12. $\dfrac{x^2-y^2}{5\,xy} \times \dfrac{x-y}{x+y} \times \dfrac{10\,x^2y^2}{(x-y)^2} \times \dfrac{1}{2\,xy}.$

13. $\dfrac{3(a-b)^2}{4(a+b)^2} \times \dfrac{7(a^2-b^2)}{9(a-b)^3} \div \dfrac{14\,ab}{8(a+b)}.$

14. $\dfrac{6\,x^2y-4\,xy^2}{45\,x^2-20\,y^2} \times \dfrac{30\,x+20\,y}{4\,x^2y^2} \times \dfrac{xy}{x+y}.$

15. $\left(x+1+\dfrac{1}{x}\right)\left(x-1+\dfrac{1}{x}\right) \div \dfrac{(x^2+x+1)(x^2-x+1)}{x^2(x^2-1)}.$

16. What is the reciprocal of s ? Of $\frac{2}{5}$? Multiply $\frac{2}{3}$ by the reciprocal of $\frac{6}{5}$.

17. Multiply $\dfrac{a^2}{b^2}$ by the reciprocal of $\dfrac{2\,a}{b}$.

18. What is the reciprocal of $x - y$? Multiply $x^2 - y^2$ by the reciprocal of $x - y$.

19. Divide $\dfrac{x^2}{x^2 - 1}$ by the reciprocal of $\dfrac{(x-1)^2}{x^4}$.

Work the following miscellaneous examples in fractions:

20. Reduce $\dfrac{x^2 - 4}{6 - 3x}$ to its lowest terms.

21. Reduce $x^2 - x + 1 - \dfrac{x-2}{x+1}$ to an improper fraction.

22. Reduce $\dfrac{a^2 - ab - 3a + 2b}{a - 1}$ to a mixed number.

Simplify the following:

23. $\dfrac{a}{b} - \dfrac{b}{a} - \dfrac{a-b}{a+b}$.

24. $\dfrac{a+1}{x^2 - 1} + \dfrac{a^2 + a}{x^2 - x}$.

25. $\left(\dfrac{9a^2}{b^2} - 1\right)\left(\dfrac{b}{3a-b} + 1\right)$.

26. $\dfrac{9a^3b^2 - 12a^2b^3}{9a^2 - 16b^2}$.

27. $\left(\dfrac{a}{b} + \dfrac{b}{a}\right) \div \dfrac{a^4 - b^4}{a^2b^2}$.

28. $2\frac{5}{8} \times 5\frac{1}{3} \div 1\frac{7}{9}$.

29. $3\frac{7}{8} + 2\frac{3}{4} - 1\frac{5}{8}$.

30. $\dfrac{x-3}{2x+6} - \dfrac{x^2 + 6}{6x^2 - 54} - \dfrac{x+3}{3x-9}$.

31. $\dfrac{4a-b}{2x+y} \cdot \dfrac{2a}{4a^2 - ab} \div \dfrac{4}{4x^2 - y^2}$.

EXRCISE 104

VERBAL PROBLEMS

1. An automobile traveled a miles the first hour, b miles the next hour, c miles the next, and d miles each of the next two hours; what was its average rate per hour during the trip?

2. A man painted an automobile in 8 days. What part of the work did he do in 1 day? In x days? In c days?

3. A room is x yards long and y yards wide. Find its area in square feet.

4. A car travels y yards in h hours. What was its rate in feet per minute ?

5. If the average weight of three objects is a pounds, and of four other objects is b pounds, what is the average weight of all seven objects ?

6. Express as a fraction 20 % of x, also 90 % of x, 120 % of x, $133\frac{1}{3}$ % of x, $87\frac{1}{2}$ % of $2\,x$, $66\frac{2}{3}$ % of $320 - x$.

7. A man walks $\dfrac{s}{n}$ miles an hour. How many hours will it take him to walk $\dfrac{a}{b}$ miles ?

8. A printing press can do a piece of work in h hours and runs 3 hours. Another press can do the same piece of work in p hours and runs c hours. What fractional part of the work has each done ? What part have both done together ?

9. A boy has c cents, his father gave him b cents, and he then spent n nickels. How many cents·did he have left ?

10. Give the product of three consecutive numbers of which m is the middle number ?

11. How many minute spaces does the hour hand of a clock move over while the minute hand moves over 60 minute spaces ? Over 48 minute spaces ? 12 spaces ? x spaces ?

EXERCISE 105

1. A boy earns e cents a day for d days. On the average he spends s cents a day. Obtain a formula for the number of dollars (D) which he has at the end of the d days. By use of this formula find how many dollars he has at the end of 20 days during which he has earned on the average 90 cents and spent 35 cents a day.

2. If coal is worth c dollars a ton, find a formula for T, the number of tons which can be obtained in exchange for f bushels of wheat worth h cents a bushel, and w bushels of corn worth y cents a bushel ?

3. By use of this formula find the number of tons of coal worth $9 a ton which can be obtained in exchange for 30 bu. of wheat at $1.35 per bushel, and 35 bu. of corn at 90 cents a bushel.

4. If the cost of an article is c and the article is to be sold at a profit of r per cent (expressed decimally), find a formula for s, the selling price.

5. A boy made a bookcase in a manual training shop. The material and his time were worth $8.50. He wishes to sell the bookcase at a profit of 20 per cent. What must he ask for the bookcase? Use formula for s, Ex. 4.

6. A boy wishes to sell a desk for $10 at a profit of 25 per cent. What is its cost? Use formula for s, Ex. 4.

7. A girl wishes to sell canned goods for $4 a dozen at a profit of $33\frac{1}{3}$ per cent. What is their cost? Use formula for s, Ex. 4.

8. A 5 and 10 cent store wishes to sell an article for 10 cents at a gain of 20 per cent. What is the cost of the article? Use formula for s, Ex. 4.

9. A 5 and 10 cent store wishes to sell an article for 5 cents at a gain of 22 per cent. Using formula for s, Ex. 4, find the cost of the article to three decimal places.

10. In the formula $V = \dfrac{\pi r^2}{2}(h + h')$, find V when $\pi = \frac{22}{7}$, $r = 12$, $h = 3.6$, and $h' = 2.7$.

11. The formula $H.P. = \dfrac{plan}{33000}$ is used in determining the horse power of a steam engine. Find H. P. when $p = 80$, $l = 4\frac{1}{2}$, $a = 2$, and $n = 400$.

CHAPTER X

FRACTIONAL AND LITERAL EQUATIONS

PART I

95. In solving an equation containing fractions we *multiply both members of the equation by the smallest number which contains all of the denominators as factors.*

Ex. 1. Solve $\dfrac{x+3}{4} - \dfrac{x+4}{6} = \dfrac{x}{2} - 2$.

The smallest number that will contain all the denominators is 12. Multiplying both members of the equation by 12, we obtain
$$3(x+3) - 2(x+4) = 6x - 24.$$
Removing parentheses,
$$3x + 9 - 2x - 8 = 6x - 24.$$
Transposing terms,
$$3x - 2x - 6x = -9 + 8 - 24.$$
$$-5x = -25.$$
$$x = 5. \quad \textit{Ans.}$$

CHECK.
$$\frac{5+3}{4} - \frac{5+4}{6} = \frac{5}{2} - 2.$$
$$2 - 1\tfrac{1}{2} = \tfrac{1}{2}.$$

Ex. 2. Solve $\dfrac{2}{x^2-4} + \dfrac{5}{x+2} - \dfrac{2}{x-2} = 0$.

Multiplying by $x^2 - 4$, the smallest number that will contain all the denominators,
$$2 + 5(x-2) - 2(x+2) = 0.$$
$$2 + 5x - 10 - 2x - 4 = 0.$$
$$3x = 12.$$
$$x = 4. \quad \textit{Ans.}$$

Let the pupil check the work.

188

EXERCISE 106

ORAL

At sight solve each of the following:

1. $\dfrac{x}{3}=5.$

2. $\dfrac{2x}{3}=4.$

3. $\dfrac{5x}{6}=-10.$

4. $2=\dfrac{3}{x}.$

5. $\dfrac{6}{x}=-2.$

6. $\dfrac{1}{x}=4.$

7. $\dfrac{3}{x}=1.$

8. $\dfrac{2x}{3}=-8.$

9. $\dfrac{2}{3x}=8.$

10. $5=\dfrac{2x}{3}.$

11. $5=\dfrac{3}{2x}.$

12. $\dfrac{3}{x}=-1.$

13. $1=-\dfrac{3}{x}.$

14. $\dfrac{1}{x}=\dfrac{2}{3}.$

15. $\dfrac{2}{x}=-\dfrac{1}{5}.$

16. $\dfrac{2}{x}=\dfrac{4}{5}.$

17. $\dfrac{2}{3}=\dfrac{x}{2}.$

18. $\dfrac{3}{4}=\dfrac{1}{x}.$

19. $\dfrac{5}{2}=\dfrac{10}{x}.$

20. $\dfrac{p}{20}=40.$

21. $\dfrac{2}{a}=8.$

22. $\dfrac{l}{5}=10.$

23. $\dfrac{2}{w}=8.$

24. $5=\dfrac{25}{t}.$

25. $0=\dfrac{x}{5}+5.$

26. $\dfrac{3x}{2}-4=0.$

27. $\dfrac{4y}{5}+6=0.$

28. $0=8-\dfrac{4x}{3}.$

EXERCISE 107

Solve and check:

1. $\dfrac{x}{8}-\dfrac{3}{2}=0.$

2. $\dfrac{x-3}{2}=4.$

3. $x+\dfrac{x}{2}=18.$

4. $\dfrac{x}{2}+\dfrac{x}{4}=15.$

5. $\dfrac{x}{2}-\dfrac{x}{3}=10.$

6. $\dfrac{3-x}{2}=\dfrac{x-2}{3}.$

7. $\dfrac{2}{x-3}=1.$

8. $\dfrac{2}{5x-1}=\dfrac{2}{3x+5}.$

9. $\dfrac{x+1}{x-2}=\dfrac{3}{2}.$

10. $\dfrac{2x-3}{4x-5}=\dfrac{4}{7}.$

11. $5+\dfrac{x}{2}=x-5.$

12. $\dfrac{x}{x-3}=\dfrac{x+1}{x-4}.$

13. $\dfrac{x+3}{2}=4-\dfrac{x}{3}.$

14. $\dfrac{x}{8}-6=1-\dfrac{x}{6}.$

15. $\dfrac{4}{9}-\dfrac{1}{6x}=\dfrac{1}{2}-\dfrac{4}{9x}.$

16. $\dfrac{x}{3}-\dfrac{3x}{5}+\dfrac{7x}{5}=\dfrac{34}{15}.$

17. $\dfrac{3x}{4}-5=\dfrac{7x}{8}+9\frac{1}{2}.$

18. $\dfrac{5x}{3}-8=\dfrac{7x}{9}+3\frac{1}{2}.$

19. $\dfrac{2x}{3}-\dfrac{2x+1}{5}=\dfrac{1}{3}.$

20. $\dfrac{2x-3}{4}+\dfrac{x+1}{6}=\dfrac{5x+2}{12}.$

21. $\dfrac{1}{2x}-\dfrac{3}{x}+\dfrac{5}{3x}=\dfrac{3}{4x}-\dfrac{19}{24}.$

22. $\dfrac{3x+5}{4}=1-\dfrac{x+4}{6}.$ **23.** $3=\tfrac{3}{5}(x-2).$

Sug. Remove the parenthesis; thus, $3=\dfrac{3x}{5}-\dfrac{6}{5}$, etc.

24. $2x-8-\tfrac{1}{7}(24-2x)=0.$ **25.** $\tfrac{3}{4}(x-1)=\tfrac{1}{3}(x-2).$

26. $\dfrac{3-2x}{8}-\dfrac{x-3}{6}-1=\dfrac{x+4}{3}+\dfrac{1}{24}.$

27. $\dfrac{3x-1}{7}-\dfrac{x+1}{6}-\dfrac{4x+1}{21}=\dfrac{3(x-1)}{4}-3.$

28. $\dfrac{2x+5}{5}-\dfrac{x+1\frac12}{10}+x=\dfrac{5x-10\frac12}{20}-\dfrac{1}{5}.$

29. $\dfrac{3-x}{10}-\tfrac{2}{3}(5+x)+\dfrac{x-3}{5}-\tfrac12(2x+5)=\dfrac{7x-4}{6}.$

30. $2(x+\tfrac13)+x\left(1-\dfrac{1}{2x}\right)=\dfrac{6x+5}{12}+\dfrac{x+5}{4}.$

31. $\dfrac{x+5}{7}-\dfrac{x+7}{5}+\dfrac{x+1}{2}-\dfrac{2x-5}{10}=\dfrac{x+22}{70}.$

32. $\tfrac{2}{3}(5x+2)-\tfrac{3}{4}(7x-2)+\tfrac12(3x-2)=x-\dfrac{x}{2}.$

33. $.5x-.4x=.3.$

34. $1.5x-5=x.$

35. $.6x-1.5=.2-.15x.$

36. $.7x+.8x=4.5.$

37. $.8x=3.6-.4x.$

38. $\dfrac{.5}{x}=3-\dfrac{.7}{x}.$

39. $.15x-.12x=.09.$

40. $\dfrac{5}{2x-1}=\dfrac{8}{3x+1}.$

41. $\dfrac{6x-5}{3x-3}=\dfrac{8x-7}{4x+4}.$

42. $\dfrac{x}{3}-\dfrac{x^2-5x}{3x-7}=\dfrac{2}{3}.$

43. $\dfrac{3}{3-x}+\dfrac{4}{3+x}=\dfrac{8x+3}{9-x^2}.$

44. $\dfrac{2x+1}{2x-1} - \dfrac{10}{4x^2-1} = \dfrac{2x-1}{2x+1}.$

45. $\dfrac{1}{x+1} + \dfrac{1}{x-1} = \dfrac{2}{x+2}.$

In the following reduce each fraction to its lowest terms and then solve :

46. $\dfrac{x}{2} = \dfrac{x^2-1}{x+1}.$

48. $\dfrac{x^2-x-2}{x-2} = 4 - 6(x-3).$

47. $\dfrac{4x+4}{x^2+x} - \dfrac{3}{4} = 0.$

49. $\dfrac{x^2-6x+9}{x-3} = 2x-7.$

Find the value of the letter in each of the following :

50. $\dfrac{b}{5} - \dfrac{b}{6} = 7.$

53. $\dfrac{a+3}{4} - 5 = 7.$

51. $\dfrac{1}{v+2} + \dfrac{7}{3(v+2)} = \dfrac{2}{3}.$

54. $\dfrac{15}{3-2s} - \dfrac{2}{6-4s} - 14 = 0.$

52. $\dfrac{1}{3(p-7)} + \dfrac{1}{6} = \dfrac{1}{2p-14}.$

55. $\dfrac{4}{2t+2} + \dfrac{31}{3t+3} = \dfrac{1}{6}.$

56. If $A = lw$, $A = 600$, and $w = 20$, find the value of l.

Do you know the meaning of this process in arithmetic in connection with the rectangle ?

57. If $C = \tfrac{5}{9}(F-32)$, find F when $C = 50$. When $C = 100$.

58. If $LW = lw$, and $L = 8$, $W = 100$, and $w = 40$, find l.

59. If $R = \dfrac{gs}{g+s}$, find s when $R = 10$ and $g = 32$.

96. A **numerical equation** is an equation in which the known quantities are expressed by figures. Thus, all the equations on page 190 are numerical equations.

A **literal equation** is an equation in which some or all of the known quantities are denoted by letters ; as by a, b, c . . ., or m, n, p . . .

The methods used in solving literal equations are the same as those used in solving numerical equations.

Ex.   Solve $a(x - a) = b(x - b)$.

$$ax - a^2 = bx - b^2.$$
$$ax - bx = a^2 - b^2.$$
$$(a - b)x = a^2 - b^2.$$
$$x = a + b. \quad \textit{Root.}$$

CHECK. $a(x - a) = a(a + b - a) = ab.$
$$b(x - b) = b(a + b - b) = ab.$$

EXERCISE 108

ORAL

At sight give the value of x in each of the following:

1. $3x = a.$ **5.** $a = 5x.$ **9.** $2b - x = 0.$ **13.** $a = \dfrac{bx}{c}.$

2. $ax = b.$ **6.** $b = px.$ **10.** $0 = x - 2a.$

3. $\dfrac{x}{a} = 1.$ **7.** $1 = rx.$ **11.** $ax = \dfrac{1}{b}.$ **14.** $ax - 1 = 0.$

                                                    15. $c - bx = 0.$

4. $\dfrac{a}{x} - 1 = 0.$ **8.** $\dfrac{x}{p} = 4.$ **12.** $\dfrac{a}{x} = p.$ **16.** $c = rx.$

At sight give the value of the underscored letters in the following:

17. $p = b\underline{x}.$ **20.** $a = l\underline{x}.$ **23.** $i = pr\underline{x}.$ **26.** $i = \underline{p}rt.$

18. $p = b\underline{r}.$ **21.** $a = l\underline{w}.$ **24.** $i = pr\underline{t}.$ **27.** $v = l\underline{w}h.$

19. $p = \underline{b}r.$ **22.** $a = \underline{l}w.$ **25.** $i = p\underline{r}t.$ **28.** $v = l\underline{w}h.$

EXERCISE 109

Solve for x and check :

1. $3x + 2a = x + 8a.$ **6.** $3cx = a - (2b - a + cx).$

2. $9ax - 3b = 2ax + 4b.$ **7.** $2ax - 3b = cx + 2d.$

3. $5ax - c = ax - 5c.$ **8.** $(x + a)(x - b) = x^2.$

4. $ax + bx = c.$ **9.** $ab(x + 1) = a^2 + b^2x.$

5. $ax + b = bx + 2b.$ **10.** $(x - 1)(x - 2) = (x - a)^2.$

11. $\dfrac{x}{a}=\dfrac{b}{2\,a}.$

12. $\dfrac{3\,x}{b}=\dfrac{5\,c}{4\,b}.$

13. $\dfrac{ax}{b}=\dfrac{c}{d}.$

14. $\dfrac{1}{x-a}=2.$

15. $\dfrac{1}{x-p}=\dfrac{2}{b}.$

16. $5=\dfrac{8\,p}{x-p}.$

17. $p+q=\dfrac{a+b}{x}.$

18. $\dfrac{a+b}{x}=\dfrac{c+d}{5}.$

19. $\dfrac{c-d}{x}=p+q.$

20. $\dfrac{a}{x}-b=\dfrac{c}{3\,x}.$

21. $\dfrac{3}{x-3\,a}=\dfrac{5}{4\,a}.$

22. $\dfrac{3}{x-a}=\dfrac{4}{a-b}.$

23. $p=\dfrac{q}{r+8x}.$

24. $a=\dfrac{x-b}{x-c}.$

25. $\dfrac{x-p}{x-q}=\dfrac{r}{8}.$

26. $\dfrac{x}{x-m}=\dfrac{m+n}{n}.$

27. $\dfrac{x-a}{x-b}=\dfrac{x-c}{x-d}.$

28. $a^2x=(a-b)^2+b^2x.$

29. $\dfrac{c}{3\,x}+5=\dfrac{a}{x}-3\,b.$

30. $\dfrac{a+x}{a-2\,x}=\dfrac{a-x}{a+2\,x}.$

31. $(a-b)x=a^2-(a+b)x.$

32. $a-\dfrac{b}{x}=\dfrac{a}{x}-b.$

33. $\dfrac{4\,x-a}{2\,x-a}-1=\dfrac{x+a}{x-a}.$

34. $\dfrac{x}{a}+\dfrac{x}{b}+\dfrac{x}{c}=d.$

Solve the following :

35. $a=x+rtx$ for x.

36. $a=p+prt$ for p.

37. $a=p(1+rx)$ for x.

38. $a=p(1+rt)$ for t.

39. $ab=cd+p$ for p.

40. $ad=cd+p$ for d.

41. Solve $C=\dfrac{E}{R}$ for E. Also for R.

Solve each of the following for the underscored letter :

42. $3\,a+bc=p\underline{q}.$

43. $4\,l\underline{m}-3\,rs=pq.$

44. $LW=l\underline{w}.$

45. Solve $\dfrac{D}{d}=q+\dfrac{r}{d}$ for r. Also for d. Also for q.

46. $a=\tfrac12\,b\underline{h}.$

47. $s=\pi r\underline{l}.$

48. $T=\pi r(r+\underline{l}).$

1. What is the cost of e eggs at c cents a dozen ?

2. If l feet of wire weighs p pounds, find the weight of y yards of the wire.

3. Multiply a by b and m by n and add the results.

4. Express the sum of a yards, b feet and i inches in inches. In feet.   In yards.

5. Indicate in two different ways that the sum of a and b is to be divided by c.

6. If q is the quotient, r the remainder, and d the divisor, find the dividend.

7. What is the average size of 7 and 11 ? of 7 and x ? of a and b ?

8. Express in symbols 10 % of x.   5 % of x.   115 % of b.

Express each of the following as an equation :

9. $\frac{2}{3}$ of x equals 12 less $\frac{1}{2}x$.

10. $\frac{1}{3}$ of x added to $\frac{1}{4}$ of x gives 42.

11. State $\dfrac{x}{3} - \dfrac{x}{4} = 28$ as a problem concerning an unknown number.

12. What per cent of 8 is x ?   Of 9 is $x + 2$?

13. A pupil has worked 8 out of x examples.   What per cent of the x examples has he worked ?   If he should work l more examples, what per cent of the examples will he then have worked ?

14. A pupil's grades in three subjects are 89, 92, 85.   If his grade in a fourth subject be represented by x, what is his average grade in all four subjects ?

15. Express in the form of a fraction the average value of $x, x + 2, x - 3$.

16. If x is added to the numerator of $\frac{3}{10}$, what does the fraction become ?

17. If x is subtracted from the denominator of $\frac{7}{11}$, what does the fraction become?

18. If x is added to both numerator and denominator of $\frac{a}{b}$, what does the fraction become?

19. If 2 gal. of alcohol is mixed with 8 gal. of water, what per cent of the solution is alcohol? If 6 more gallons of water are then added, what per cent of the whole is the alcohol?

20. A given mixture consists of 4 qt. of alcohol and 5 qt. of water. If x more quarts of water be added, what fractional part of the whole will the alcohol be?

21. A given alloy consists of 3 lb. of tin and 20 lb. of copper. If x pounds of copper be added, what part of the whole will the tin be?

22. $\frac{58}{7} = 8 + \frac{2}{7}$. Express in like manner the statement that $50 \div x$ gives a quotient 7 and a remainder 2.

23. Also that x divided by 7 gives 6 for a quotient and 3 for a remainder.

24. After one courier has travelled 6 hours, another courier starts after him and overtakes him in x hours. How many hours does the first courier travel? If he travels at the rate of 5 miles an hour, how many miles does he go? If the second courier travels at the rate of 7 miles an hour, how many miles does he go?

25. Express as a common fraction the annual income from x dollars invested at 5 per cent. Also the income from $\frac{2}{3}$ of x dollars invested at 4 per cent.

26. If $\frac{4}{5}$ of x dollars is invested at 5 %, and $\frac{1}{5}$ of x dollars at 6 %, express the total annual income from these investments.

EXERCISE 111

1. One third of a certain number is 68. Find the number.

2. A man sold two thirds of his farm for $6400. Find the value of his whole farm.

3. 34 diminished by $\frac{1}{8}$ of a certain number gives 23. Find the number.

4. What number increased by $\frac{1}{8}$ of itself and then by 20 equals 60?

SUG.   NUMBERS DEALT WITH   SYMBOLS FOR NUMBERS

(1) The unknown number $= x.$

(2) $\frac{1}{8}$ of the unknown number $= \frac{1}{8}x.$

(3) A known number $= 20.$

(4) Another known number $= 60.$

RELATION OF EQUALITY AMONG NUMBERS

The sum of the first three numbers equals the last, etc.

5. What number diminished by $\frac{1}{4}$ of itself and then increased by 30 equals 90?

6. What sum of money diminished by $\frac{1}{5}$ of itself and then by \$30 equals \$160.60?

7. $\frac{2}{3}$ of a number added to 5 times the number gives 340. Find the number.

8. The difference between $\frac{2}{3}$ and $\frac{1}{2}$ of a certain number is 14. Find the number.

9. Two thirds of a certain number is 12 more than one third of the number. Find the number.

10. A man left \$24,000 to his two children. As his daughter had cared for him in his old age, he left his son only $\frac{2}{3}$ as much as his daughter. What did each receive?

11. A farm is worked on shares. As the owner of the farm supplies the tools and fertilizers, the tenant receives only $\frac{3}{4}$ as large a share of the profits as the owner. If one year's profits are \$4410, what does each receive?

12. Separate 126 into two parts such that one of them is $\frac{1}{5}$ as large as the other. $\frac{2}{5}$ as large.

13. Find two consecutive numbers such that one seventh of the greater number exceeds one ninth of the less by 1.

14. A macadam road cost $18,000. The county paid $\frac{1}{2}$ as much of the cost as the township, and the state paid $\frac{1}{6}$ as much as the township. How much did each pay?

15. A certain kind of concrete contained $\frac{1}{2}$ as much sand as gravel and $\frac{1}{2}$ as much cement as sand. How many pounds of each were in $1\frac{3}{4}$ tons of concrete?

16. If a 12-year-old boy and a 16-year-old boy together earn $48 in mowing lawns, and the younger boy receives only half as much as the other, how much does each boy receive?

17. Separate $6800 into three parts such that the second part equals $\frac{1}{2}$ of the first part, and the third part equals $\frac{1}{5}$ of the first part.

18. A farmer obtained 2720 pounds of cream in one month by the use of a separator. This is $\frac{1}{5}$ more than he would have obtained if his milk had been skimmed by hand. How much would he have obtained by the latter process?

19. A cubic foot of water and a cubic foot of alcohol together weigh 112.5 lb. The alcohol weighs $\frac{4}{5}$ as much as the water. What is the weight of a cubic foot of each?

20. Two men kept a store for a year and made $4800. The owner of the store building received 40% more of the profits than the other. How much did each receive?

21. In building a macadam road the county pays twice as much as the state, and the township 50% more than the state. What does each pay if the road costs $18,000?

22. A boy's grade in a subject is 89. What is his grade in another subject if his average in the two subjects is 92?

SUG.	NUMBERS DEALT WITH	SYMBOLS
	Grade in first subject	= 89.
	Grade in second subject	= x.
	Average in the two subjects	= 92.

EQUALITY AMONG NUMBERS

(Sum of the first two numbers) ÷ 2 = third number, etc.

23. A boy's grades in two subjects are 84 and 88. What must his grade in a third subject be in order that his average shall be 89?

24. A girl's grades are, arithmetic 87, reading 92, and geography 85. What grade must she have in spelling to make her general average 90?

25. The profits in a given business for two years were $2200 and $2800. What must the profits for a third year be in order that the average for the three years shall be $3000?

26. Three loads of hay weigh 1950, 1800, 2100 lb. respectively. What must a fourth load weigh in order that the average weight of all four loads shall be 2000 lb.?

27. What number must be added to the numerator of $\frac{11}{45}$ to make the value of the resulting fraction $\frac{4}{5}$?

28. What number must be subtracted from the numerator of $\frac{27}{35}$ to make the resulting fraction equal to $\frac{4}{7}$?

29. What number must be added to the denominator of $\frac{3}{8}$ to make the resulting fraction equal to $\frac{1}{5}$?

30. By how much must the denominator of $\frac{9}{20}$ be increased in order to make the resulting fraction equal to $\frac{3}{8}$?

31. A given mixture of alcohol and water contains 18 qt., of which 4 qt. are alcohol. How much water must be added in order that $\frac{1}{6}$ of the resulting mixture shall be alcohol?

Sug.	NUMBERS DEALT WITH	SYMBOLS
(1)	Number of quarts in the mixture	= 18.
(2)	Number of quarts of alcohol	= 4.
(3)	Number of quarts of water to be added	= x.
(4)	Number of quarts in final mixture	= $18 + x$.
(5)	Fraction of alcohol in resulting mixture	= $\frac{1}{6}$.

RELATION AMONG NUMBERS

2d number ÷ 4th = last number, etc.

32. A mixture contains 9 pt. of which 4 pt. are alcohol. How many pints of water must be added in order that 10 % of the resulting mixture shall be alcohol?

33. A given mixture consists of 4 gallons of acid and 9 gallons of water. How many gallons of water must be added in order that the acid shall be $\frac{1}{6}$ of the mixture?

34. 8 bushels of oats have been mixed with 15 bushels of corn. How many bushels of corn must be added in order that $\frac{1}{8}$ of the resulting mixture shall be oats?

35. An alloy of tin and copper weighs 20 lb. and contains 3 lb. of tin. How many pounds of copper must be added in order that 5 % of the resulting alloy may be tin?

36. If 100 lb. of sea water contains $2\frac{1}{2}$ lb. of salt, how much fresh water must be added to it in order that 100 lb. of the mixture shall contain 1 lb. of salt?

37. If 100 lb. of sea water contains $2\frac{1}{2}$ lb. of salt, how much fresh water must be added to the sea water in order that 20 lb. of the mixture shall contain 4 oz. of salt?

38. If 100 lb. of sea water contains $2\frac{1}{2}$ lb. of salt, how much water must be evaporated from the salt water in order that 8 lb. of the water left shall contain 1 lb. of salt?

39. A man travels 5 mi. an hour for 6 hours, when another man starts at the same place, following at the rate of 7 mi. an hour. When will the second overtake the first?

SUG.   NUMBERS DEALT WITH   SYMBOLS

Number of hours first courier goes	$= x + 6.$
Number of hours second courier goes	$= x.$
Number of miles per hour first courier goes	$= 5.$
Number of miles per hour second courier goes	$= 7.$
Entire number of miles first courier goes	$= 5(x + 6).$
Entire number of miles second courier goes	$= 7x.$

EQUALITY AMONG NUMBERS

Last number = next to last number.

40. In a certain battle an officer sent an order by a messenger who traveled 8 mi. per hour on a bicycle. Three quarters of an hour later he sent a reversal of his previous order by a messenger on a motor cycle traveling 30 mi. per hour. In how many hours did the latter messenger overtake the former?.

41. A courier who travels $5\frac{1}{2}$ mi. an hour was followed after 8 hr. by another, who went $7\frac{1}{3}$ mi. an hour. In how many hours will the second overtake the first?

42. A train running 40 mi. an hour left a station 45 min. before a second train running 45 mi. an hour. In how many hours will the second train overtake the first?

43. A certain number when divided by 7 gives 6 for a quotient and 3 for a remainder. Find the number.

NUMBERS DEALT WITH	SYMBOLS
(1) The required number	$= x.$
(2) The divisor	$= 7.$
(3) The quotient	$= 6.$
(4) The remainder	$= 3.$

EQUALITY AMONG NUMBERS

(Difference between the first and last numbers) ÷ by second number = third number.

44. A certain number when divided by 12 gives 11 for a quotient and 5 for a remainder. Find the number.

45. $\frac{4}{5}$ of a man's property is invested at 5 % and $\frac{1}{5}$ at 6 %. His annual income is $ 780. How many dollars is he worth?

46. My annual income is $ 910. If $\frac{1}{4}$ of my property is invested at 5 %, $\frac{2}{3}$ at 4 %, and the rest at 6 %, find the amount of my property.

47. A given mixture is composed of 8 gallons of acid and 15 gallons of water. How many gallons of water must be added in order that 20 per cent of the resulting fluid shall be acid?

48. Three of a pupil's grades are 88, 72, 92. What must be his grade· in a fourth subject in order that his average in the four subjects shall be 85 ?

49. What number when divided by 12 gives 13 for a quotient and 3 for a remainder ?

50. A train running 40 mi. an hour left a station 30 min. before a second train running 45 mi. an hour. In how many hours will the second train overtake the first?

51. A certain room is 20 ft. long and 12 ft. wide. The walls and ceiling of the room together have an area of 752 sq. ft. How high is the ceiling ?

52. Find three consecutive numbers such that if they be divided by 2, 3, and 4 respectively, the sum of the quotients will equal the next higher consecutive number.

53. A woman can write 15 words per minute with a pen, and a girl can write 40 words per minute on the typewriter. The woman has a start of 3 hours in copying a certain manuscript. How long before the girl using the typewriter will overtake the woman ?

EXERCISE 112

1. The formula for the area of a rectangle is $a = lw$. Solve this formula for l. Express your answer in words as a rule.

In order to keep firmly in mind the fact that area is a number, it is well often to state the formula $a = lw$ as follows: (the area-number of a rectangle) equals (the length-number) $\times$ (the width-number). State $l = \dfrac{a}{w}$ similarly.

2. Solve $a = lw$ for w, and similarly state the result as a rule. (Thus, width-number equals area-number, etc.)

3. A certain cantonment was to be in the shape of a rectangle, was to contain 20.4 sq. mi., and have a frontage of 4 miles. How deep must the camp be? Solve by use of the formula $l = \dfrac{a}{w}$.

4. Solve $v = lwh$ for h, and state the result as above in Exs. 1 and 2. In stating the rules in Exs. 4–7, use the form of statement shown in Ex. 1.

5. Solve $v = lwh$ for l, and state the result as a rule.

6. Solve $p = br$ for b, and state the result as a rule.

7. Solve $p = br$ for r, and state the result as a rule.

8. Hence $p = br$ takes the place of what three rules? Count up the number of letters in the three rules taken together. Count up the number of letters and symbols in $p = br$. Compare the two results.

9. In a certain manual training school, boys make articles which they wish to sell at a profit of 20 per cent. Obtain a formula for the selling price, s, when c denotes the cost price.

10. Solve the formula $s = \dfrac{6\,c}{5}$ for c. By use of the formula obtained, find the cost of an article when the selling price is to be $18.

11. In division, the relation of the dividend (D), divisor (d), quotient (q), and remainder (r) is expressed by the formula $D = dq + r$. Solve this formula for r. Express the result as a rule.

12. Solve $D = dq + r$ for q and express the result as a rule. Give a numerical application of this rule.

13. The formula for the distance (d), in terms of the rate (r), and time (t) is $d = rt$. Solve this formula for t and express the result as a rule.

14. When a certain tree at a distance of $\frac{3}{4}$ of a mile was struck by lightning, the sound was heard in $3\frac{2}{3}$ seconds. From these facts deduce the velocity of sound, by use of the formula $r = \dfrac{d}{t}$.

EXERCISE 118

REVIEW

Factor the following:

1. (1) $3 x^3 - 12 x^2y + 12 xy^2$.   (3) $x^2 - 7 x + 10$.

(2) $4 a^2 - 16 b^2$.   (4) $x^5 - a^4x$.

2. (1) $a^2x^2 - 8 ax + 16$.   (3) $9 x^4 - 42 x^3y^2 + 49 x^2y^4$.

(2) $3 x^4 - 3 a^2b^2c^2$.   (4) $256 x^2 - 81 y^2$.

3. If the product is p, the multiplier m, and the multiplicand M, then $p = mM$. From this formula find M in terms of the other letters. State the result as a rule.

4. Also solve $p = mM$ for m and state the result as a rule.

Simplify:

5. $\dfrac{x}{2 y} - \dfrac{x - y}{2(x + y)}$.

7. $\dfrac{6 x + 6}{x - 2} \cdot \dfrac{x^2 - 4}{3(x^2 - 1)}$.

6. $\dfrac{1}{2 p - 2} - \dfrac{1}{3 p + 3} - \dfrac{1}{p^2}$.

8. $\dfrac{x^2y - y^3}{10 xy} \div \dfrac{x^2 - xy}{5 x^2y}$.

9. By aid of the principles of short multiplication, compute in a short way

(1) 148×152.   (3) 996^2.   (5) $11\frac{3}{4} \times 12\frac{1}{4}$.

(2) $7\frac{1}{2} \times 8\frac{1}{2}$.   (4) 1004^2.   (6) 127×133.

10. By use of the formula $p = br$, find the rate of income from a $4\frac{1}{4}$ per cent liberty bond bought at 95.

11. What number diminished by $\frac{1}{4}$ of itself and then increased by 30 equals 90?

Solve and check the following:

12. $\dfrac{1}{5 x} - \dfrac{3}{10 x} = \dfrac{2}{15 x} - \dfrac{7}{12}$.

13. $\frac{1}{4}(x + 1) - \frac{1}{3}(3 - x) = \frac{1}{2}(x - 1)$.

14. $3 - \dfrac{4}{y+1} = 1.$

16. $\dfrac{ax}{b} + \dfrac{b}{a} = \dfrac{a}{b} - \dfrac{bx}{a}$

15. $\dfrac{x+a}{x+b} = \dfrac{3}{4}.$

17. $\dfrac{l+x}{l-x} = \dfrac{x^2}{l^2 - x^2}.$

18. In $p = br$, let $b = 400$ and $p = 32$ and find r. Write this as a problem in percentage.

19. Two men manage a store, and as one of them owns the building, the other receives only $\frac{3}{5}$ as large a share of the profits as the owner of the store. If the profits for one year are $6600, what does each receive?

20. Write $d = rt$ as a rule. Also $r = \dfrac{d}{t}.$ Also $t = \dfrac{d}{r}.$ Count the number of letters in these three rules. Count the number of symbols in $d = rt$. Compare the two results. What does this comparison teach us as to the value of algebra?

21. What three rules are represented by the formula $a = lw$?

22. A dealer in fruit found that when the price of a certain kind of fruit was 20 cents a quart, he sold 5 qt. per day; when it was 16 cents, he sold 20 qt. per day; when it was 12 cents, he sold 45 qt. per day; when it was 8 cents, he sold 150 qt. per day. Make a graph of these facts.

From the graph determine approximately the number of quarts sold per day when the price was 10 cents a quart. Also when it was 15 cents a quart.

23. If the circumference of a circle is denoted by c and the radius by r, then $c = 2\pi r$. Solve this formula for r, and state the result as a rule. Make up and solve a numerical example by use of your rule.

24. A certain number when divided by 8 gives 9 for a quotient and 3 for a remainder. Find the number.

25. A pupil's grades in two studies are 72 and 94. What grade must he get in a third subject in order to make an average of 85 in all three subjects?

26. If $K = \frac{1}{2}h(b + b')$, $K = 280$, $h = 12$, and $b = 10$, find b'

27. If $V = \pi r^2 h$, $V = 1540$, $\pi = \frac{22}{7}$, and $r = 7$, find h.

28. If $T = \pi r(r + l)$, $T = 1144$, $\pi = \frac{22}{7}$, $r = 14$, find l.

PART II

97. Examples in the Solution of Equations.

Ex. 1.   Solve $.2\,x + .05\,x = 1.5$

$$.25\,x = 1.5.$$

$$x = 6. \quad \textit{Ans.}$$

Ex. 2.   Solve $\dfrac{4}{1+x} + \dfrac{x+1}{1-x} - \dfrac{x^2-3}{1-x^2} = 0.$

Multiplying by the L. C. D., $1 - x^2$,

$$4(1-x) + (x+1)^2 - x^2 + 3 = 0.$$
$$4 - 4\,x + x^2 + 2\,x + 1 - x^2 + 3 = 0.$$
$$-2\,x = -8.$$
$$x = 4. \quad \textit{Root.}$$

Let the pupil check the work.

EXERCISE 114

Solve and check :

1. $.5\,x + .3\,x = 1.6.$
2. $.8\,x = 2.4 - .4\,x.$
3. $.7\,x - .04\,x = .99.$
4. $1.4\,y - .06\,y = .67.$
5. $.28\,p + .3\,p = .093.$
6. $3.12\,x + 2.18\,x = 10.6.$
7. $5.46\,y + .04\,y = 7.7.$
8. $3.2\,x - .6\,x = 1.2\,x - 14.$

Solve to the nearest thousandth each of the following :

9. $.6\,x - 1.5 = .2 - .15\,x.$
10. $.7\,p - .32\,p = .5.$
11. $\dfrac{1.5x - 1.6}{12} = \dfrac{3.5x - 2.4}{8}.$
12. $1.82\,x - 4.27 = 6.53 - 2.47\,x.$
13. $1.72\,x - 1.74 = 2.89 + 4.26.$
14. $3(2\,x + 4) = 2.5\,x - .7(x - 1).$
15. $2.3(3.5\,x - 1.5) - 1.5(2.5\,x + 1.5) = 0.$
16. Is 1.7 the root of the equation $2\,x + 7 = 10.2$?

Solve and check the following :

17. $\dfrac{x^2 - x + 1}{x - 1} = 2\,x - \dfrac{x^2 + x + 1}{x + 1}.$

18. $\dfrac{x - 3}{2\,x + 6} - \dfrac{\frac{1}{6}x^2 + 1}{x^2 - 9} = \dfrac{x + 3}{3\,x - 9}.$

19. $\dfrac{3}{x-1}+\dfrac{4}{x+1}-\dfrac{5}{2x-2}=\dfrac{11}{3x+3}+\dfrac{4}{1-x^2}.$

20. $\dfrac{x+1}{2x-3}-\dfrac{x^2+7}{4x^2-9}=\dfrac{2}{2x+3}-\dfrac{x-1}{6-4x}.$

21. $\dfrac{3x^2-5}{3x-6}-\dfrac{7}{6x+12}-2-x=\dfrac{7}{2x^2-8}.$

22. $\dfrac{3-2x}{4}+\dfrac{x}{6}-\dfrac{1-6x}{15-7x}=\dfrac{2-3x}{9}.$

Sug. In solving this equation, it will save labor to multiply both sides first by 36 (that is, by the smallest number that will contain 4, 6, and 9 as factors), to collect terms, and then multiply by $15-7x$.

23. $\dfrac{5x+13}{12}=\dfrac{2x+5}{6}+\dfrac{23}{4x-36}-\dfrac{5-\frac{1}{4}x}{3}.$

24. $\dfrac{3x-1}{30}+\dfrac{4x-7}{15}=\dfrac{x}{4}-\dfrac{2x-3}{12x-11}+\dfrac{7x-15}{60}.$

EXERCISE 115

Solve for x and check :

1. $\dfrac{a}{x}-5=\dfrac{a}{3x}+7.$

2. $\dfrac{a}{x}-3b=\dfrac{c}{3x}+5.$

3. $\dfrac{x}{r}=\dfrac{1}{rs}-\dfrac{x}{s}.$

4. $\dfrac{x}{p}-\dfrac{2x}{q}=\dfrac{q^2-4p^2}{pq}.$

5. $\dfrac{3}{4}\left(\dfrac{x}{a}-1\right)=\dfrac{2}{3}\left(\dfrac{x}{a}+1\right).$

6. $\dfrac{2a}{3}\left(\dfrac{x}{a}-a\right)=\dfrac{3a}{5}\left(\dfrac{x}{a}+a\right).$

7. $\dfrac{ax-b}{ab}+\dfrac{bx-c}{bc}+\dfrac{cx-a}{ac}=0.$

8. $\dfrac{ax}{3a+b}+\dfrac{bx}{3a-b}=\dfrac{3a^2x+b^2}{9a^2-b^2}.$

9. $\dfrac{a^2-x}{c}-\dfrac{b^2-x}{a}-\dfrac{c^2-x}{b}=\dfrac{a^2}{c}-\dfrac{b^2}{a}.$

10. $\dfrac{5a^2-7x}{3ab}+\dfrac{ab^2+10x}{5ac}=\dfrac{10c^2+3x}{6bc}+\dfrac{5(a-c)}{3b}+\dfrac{b^2}{5c}.$

11. $\dfrac{x-p}{p-q}+\dfrac{x+3\,q}{p+q}=3.$

13. $\dfrac{lm-p^2x}{l-m}+\dfrac{l}{3\,p}=\dfrac{px}{3\,m}.$

12. $\dfrac{ab-c^2x}{a-b}-\dfrac{a}{3\,c}=\dfrac{x}{3\,b}.$

14. $\dfrac{x}{a}-\dfrac{x}{a-b}=\dfrac{1}{a+b}-\dfrac{x}{b}.$

Solve the following:

15. $R=\dfrac{gs}{g+s}$ for s.

16. $R=\dfrac{gs}{g+s}$ for g.

17. $\dfrac{1}{f}=\dfrac{1}{p}+\dfrac{1}{p'}$ for f.

18. $\dfrac{1}{f}=\dfrac{1}{p}+\dfrac{1}{p'}$ for p.

19. $C=\tfrac{5}{9}(F-32)$ for F.

20. $T=2\,\pi r(r+h)$ for h.

21. $a=\tfrac{1}{2}h(b_1+b_2)$ for h.

22. $a=\tfrac{1}{2}h(b_1+b_2)$ for b_1.

EXERCISE 116
VERBAL PROBLEMS

1. If e is an even number, give the next two smaller odd numbers.

2. A train runs m miles an hour. How far will it go in c/r hours?

3. If i inches are cut from one end of a string which is p inches long, and c inches are then cut from the other end, how much of the string will be left?

4. If s is odd, give the next smaller even number.

5. If apples are sold d for a dime, how many can be bought for c cents?

6. If a man walks y yards in m minutes, how long will it take him to walk x/y miles?

7. If r per cent of p is x, find r per cent of y.

8. What must be added to x to make the result equal to $\dfrac{p}{q}$?

9. A man has hay enough to last c cows y days. How long will the hay last n cows?

10. If 2 oranges cost c cents, what will 1 dozen oranges cost?

11. If a pupil's average grade in three subjects is 87 and his grade in a fourth subject is x, represent his average grade in all four subjects in the form of a fraction.

12. If x be added to both numerator and denominator of $\frac{19}{55}$, what will the resulting fraction be?

13. A baseball team has won 25 games out of 36. If it should play x more games and win them all, how many games will it have played? How many will it have won? What would be its average of games won (expressed as a fraction)?

14. If a baseball team has won 25 games out of 36, and should play 12 more games and win x of them, what would be its average of games won?

15. If a baseball team has won 19 games out of 36 and should play x more and win $\frac{3}{4}$ of them, what would be its average of games won?

16. An alloy of silver and copper weighs 120 lb. and contains 8 lb. of silver. If x pounds of silver be added to it, what will the resulting mass weigh? What fraction of it will be silver?

17. A given mixture of alcohol and water contains 13 qt., of which 5 quarts are alcohol. If x quarts of alcohol are added, what fractional part of the resulting fluid would the alcohol be?

18. If a boy can do a given piece of work in 15 days, what fractional part of it can he do in 1 day? In x days? If a man can do the same piece of work in 9 days what part of it can he do in 1 day? In x days? How much will the man and boy do together in x days?

19. If sound travels at the rate of 1100 ft. per second, how long will it take it to go 5000 ft.? 1000 yd.? If a bullet travels x ft. per second, how long will it take it to go 3000 ft.?

20. If a bullet travels x feet per second and sound 1100 ft. per second, how many seconds in all will it take a bullet to travel to a target 3000 ft. distant and for the sound of the impact on the target to return?

21. Express the following as an equation: x exceeds the sum of a third and fourth parts of itself by 12.

EXERCISE 117

1. Find the number the sum of whose third, fourth, and fifth parts is 94.

2. A certain number exceeds the sum of its third, fourth, and tenth parts by 38. Find the number.

3. What number increased by .15 of itself equals 690 ?

4. A number increased by .06 of itself and then by $100 equals $312. Find the number.

5. What number increased by $66\frac{2}{3}$ per cent of itself equals 205 ?

6. A boy's average grade in three subjects is 88; what must his grade in a fourth subject be in order that his average in all four subjects shall be 90 ?

SUG.	NUMBERS DEALT WITH	SYMBOLS
(1)	Average grade in three subjects	= 88.
(2)	Grade in fourth subject	= x.
(3)	Average grade in all four subjects	= 90.

RELATION OF EQUALITY

(Sum of second number and 3 times first number) ÷ 4 = last number, etc.

7. The average weight of 4 loads of coal is 1950 lb. What must be the weight of a fifth load in order that the average weight of all 5 loads shall be 2000 lb.?

8. The average wheat crop of the United States for four years was 660 millions of bushels. What would the crop for the fifth year need to be in order to make the average for the five years 700 million bushels ?

9. A physician has 60 pints of a 5 per cent solution of acid. How many pints of water must be added to it to make it a 2 per cent solution ?

10. A given alloy of tin and copper weighs 120 lb. and is 30 per cent tin. How much copper must be added to it in order that 10 per cent of the resulting alloy shall be tin ?

11. A given mixture of alcohol and water contains 20 gallons and is 40 per cent pure (*i.e.* 40 per cent of the mixture is alcohol and the rest water). How many gallons of water must be added to the mixture to make it 10 per cent pure?

12. What number must be added to both numerator and denominator of $\frac{12}{55}$ in order that the value of the resulting fraction shall be $\frac{1}{2}$?

13. What number must be subtracted from both numerator and denominator of $\frac{34}{47}$ in order that the value of the resulting fraction shall be $\frac{2}{3}$?

14. A baseball nine has played 36 games of which it has won 25. How many games must it win in succession to bring its average of games won up to .75?

15. A baseball nine has won 25 games out of 36 played. It still has 12 games to play. How many of these will it need to win in order to bring its average of games won up to .75?

16. A baseball nine has won 19 games out of 36 games played. If after this it should win $\frac{3}{4}$ of the games played, how many games would it need to play to bring its average of games won up to .66$\frac{2}{3}$?

17. A pupil has worked 15 problems. If he should work 9 more problems and get 8 of them right, his average would be .75. How many problems has he worked correctly thus far?

18. A mass of copper and silver alloy weighs 120 lb. and contains 8 lb. of copper. How much copper must be added to the mass in order that 100 lb. of the resulting alloy shall contain 10 lb. of the copper?

19. A mass of copper and silver alloy weighs 120 lb. and contains 8 lb. of silver. How much silver must be added to the mass in order that 1 lb. of the resulting alloy shall contain $2\frac{1}{5}$ oz. of silver?

20. How much water must be added to 50 gallons of milk containing 5% of butter fat to make a mixture containing 3% of butter fat?

Sug. The 50 gal. of milk contain $50 \times .05$ or 2.5 gal. butter fat.

21. A certain kind of cream is $\frac{2}{5}$ butter fat, and a certain kind of milk is 3% butter fat. How many gallons of the cream must be added to 40 gallons of milk to make a mixture which is 5% butter fat?

22. If a boy can do a piece of work in 15 days which a man can do in 9 days, how long would it take both working together to do the piece of work?

Sug.   NUMBERS DEALT WITH   SYMBOLS

(1) Number of days takes boy alone $= 15.$
(2) Number of days takes man alone $= 9.$
(3) Number of days takes both together $= x.$
(4) Part boy does in 1 day $= \frac{1}{15}.$
(5) Part man does in 1 day $= \frac{1}{9}.$
(6) Part both together do in 1 day $= \frac{1}{x}.$

RELATION OF EQUALITY AMONG NUMBERS

Sum of (4) and (5) $=$ (6).

23. A can spade a garden in 3 days, B in 4 days, and C in 6 days. How many days will they require working together?

24. A and B together can mow a field in 4 days, but A alone could do it in 12 days. In how many days can B mow it?

25. One pipe can fill a given tank in 48 min. and another can fill it in 1 hr. and 12 min. How long will it take he pipes together to fill the tank?

26. Two inflowing pipes can fill a cistern in 27 and 54 min. respectively, and an outflowing pipe can empty it in 36 min. All pipes are open and the cistern is empty. In how many minutes will it be full?

Sug. Since emptying is the opposite of filling, we may consider that a pipe which empties $\frac{1}{36}$ of a cistern in a minute will *fill* $-\frac{1}{36}$ of it each minute.

27. A tank has four pipes attached, two filling and two emptying. The first two can fill it in 40 and 64 min. respectively, and the other two can empty it in 48 and 72 min. respectively. If the tank is empty and the pipes all open, in how many minutes will it be full?

28. A rifle ball is fired at a target 1100 yd. distant and $4\frac{1}{2}$ sec. after firing the shot the marksman heard the impact of the bullet on the target. If the bullet traveled at the rate of 2200 ft. per second, what was the rate at which the sound of the impact traveled back to the marksman?

29. A rifle ball is fired at a target 1000 yd. distant, and 4 sec. after firing the shot, the marksman heard the im pact of the bullet on the target. If sound traveled at the rate of 1100 ft. per second, at what rate did the bullet travel?

30. What number increased by $87\frac{1}{2}$ per cent of itself equals 195?

31. If a certain number be added to the numerator of $\frac{19}{36}$ and also subtracted from the denominator, the value of the fraction will become $\frac{2}{3}$. Find the number.

32. A given mixture of alcohol and water contains 24 gallons and is 15 per cent pure. How much alcohol must be added to it to make it 20 per cent pure?

33. A pupil's average grade in five subjects is 84. What must be his grade in a sixth subject in order that his average in all six subjects shall be 86?

34. Divide the number 54 into 4 parts, such that the first increased by 2, the second diminished by 2, the third multiplied by 2, and the fourth divided by 2, will all produce equal results.

Sug. First part $= x - 2$.
Second part $= x + 2$, etc.

35. In the United States the gold dollar is 90 % gold and 10 % copper. If a mass of gold and copper weighing 24 lb. is 75 % gold, how many pounds of gold must be added to it to make it ready for coinage into gold dollars?

36. My annual income is $ 910. If $\frac{1}{4}$ of my property is invested at 5 %, $\frac{2}{3}$ at 4 %, and the rest at 6 %, find the amount of my property.

37. If one pipe can fill a swimming tank in 1 hr. and another can fill it in 36 min., how long will it take the two pipes together to fill the tank?

38. A baseball player who has been at the bat 150 times has a batting average of .240. How many more times must he bat in order to bring his average up to .250, provided that in the future his base hits equal half the number of times he bats?

39. A girl has worked a certain number of problems and has $\frac{2}{3}$ of them right. If she should work 9 more problems and get 8 of them right, her average would be .75. How many problems has she worked?

40. How much water must be added to 5 gallons of alcohol which is 90 % pure, in order to make a mixture which is 80 % pure?

REVIEW

Factor:

1. (1) $\frac{1}{4}ax + \frac{1}{4}ay$. (4) $4x^2 - a^2 - 2ab - b^2$.

(2) $x^5 - 256x$. (5) $a^6 - 8a^4 + 16a^2$.

(3) $x^3 + 6x^2 - 135x$. (6) $x^2 - .7x + .12$.

2. (1) $a^2 + b^2 - x^2 + 2ab$. (3) $5(a+b)x - 7(a+b)y$.

(2) $\pi a^2 - \pi b^2 + \pi c^2$. (4) $(a+x)^2 - 4(a+x) + 4$.

3. Compute in a short way by the aid of factoring:

(1) $89.8^2 - 10.2^2$. (2) $3.1416 \times 169 - 3.1416 \times 144$

4. Simplify $\dfrac{5}{4(x+y)} + \dfrac{3}{x-y} - \dfrac{13x+7y}{4(x^2-y^2)}$.

5. Simplify $\dfrac{20a+5}{6a+6} \cdot \dfrac{16a-4}{5a+5} \div \dfrac{16a^2-1}{a^2+2a+1}$.

6. Simplify $\left(a - \dfrac{b^2}{a}\right)\left(\dfrac{a}{b} + \dfrac{b}{a}\right) \div \dfrac{a^4 - b^4}{ab}$.

Solve for x and check:

7. $\dfrac{2}{x-2} - \dfrac{5}{x+2} = \dfrac{2}{x^2-4}$.

8. $\dfrac{x-b}{x-2a} - \dfrac{x+b}{x+2a} - \dfrac{4a^2-b^2}{x^2-4a^2} = 0$.

9. $6 + \dfrac{1}{4}\left(x - 9 - \dfrac{x-3}{5}\right) + 3 + \dfrac{x-5}{5} = \dfrac{x}{2}$.

Find the value of x in the shortest way, when

10. $\frac{44}{7}x = \frac{44}{7} \times 19 + \frac{44}{7} \times 41$.

11. $3.1416 x = 3.1416(723) - 3.1416(476)$

12. If $\dfrac{3}{x} - \dfrac{2}{y} = 4$, find the value of x when $y = \frac{1}{4}$. Also when $y = -\frac{3}{4}$. When $y = \frac{4}{5}$.

Solve and check:

13. $\dfrac{3px - 3qx}{x^2 - q^2} - \dfrac{p - 2q}{x + q} + \dfrac{p - q}{q - x} = 0$.

14. $.6x + .05 = .2 - .15x$.

15. $\dfrac{1.5\,x - 1.35}{12} - \dfrac{3.5\,x - 2.4}{8} = 0.$

16. If a bushel of wheat occupies $1\frac{1}{4}$ cu. ft., find a formula for a number of bushels of wheat (W) in a rectangular bin, l feet long, w feet wide, and h feet high. Then solve this formula for h.

17. A rectangular bin is to contain 1000 bushels of wheat. If the bin is to be 24 ft. long and 5 ft. wide, find its depth by use of the formula obtained last in Ex. 16.

18. A number exceeds the sum of its third and fourth parts by 45. Find the number.

19. Copy the following and fill in the vacant place :

SUBJECT	RULE	FORMULA
Percentage		$r = \dfrac{p}{b}$
Volume of box		$h = \dfrac{v}{lw}$
Interest	Amount = principal + interest	
Division		$D - r = dq$
Motion	Rate = distance ÷ time	
Interest		$p = \dfrac{a}{1 + rt}$

20. Solve $a = p(1 + rt)$ for r and state the result as a rule.

21. Solve $a = p(1 + rt)$ for t and state the result as a rule.

22. Any number multiplied by the reciprocal of itself gives what? Illustrate by the multiplication of some monomial by its reciprocal.

23. Write an algebraic expression which must represent an even number.

24. Write an expression which must represent an odd number.

25. Divide $a^3 - b^3$ by $a - b$. What then are two factors of $a^3 - b^3$? From these results form a rule for factoring the difference of two cubes. By use of your rule factor $x^3 - 8\, y^3$.

26. In like manner by dividing $a^3 + b^3$ by $a + b$ obtain a rule for factoring the sum of two cubes.

27. In $5.27\, x - 2.16 = 3.72 + 8\, x$ find the value of x to the nearest thousandth.

28. The average profit in a certain business for three years was $2400. What must the profit for a fourth year be in order that the average for the four years shall be $2500?

29. Solve $\dfrac{1}{3}\left(\dfrac{x}{b} - 1\right) = \dfrac{1}{2}\left(\dfrac{x}{b} + 1\right)$ for x.

30. Explain the difference between the Centigrade and Fahrenheit scales in measuring temperatures. If C is the Centigrade temperature equivalent to a Fahrenheit temperature F, then $C = \frac{5}{9}(F - 32)$. By use of this formula find the value of C when $F = 212°$.

31. If iron melts at 2700° F., what is the Centigrade temperature at which it melts?

32. Solve $C = \frac{5}{9}(F - 32)$ for F. To what use can this result be put?

33. If tin melts at 228° C., find the Fahrenheit temperature at which tin melts.

34. If copper melts at 1091° C., find the Fahrenheit temperature at which copper melts.

35. The adjoining table gives the number of miles of railroad in the United States in the years as specified. Make a graph of these facts.

Year	Miles of Railroad	Year	Miles of Railroad
1830	23	1880	93,267
1840	2,818	1890	167,191
1850	9,021	1900	198,964
1860	30,626	1910	249,992
1870	52,922	1920	253,152

From the graph determine approximately the number of miles of railroad in the United States in the year 1875. In 1895.

CHAPTER XI

SIMULTANEOUS EQUATIONS

PART I

98. Simultaneous Equations are a set or system of equations in which more than one unknown quantity is used, and the same symbol stands for the same unknown number.

Thus, in the group of three simultaneous equations,

$$x + y + 2z = 13,$$
$$x - 2y + z = 0,$$
$$2x + y - z = 3,$$

x stands for the same unknown number in all of the three equations, y for another unknown number, and z for still another.

99. Independent Equations are those which cannot be derived one from the other.

Thus, $\left. \begin{array}{l} x + y = 10 \\ 2x = 20 - 2y \end{array} \right\}$ are not independent equations, since by transposing $2y$ in the second equation and dividing it by 2, we may convert the second equation into the first.

But $\left. \begin{array}{l} 3x - 2y = 5 \\ 5x + y = 6 \end{array} \right\}$ are independent equations, since neither one of them can be converted into the other.

100. Elimination is the process of combining two equations containing two unknown quantities so as to form a single equation with only one unknown quantity. Or, in general, elimination is the process of combining several simultaneous equations so as to form equations one less in number and containing one less unknown quantity.

There are two principal methods of elimination: I, *addition and subtraction;* II, *substitution.*

These methods are presented to best advantage in connection with illustrative examples.

101. I. Elimination by Addition and Subtraction.

Ex.   Solve
$$\begin{cases} 12\,x + 5\,y = 75. & (1) \\ 9\,x - 4\,y = 33. & (2) \end{cases}$$

In order to make the coefficients of y in the two equations alike, we multiply equation (1) by 4, and (2) by 5,

$$48\,x + 20\,y = 300. \quad (3)$$
$$45\,x - 20\,y = 165. \quad (4)$$

Add equations (3) and (4), $93\,x = 465.$

Divide by 93, $\qquad\qquad x = 5.$ *Root.*

Substitute for x its value 5, in equation (1),

$$60 + 5\,y = 75,$$
$$\therefore\ y = 3. \quad Root.$$

Cʜᴇᴄᴋ. $\quad 12\,x + 5\,y = 12 \times 5 + 5 \times 3 = 75.$
$$9\,x - 4\,y = 9 \times 5 - 4 \times 3 = 33.$$

Since y was eliminated by adding equations (3) and (4), the above process is called elimination by *addition.*

The same example might have been solved by the method of subtraction.

Thus, multiply equation (1) by 3, and (2) by 4,

$$36\,x + 15\,y = 225. \quad (5)$$
$$36\,x - 16\,y = 132. \quad (6)$$

Subtract (6) from (5), $\quad 31\,y = 93,$
$$y = 3,$$
$$\text{and}\quad x = 5.$$

It is important to select, in every case, the smallest multipliers that will cause one of the unknown quantities to have the same coefficient in both equations.

Thus, in the last solution given above, instead of multiplying equation (1) by 9, and (2) by 12, we divide these multipliers by their common factor, 3, and get the smaller multipliers, 3 and 4.

Hence, in general,

Multiply the given equations by the smallest numbers that will cause one of the unknown quantities to have the same coefficient in both equations:

If the equal coefficients have the same sign, subtract the corresponding members of the two equations; if the equal coefficients have unlike signs, add.

EXERCISE 119

Solve and check :

1. $4x + 3y = 11.$
$5x - 2y = 8.$

2. $7x + 5y = 38.$
$4x - 3y = 10.$

3. $5x + 7y = 23.$
$4x - y = 25.$

4. $5x - 2y = 11.$
$3x + 5y = -12.$

5. $8x + 3y = 43.$
$3x + 5y = 20.$

6. $7x + 2y = -8.$
$5x + 3y = -1.$

7. $9x - 8y = 5.$
$15x + 12y = 2.$

8. $13x + 6y = 29.$
$7x - 6y = 11.$

9. $11x + 9y = -19.$
$5x + 6y = -8.$

10. $5p + 4q = 17.$
$3p - 2q = 8.$

11. $4l + 5m = 10.$
$7l + 3m = 6.$

12. $3A - 2B = 1.$
$A + B = 2.$

13. $5r - 9s + 17 = 0.$
$14r + 15s - 73 = 0.$

14. $5x - 29 = 2y.$
$3y + 2x - 23 = 0.$

15. $3x - 2y - 1 = 0.$
$x + y = 2.$

16. $x + 5y = -3.$
$7x + 8y = 6.$

17. If $x = 3$ and $y = 4$, does $12x - 5y = 14$?

18. Make up and solve a pair of simultaneous equations in which $x = 7$ and $y = 2$.

19. Make up and solve a pair of simultaneous equations in which the roots are 5 and -3.

102. II. Elimination by Substitution.

Ex.   Solve $5x + 2y = 36.$ **(1)**

$\qquad\qquad\qquad\qquad 2x + 3y = 43.$ **(2)**

From (1) we obtain $5x = 36 - 2y,$

$$\therefore x = \frac{36 - 2y}{5}. \quad (3)$$

In equation (2) substitute for x its value given in (3),

$$2\left(\frac{36 - 2y}{5}\right) + 3y = 43.$$

$$\frac{72 - 4y}{5} + 3y = 43.$$

$$72 - 4y + 15y = 215.$$

$$11y = 143.$$

$$y = 13. \quad \textit{Root.}$$

Substitute for y in (3), $x = \dfrac{36 - 26}{5} = 2.$ *Root.*

Let the pupil check the work.

Hence, in general,

In one of the given equations obtain the value of one of the unknown quantities in terms of the other unknown quantity; Substitute this value in the other equation and solve.

EXERCISE 120

Solve by the method of substitution and check :

1. $3x + 2y = 12.$
$\quad y = 2x - 1.$

2. $5x + 3y = 29.$
$\quad y = 7 - x.$

3. $3x - 5y = 1.$
$\quad y = 2x - 3.$

4. $8x - 3y = 30.$
$\quad y = x - 5.$

5. $x = 2y - 11.$
$\quad 3x + 7y = 19.$

6. $9x + 5y = -28.$
$\quad x = 3y + 4.$

7. $5x - 8y = -1.$
$3y = 5x - 9.$

8. $11x - 5y = 21.$
$5x = 9 + 2y.$

9. $7x + 4y = 40.$
$2y = x + 2.$

10. $y - 3x = -2.$
$2x + 5y = 7.$

11. $3x - 4y = 1.$
$4x - 5y = 1.$

12. $2x + 3y = 1.$
$3x + 4y = 2.$

Find out which of the following sets of equations are worked more readily by the method of addition and subtraction, and which by the method of substitution, and work each example accordingly :

13. $x = 3y - 5.$
$2x + 5y = 12.$

14. $4x + 3y = 1.$
$2x - 6y = 3.$

15. $2x - 6y = 3.$
$x = 3 - 2y.$

16. $3x - 5y = 19.$
$4x + 7y = -2.$

17. $y = 2x + 3.$
$3y + x = 2.$

18. $x = 2y + 6.$
$5x + 3y = 4.$

19. $2x + 7y + 1 = 0.$
$5x + 8y - 7 = 0.$

20. $x - 3 = 0.$
$2y + 3x = 5.$

21. $2x + 3y = 1.$
$2x + 4y = 2.$

22. $y - 3x = 9.$
$2x + 7y = -6.$

23. $7x + 8y = 19.$
$5x + 6y = 13\frac{1}{2}.$

24. $x = 2y - 3.$
$y = 5x - 21.$

25. $y = 3.$
$2x = 3y - 17.$

26. $y = 3x.$
$4x + 5y = 38.$

27. Make up and solve an example in simultaneous equations which is solved more readily by the method of addition and subtraction than by the method of substitution.

28. Make up and solve an example of which the reverse of Ex. 27 is true.

EXERCISE 121

Solve and check:

1. $\dfrac{x}{2} + \dfrac{y}{3} = 10.$

$\dfrac{x}{3} + \dfrac{y}{2} = 10.$

2. $\dfrac{x}{2} - \dfrac{3\,y}{4} = -12.$

$\dfrac{2\,x}{3} + \dfrac{y}{5} = 8.$

3. $\dfrac{3\,x}{4} + \dfrac{5\,y}{2} = 8.$

$\dfrac{y}{2} = \dfrac{3\,x}{2} - 5.$

4. $\dfrac{2\,A}{5} - \dfrac{3\,B}{4} = 3\tfrac{2}{3}.$

$\dfrac{3\,A}{10} - \dfrac{B}{2} = 2\tfrac{1}{2}.$

5. $\dfrac{7\,x}{2} - \dfrac{5\,y}{4} = -12.$

$\dfrac{y}{4} = \dfrac{3\,x}{2} + 4.$

6. $\tfrac{1}{2}\,x + \tfrac{1}{3}\,y = 5.$
$\tfrac{1}{3}\,x + \tfrac{1}{2}\,y = 5.$

7. $\tfrac{1}{4}\,r - \tfrac{1}{6}\,s = 1.$
$\tfrac{1}{3}\,r - \tfrac{1}{4}\,s = 1.$

8. $.2\,x - .7\,y = .9.$
$.5\,x + .3\,y = .2.$

9. $.4\,l + .3\,m = .1.$
$.2\,l - .6\,m = .3.$

10. $\dfrac{x+1}{3} - \dfrac{y}{4} = -1.$

$\dfrac{y}{3} - \dfrac{3\,x+1}{4} = 0.$

11. $\dfrac{2\,x-y}{5} + \dfrac{3\,x+2\,y}{11} = 2.$

$-\dfrac{2\,x}{3} + \dfrac{4\,x+y-1}{4} = 1.$

12. $.3\,x + .4\,y = 1.1.$
$.5\,x + .2\,y = .9.$

13. $5\,r = 2\,s.$
$s - 8 - 3(r-3) = 0.$

14. $\dfrac{4\,x+5\,y}{40} = x - y.$

$\dfrac{2\,x-y}{3} + 2\,y = \tfrac{1}{2}.$

15. $\dfrac{3\,x}{2} - \dfrac{y+3}{4} = 0.$

$\dfrac{7\,x-4}{5} - \dfrac{y+3}{6} = 0.$

16. $2(x+4) - 3(y-5) = 0.$
$3(x+1) - y = 0.$

17. $\dfrac{x+1}{2} - \dfrac{5\,x+y}{4} = -4.$

$y = x - 2.$

18. $\dfrac{3\,x+2}{10} - \dfrac{y+2}{4} = 0.$

$\dfrac{4}{7}(x+1) = \dfrac{2\,y}{3}.$

19. $\dfrac{3x-y}{5} - \dfrac{4x-3y}{6} + 1 = 0.$

$y = x - 8.$

20. $\dfrac{2y-x}{3} - \dfrac{6x-y}{2} + 3 = 0.$

$y = 2x.$

21. $\dfrac{3x+2y}{5} = 3.$

$\dfrac{7x-3y}{4} = 3.$

22. $x + y = \frac{1}{6}(y - 3).$

$\dfrac{2x+5}{y+x} = -9.$

103. Literal Equations.

Ex. 1.   Solve $\qquad$ $ax + by = c.$ $\qquad$ (1)

$\qquad\qquad\qquad\qquad px + qy = r.$ $\qquad$ (2)

Multiply (1) by q, $\qquad$ $aqx + bqy = cq.$ $\qquad$ (3)

Multiply (2) by b, $\qquad$ $bpx + bqy = br.$ $\qquad$ (4)

Subtract (4) from (3), $\quad$ $(aq - bp)x = cq - br.$ $\quad$ (5)

$$x = \frac{cq - br}{aq - bp}. \quad Ans.$$

Multiply (1) by p, $\qquad$ $apx + bpy = cp.$ $\qquad$ (6)

Multiply (2) by a, $\qquad$ $apx + aqy = ar.$ $\qquad$ (7)

Subtract (7) from (6), $\quad$ $(bp - aq)y = cp - ar.$

$$y = \frac{cp - ar}{bp - aq} \quad Ans.$$

Let the pupil check the work.

In solving simultaneous literal equations, observe that if the value obtained for the first unknown is a fraction containing a binomial term (or the value is complex in other ways), it is better not to find the value of the other unknown as in numerical equations, *i.e.* by substituting the value found in one of the original equations and reducing. A better method is to take both of the original equations and eliminate anew. See the solution of the preceding example.

Ex. 2.   Solve $ax + by = c.$ (1)

$$y = x + 1.$$ (2)

Substitute from (2) in (1),

$$ax + b(x + 1) = c$$ (3)

$$ax + bx + b = c$$

$$(a + b)x = c - b$$

$$x = \frac{c - b}{a + b}. \quad Ans.$$

$$y = x + 1 = \frac{c - b}{a + b} + 1 = \frac{a + c}{a + b}. \quad Ans.$$

EXERCISE 123

Solve and check:

1. $x + 3y = 7a.$
   $2x + y = 7a.$

2. $3x + 2y = 6a + 4b.$
   $5x - y = 10a - 2b.$

3. $x + ay = 2a.$
   $bx + y = ab + 1.$

4. $ax + by = 2ab.$
   $y = x - a.$

5. $2ax + 3by = 4ab.$
   $5ax + 4by = 3ab.$

6. $ax + by = c.$
   $mx + py = d.$

7. $bx + ay = a + b.$
   $ab(x - y) = a^2 - b^2.$

8. $ax + ay = b^2.$
   $y = x - a.$

9. $ax + by = 2.$
   $3ax - 2by = 1.$

10. $px + qy = p^2 - 4q^2.$
    $y = 2x.$

11. $cx + y = c^2 - d^2.$
    $y = dx.$

12. $px + qy = 2pq.$
    $x - y = p - q.$

13. $ax + by = c.$
    $px - qy = r.$

14. $ax + by = c.$
    $y = x + 2.$

15. $\frac{x}{a} + \frac{y}{b} = c.$
    $y = x + 1.$

16. $\frac{2x}{c} + \frac{3y}{d} = 13.$
    $\frac{4x}{c} - \frac{6y}{d} = -10.$

17. $\frac{a}{b + y} = \frac{b}{a - x}.$
    $a(a - y) = b(b + x).$

18. Make up and solve a pair of simultaneous equations in which $x = 3\,a$ and $y = -2\,a$ are the roots.

104. Three or More Simultaneous Equations.

Ex.   Solve
$$\begin{cases} 3\,x + 4\,y - 5\,z = 32. & (1) \\ 4\,x - 5\,y + 3\,z = 18. & (2) \\ 5\,x - 3\,y - 4\,z = 2. & (3) \end{cases}$$

If we choose to eliminate z first, multiply (1) by 3, and (2) by 5,
$$9\,x + 12\,y - 15\,z = 96. \qquad (4)$$
$$20\,x - 25\,y + 15\,z = 90. \qquad (5)$$
Add (4) and (5), $\overline{\qquad 29\,x - 13\,y = 186.} \quad (6)$

Also multiply (2) by 4, (3) by 3,
$$16\,x - 20\,y + 12\,z = 72. \qquad (7)$$
$$15\,x - 9\,y - 12\,z = 6. \qquad (8)$$
Add (7) and (8), $\overline{\qquad 31\,x - 29\,y = 78.} \quad (9)$

We now have a pair of simultaneous equations,
$$\begin{cases} 29\,x - 13\,y = 186. \\ 31\,x - 29\,y = 78. \end{cases}$$

Solving these, obtain $\left. \begin{array}{l} x = 10. \\ y = 8. \end{array} \right\}$ *Roots.*

Substitute for x and y in equation (1),
$$30 + 32 - 5\,z = 32$$
$$z = 6. \quad \textit{Root.}$$

CHECK. $3\,x + 4\,y - 5\,z = (3 \times 10) + (4 \times 8) - (5 \times 6) = 32.$
$4\,x - 5\,y + 3\,z = (4 \times 10) - (5 \times 8) + (3 \times 6) = 18.$
$5\,x - 3\,y - 4\,z = (5 \times 10) - (3 \times 8) - (4 \times 6) = 2.$

EXERCISE 123

Solve and check :

1. $x + y + z = 6.$
$3\,x + 2\,y + z = 10.$
$3\,x + y + 3\,z = 14.$

2. $x + 2\,y + z = 7.$
$2\,x + y + 3\,z = 14.$
$x + 3\,y - z = 2.$

3. $3\,x - y - 2\,z = 11.$
$4\,x - 2\,y + z = -2.$
$6\,x - y + 3\,z = -3.$

4. $5\,x - 6\,y + 2\,z = 5.$
$8\,x + 4\,y - 5\,z = 5.$
$9\,x + 5\,y - 6\,z = 5.$

5. $A + 3B = 4 - 2C.$
$3A + 4C = 10 - 6B.$
$9B + 8C = 15 - 4A.$

6. $y = 3z - 2x.$
$x = 3 + 2y - 4z.$
$z = 1 + 2x - 2y.$

7. $2x + 3y = 7.$
$3y + 4z = 9.$
$5x + 6z = 15.$

8. $2x + 4y + 3z = 6.$
$6y - 3x + 2z = 7.$
$3x - 8y - 7z = 6.$

9. $x + 3y + 3z = 1.$
$3x - 5z = 1.$
$9y + 10z + 3x = 1.$

10. $x + y = 4.$
$x + z = 6.$
$y + z = 8.$

11. $l - 3m = 1.$
$2m = p - 5.$
$l = 3p + 1.$

12. $2p + q - r = 11.$
$5p - 2q = 4.$
$4p - 3r = 20.$

13. $x + y + z = -1.$
$x = 10z - 7.$
$y = 2x - 11.$

14. $3x - \frac{1}{4}y + z = 7\frac{1}{2}.$
$2x - \frac{1}{3}(y - 3z) = 5\frac{1}{3}.$
$2x - \frac{1}{2}y + 4z = 11.$

15. Make up and solve a set of three simultaneous equations in which $x = 2$, $y = 3$, and $z = 4$.

16. Make up and solve a set of equations in which $x = 3$, $y = -2$, and $z = -1$.

EXERCISE 124

VERBAL PROBLEMS

1. A man walks the length of a block and halfway back. He finds that he has then gone f feet. How long is the block?

2. How many pounds are in x tons and p pounds?

3. What is the amount of d dollars at p per cent for y years at simple interest?

4. If a stream flows a miles an hour and a motor boat runs b miles an hour, how fast can the boat run downstream? Upstream?

5. If a boy goes to a store and buys x articles at c cents each, how many cents change will he receive for a dollar? Also express the change received as a fraction of a dollar.

6. The fractional part of a piece of work which is completed is a/b. What part of the work is yet to be done?

7. Two men, starting at the same place, walk in opposite directions, one at the rate of a miles an hour, the other at b miles an hour. How far apart will they be in h hours. Answer the same if the men should walk in the same direction.

8. The sum of two numbers is 20. The greater of the numbers is x. Find the quotient of the greater number divided by the less. What per cent is the greater number of the less? The less of the greater?

9. Express in cents the value of a mixture composed of x bushels worth 40 cents a bushel, and y bushels worth 65 cents a bushel.

10. Draw a rectangle and mark its length by x and its width by y. What expression represents its perimeter? Its area? Also draw another rectangle of greater length and less width. If the length of the second rectangle exceeds that of the first by 3, mark the length of the second rectangle with the proper expression. If the width of the second rectangle is less than the width of the first by 1, mark the width of the second rectangle with the proper expression. What is the area of the second rectangle?

11. Express in algebraic language the annual income from x dollars at 5 %. Also the income from y dollars at 4 %.

12. What is the entire annual income from x dollars at 5 %, y dollars at 6 %, and z dollars at 7 %?

13. $\frac{67}{9} = 7 + \frac{4}{9}$. Express in like manner the statement that x divided by y gives 3 for a quotient and 5 for a remainder.

If x and y represent two numbers of which x is the greater, convert the following statements into equations:

14. The difference between two numbers is 5.

15. One number exceeds another number by 5.

16. Three times the smaller of two numbers exceeds twice the larger by 8.

17. The difference between two numbers equals one fifth of their sum.

105. In the Solution of Problems Involving Two or More Unknown Quantities it is necessary to *obtain as many independent equations as there are unknown quantities involved in the equations* and to eliminate.   (See § 99, p. 217.)

Ex. Find a fraction such that if 2 is added to both numerator and denominator, the fraction becomes $\frac{1}{2}$; but if 7 is added to both numerator and denominator, the fraction becomes $\frac{2}{3}$.

NUMBERS DEALT WITH	SYMBOLS
(1) Unknown fraction	$=\frac{x}{y}.$
(2) First number added to numerator and denominator	$=2.$
(3) Resulting fraction	$=\frac{x+2}{y+2}.$
(4) Second number added to numerator and denominator	$=7.$
(5) Resulting fraction	$=\frac{x+7}{y+7}$
(6) Value of first resulting fraction	$=\frac{1}{2}.$
(7) Value of second resulting fraction	$=\frac{2}{3}.$

EQUALITIES AMONG NUMBERS

$$(3) = (6), \text{ and } (5) = (7),$$

or $\quad \dfrac{x+2}{y+2}=\dfrac{1}{2}, \ \dfrac{x+7}{y+7}=\dfrac{2}{3},$

whence we obtain $\quad x = 3, \text{ and } y = 8,$

or $\quad \dfrac{x}{y}=\dfrac{3}{8}.$ *Ans.*

EXERCISE 125

1. Find two numbers whose sum is 23 and whose difference is 5.

2. Separate $24.80 into two parts such that the smaller part equals $\frac{1}{3}$ of the larger part.

3. Separate $28\frac{1}{2}$ into two parts such that one shall exceed the other by $2\frac{3}{4}$.

4. Separate 180 into two parts such that one part exceeds three times the other by 40.

5. The difference between two numbers is 6, and if 3 is added to the larger, the sum will be double the less. Find the numbers.

6. Separate 95 into two parts such that one part exceeds $\frac{1}{2}$ the other part by 20.

7. The half of one number plus the third of another number equals 13, while the sum of the numbers is 33. Find the numbers.

8. Separate $1000 into two parts such that one part equals four times the sum of $150 and the other part.

9. The difference of two numbers is 9. 3 increased by $\frac{5}{11}$ of the less of the two numbers equals $\frac{2}{7}$ of the greater. Find the numbers.

10. A farmer paid $94 for a horse and cow. What did each cost, if the horse cost $13 more than twice as much as the cow?

11. A cubic foot of iron and a cubic foot of aluminum together weigh 618 lb. If the weight of the iron is 14 lb. less than three times the weight of the aluminum, find the weight of each.

12. A baseball nine has played 54 games, and the number of games it has won is 3 less than twice the number it has lost. How many has it lost?

13. Twice the difference of two numbers is 6, and $\frac{1}{6}$ of their sum is $3\frac{1}{2}$. What are the numbers?

14. 2 lb. of flour and 5 lb. of sugar cost 31 cents, and 5 lb. of flour and 3 lb. of sugar cost 30 cents. Find the value of a pound of each.

15. A man hired 4 men and 3 boys for a day for $18 ; and for another day, at the same rate, 3 men and 4 boys for $17.   How much did he pay each man and each boy per day ?

16. One woman buys 4 yd. of silk and 7 yd. of satin, and another woman at the same rate buys 5 yd. of silk and 5½ yd. of satin.   Each woman pays $17.70.   What is the price of a yard of each material ?

17. 1 cu. ft. of iron and 1 cu. ft. of lead together weigh 1180 lb.; also the weight of 3 cu. ft. of iron exceeds the weight of 2 cu. ft. of lead by 40 lb.   What is the weight of 1 cu. ft. of each of these materials?

18. In an athletic meet, the winning team had a score of 26 points and the second team had a score of 21 points. If the winning team took first place in 7 events and second place in 5 events, while the second team took 6 firsts and 3 seconds, how many points does a first place count ? A second place?

19. In an athletic meet, the three winning teams made scores as follows :

Team	1sts	2ds	3ds	Total Score
A	5	2	2	33
B	3	3	1	25
C	1	4	6	23

What did each of the first three places in an event count in this meet ?

20. Make up and work an example similar to Ex. 18.

21. Two partners agree to divide their profits each year in such a way that one partner receives $1000 more than ⅔ of what the other receives.   If the profits for a given year are $10,000, what does each partner receive?

22. Separate 240 into two parts such that twice the larger part exceeds five times the smaller by 10.

23. If the cost of a telegram of 14 words between two cities is 62 cents, and one of 17 words is 71 cents, what is the charge for the first 10 words in a message and for each word after that?

24. Make up and work an example similar to Ex. 23 concerning telegraph rates between two cities near your home.

25. A farmer one year made a profit of $1640 on 20 acres planted with wheat and 30 acres planted with potatoes. The next year, with equally good crops, he made a profit of $1210 on 30 acres planted with wheat and 20 acres planted with potatoes. How much per acre on the average did he make on each crop?

26. In three successive years, the farmer raised crops with profits as follows:

(1) 20 A. wheat, 30 A. corn, 40 A. potatoes; profits $1720.
(2) 30 A. wheat, 40 A. corn, 20 A. potatoes; profits $1520.
(3) 40 A. wheat, 20 A. corn, 30 A. potatoes; profits $1440.

What were his average profits per acre for each kind of crop?

27. The freight charges between two cities on 400 lb. of first-class freight and 600 lb. of second-class freight were $14.24, while the charges on 500 lb. of first-class freight and 800 lb. of second-class were $18.48. What was the rate per 100 lb. on each class?

28. The freight charges on shipments between two places were as follows: 800 lb. of 4th class + 500 lb. of 5th class + 700 lb. of 6th class, $17.11; 1000 lb. of 4th class + 600 lb. of 5th class + 800 lb. of 6th class, $20.66; 600 lb. of 4th class + 1000 lb. of 5th class + 900 lb. of 6th class, $20.52. Find the rate per 100 lb. for each of the classes.

29. If a bushel of oats is worth 80 cents and a bushel of corn is worth $1.10, how many bushels of each must a miller use to produce a mixture of 100 bu. worth 96 cents a bushel?

Sug.	NUMBERS DEALT WITH	SYMBOLS
(1)	Number of bushels of oats	$= x.$
(2)	Number of bushels of corn	$= y.$
(3)	Number of cents in value of oats	$= 80\,x.$
(4)	Number of cents in value of corn	$= 110\,y.$
(5)	Number of bushels in mixture	$= 100.$
(6)	Number of cents in value of mixture	$= 9600.$

EQUALITIES AMONG NUMBERS

$(1) + (2) = (5)$ or $x + y = 100.$
$(3) + (4) = (6)$ or $80\,x + 110\,y = 9600$, etc.

30. How many pounds of 20 cent coffee and how many pounds of 32 cent coffee must be mixed together to make 60 lb. worth 28 cents a pound?

31. If a rectangle were 3 in. longer and 1 in. narrower, it would contain 5 sq. in. more than it does now; but if it were 2 in. shorter and 2 in. wider, its area would remain unchanged. What are its dimensions?

Sug. Draw a diagram for each rectangle considered in the problem. (See Ex. 21, p. 124.) Then make a list of numbers as usual, etc.

32. If a rectangle were made 3 ft. shorter and $1\frac{1}{2}$ ft. wider, or if it were 7 ft. shorter and $5\frac{1}{4}$ ft. wider, its area would remain unchanged. What are its dimensions?

33. If 1 be added to the numerator of a certain fraction, the value of the fraction becomes $\frac{1}{3}$; but if 1 be subtracted from its denominator, the value of the fraction becomes $\frac{1}{4}$. Find the fraction.

34. There is a fraction such that if 4 be added to its numerator the fraction will equal $\frac{4}{5}$; but if 3 be subtracted from its denominator the fraction will equal $\frac{2}{3}$. What is the fraction?

35. Find two fractions, with denominators 4 and 6, respectively, such that their sum is $3\frac{11}{12}$, but when their numerators are interchanged, their sum becomes $3\frac{7}{12}$.

36. The sum of two numbers is 97, and if the greater is divided by the less, the quotient is 5 and the remainder 1. Find the numbers.

37. Divide the number 100 into two such parts that the greater part will contain the less 3 times with a remainder of 16.

38. The difference between two numbers is 40, and the less is contained in the greater 3 times with a remainder of 12. Find the numbers.

39. A man has $ 5050 invested, part at 4 %, and the rest at 5 %. How much has he invested at each rate if his annual income is $ 220 ?

40. A man wishes to invest part of $12,000 at 5 % and the rest at 4 % so that he may obtain an income of $500. How much must he invest at each of the rates named ?

41. 7 lb. of sugar and 3 lb. of rice together cost 57 cents ; also 5 lb. of sugar and 6 lb. of rice cost 60 cents. Find the cost of a pound of each.

42. If 2 be added to the numerator of a given fraction, the value of the fraction becomes $\frac{3}{4}$. But if 6 be added to the denominator of the fraction, the value of the fraction becomes $\frac{1}{2}$. Find the fraction.

43. A man has $18,000 invested, some at 5 % and the rest at 6 %. His annual income from the two investments is $ 930. How much has he invested at each of the two rates?

44. In an orchard of 100 trees, the apple trees are 5 more than $\frac{2}{3}$ of the number of pear trees. How many trees are there of each kind ?

45. How many pounds of 18 cent coffee must be combined with how many pounds of 30 cent coffee to make a mixture of 100 lb., worth 20 cents a pound ?

46. If the length of a given building lot were diminished by 20 ft. and the width were increased by 10 ft. the area of the lot would be increased by 400 sq. ft. But if the width were diminished by 5 ft. and the length increased by 20 ft. the area would be diminished by 100 sq. ft. Find the dimensions of the lot.

47. The sum of two numbers is 98. If the greater be divided by the less, the quotient is 7 and the remainder 10. Find the numbers.

PART II

106. Example in Solving Simultaneous Equations.

Ex.   Solve $.7\,x + .05\,y = .34.$ (1)

$$12\,x - .3\,y = 4.44.$$ (2)

Multiply (1) by 100, that is, by the smallest number that will make the coefficients of both x and y integral.

$$70\,x + 5\,y = 34.$$ (3)

Similarly multiply (2) by 10,

$$120\,x - 3\,y = 44.4.$$ (4)

Multiply (3) by 3, $210\,x + 15\,y = 102.$ (5)

Multiply (4) by 5, $600\,x - 15\,y = 222.$ (6)

Add (5) and (6), $\overline{810\,x \qquad\quad = 324.}$

$$\left.\begin{array}{l} x = .4. \\ y = 1.2. \end{array}\right\} \quad Ans.$$

Let the pupil check the work.

Solve and check:

1. $1.2\,x - 3.2\,y = 1.76.$
   $3.5\,x - 4.5\,y = 6.1.$

2. $.3\,x + .5\,y = .21.$
   $.7\,x - .2\,y = .08.$

3. $3p - 2q = 2.$
$2q + 7 = 5p - 3q.$

4. $.3x + .5y = .08.$
$.07x - .05y = .002.$

5. $.4x + 5y = .014.$
$.3x - y = .001.$

6. $2x + 1.5y = 10.$
$.3x - .05y = .4.$

Solve to the nearest hundredth:

7. $.03x + .5y = .4.$
$.04x - .2y = .3.$

8. $7x + 3y = 9.$
$12x - 4.8y = 13.$

9. $3(.1x + .004) = 4(.01 - .03y).$
$.35x = 1.8y + .003.$

Solve and check:

10. $\dfrac{7}{y+3} - \dfrac{3}{x+4} = 0.$
$y(x - 2) - x(y - 5) + 13 = 0.$

11. $\frac{2}{3}(x + 3y) - \frac{1}{2}(x + 2y) = \frac{7}{12}.$
$3y - \frac{2}{3}(x + 4y + \frac{5}{6}) = 0.$

12. $(x - 5)(y + 3) = (x - 1)(y + 2).$
$xy + 2x = x(y + 10) + 72y.$

13. $\dfrac{x-2}{5} + \dfrac{x+10}{3} + \dfrac{10-y}{4} = 13.$
$\dfrac{2y+6}{3} - \dfrac{4x+y+6}{8} + 4 = 0.$

14. $\frac{1}{3}x + \frac{1}{9}y + \frac{1}{6}z = 2.$
$\frac{1}{2}x + \frac{1}{3}y + \frac{1}{8}z = 9.$
$\frac{1}{6}x + \frac{1}{2}y + \frac{1}{3}z = 3.$

15. $2x + 2y - z = 2a.$
$3x - y - z = 4b.$
$5x + 3y - 3z = 2(a + b).$

16. $\dfrac{2x}{3} + \dfrac{3y}{4} - \dfrac{4z}{5} = 18.$
$\dfrac{5x}{6} - \dfrac{5y}{8} + \dfrac{3z}{4} = -5.$
$\dfrac{3x}{2} - \dfrac{7y}{5} + \dfrac{3z}{10} = -41.$

17. $x + y + 2z = 2(a + b).$
$x + z + 2y = 2(a + c).$
$y + z + 2x = 2(b + c).$

18. $3x + 2y = \frac{13}{6}a.$
$6z - 2x = \frac{5}{3}b.$
$5y - 13z + x = 0.$

19. $x - y = 2n.$
$mx - ny = m^2 + n^2.$

20. $\dfrac{x+p}{y-q} = \dfrac{q}{p}.$
$x + y = 2q.$

21. $ax - y = c$

$y = - dx.$

22. $\dfrac{x}{a} + \dfrac{y}{b} = c.$

$\dfrac{3\,x}{b} - \dfrac{5\,y}{a} = 7\,c.$

23. Make up and solve a pair of simultaneous equations in which $x = 1.4$ and $y = - .7.$

24. Make up and solve a set of simultaneous equations in which $x = \frac{1}{2}$, $y = \frac{3}{4}$, and $z = - \frac{1}{4}.$

107. Example in Solving Simultaneous Literal Equations.

Ex. Solve $(a + b)x - (a - b)y = a^2 + b^2.$ (1)

$ax - by = 2\,ab.$ (2)

Multiply (1) by b, $(ab + b^2)x - (ab - b^2)y = a^2b + b^3.$ (3)

Multiply (2) by $(a - b)$,

$(a^2 - ab)x - (ab - b^2)y = 2\,a^2b - 2\,ab^2.$ (4)

Subtract (4) from (3),

$(- a^2 + 2\,ab + b^2)x = - a^2b + 2\,ab^2 + b^3.$

$\left. \begin{array}{l} \therefore\ x = b. \\ y = - a. \end{array} \right\}\ Ans.$

Let the pupil check the work.

108. The Use of $\dfrac{1}{x}$ and $\dfrac{1}{y}$ as Unknown Quantities enables us to solve certain equations which would otherwise be difficult of solution.

Ex. 1. Solve $\begin{cases} \dfrac{5}{x} + \dfrac{13}{y} = 49. & (1) \\[2mm] \dfrac{7}{x} + \dfrac{3}{y} = 23. & (2) \end{cases}$

Multiply (1) by 7, and (2) by 5,

$\begin{cases} \dfrac{35}{x} + \dfrac{91}{y} = 343. & (3) \\[2mm] \dfrac{35}{x} + \dfrac{15}{y} = 115. & (4) \end{cases}$

Subtract (4) from (3), $\dfrac{76}{y} = 228$, $\therefore\ \dfrac{1}{y} = 3$, or $y = \dfrac{1}{3}.$ *Root.*

Substitute the value of y in (2), hence, $x = \frac{1}{2}$. *Root.*
Let the pupil check the work.

EXERCISE 127

Solve and check :

1. $(a - 2b)x - 3y = p.$
  $y = bx.$

2. $\dfrac{x}{a+b} + \dfrac{y}{a-b} = 2.$
  $x - y = 2b.$

3. $(a - 2b)x + 3by = c.$
  $y = \frac{1}{2}x.$

4. $(a+1)x - by = a + 2.$
  $(a-1)x + 3by = 9a.$

5. $\dfrac{(a-b)x + (a+b)y}{a^2 + b^2} = 1.$
  $ax - 2by = a^2 - 2b^2.$

6. $(a+b)x + cy = 1.$
  $cx + (a+b)y = 1.$

7. $(a-b)x - (a+b)y = a^2 + b^2.$
  $bx + ay = 0.$

8. $\dfrac{4}{x} + \dfrac{7}{y} = 3.$
  $\dfrac{3}{x} + \dfrac{5}{y} = 2.$

9. $\dfrac{1}{x} - \dfrac{2}{y} = 7.$
  $\dfrac{3}{x} + \dfrac{4}{y} = 1.$

10. $\dfrac{4}{x} + \dfrac{5}{y} = 16.$
  $\dfrac{7}{x} - \dfrac{3}{y} = \dfrac{9}{2}.$

11. $\dfrac{10}{x} + \dfrac{6}{y} = 4.$
  $\dfrac{15}{x} + \dfrac{7}{y} = \dfrac{26}{3}.$

12. $\dfrac{5}{x} - \dfrac{3}{y} = 7.$
  $\dfrac{15}{y} + \dfrac{60}{x} = 16.$

13. $\dfrac{1}{x} + \dfrac{1}{y} = \dfrac{1}{n}.$
  $\dfrac{1}{x} - \dfrac{1}{y} = n.$

14. $\dfrac{a}{x} + \dfrac{b}{y} = \dfrac{a-b}{a}.$
  $\dfrac{b}{x} + \dfrac{a}{y} = \dfrac{b-a}{a}.$

15. $\dfrac{1}{x} + \dfrac{1}{y} + \dfrac{1}{z} = 2.$
  $\dfrac{2}{x} - \dfrac{1}{y} + \dfrac{1}{z} = 7.$
  $\dfrac{3}{x} + \dfrac{2}{y} + \dfrac{5}{z} = 14.$

16. $\dfrac{3}{x} - \dfrac{1}{y} = 3\frac{1}{2}.$
  $\dfrac{5}{y} + \dfrac{3}{z} = -7.$
  $\dfrac{2}{x} - \dfrac{1}{z} = 0.$

Verbal Problems

1. Give five consecutive odd numbers of which the middle one is $2x+1$.

2. How many quarts are there in p per cent of g gallons?

3. Ribbon which cost a cents a yard is sold for b cents a yard. If, when sold thus, it is sold at a gain, what is the per cent of gain? If, when sold thus, it is sold at a loss, what is the per cent of loss?

4. One factor of $\frac{c}{x}$ is $\frac{a}{c}$. Find the other factor.

5. If p pencils are sold for c cents, how many pencils can be bought for x cents?

6. If $x+3$ stands for 8, what does $2x-1$ stand for?

7. A line is divided into n parts, and each part is also divided into c smaller parts. State the total number of parts into which the line has been divided. What part of the whole line is made by x of these parts?

8. A pipe can fill a tank $\frac{2}{3}$ full in m hours. What part of the tank can it fill in h hours? In c minutes?

9. A rectangle is x inches long and y inches wide. If a strip i inches wide is cut lengthwise from the rectangle, find the area remaining. If the strip is cut from the narrow side of the rectangle, find the area remaining.

10. If x men each contribute y dollars to a certain fund, how much do they all contribute together? If the number of men be diminished by 5, and the number of dollars contributed by each be increased by 2, how many dollars are contributed?

11. If a pipe can fill a given tank in x hours, what part of the tank will it fill in 4 hours? If another pipe can fill the tank in y hours, what part of the tank will it fill in 5 hours?

12. If $74 is paid in 7 ten-dollar bills, and 4 one-dollar bills, how many bills in all are used? If $74 were paid entirely in one-dollar bills, how many bills would be used?

13. 37 means $10 \times 3 + 7$. Express a number which contains b tens and 5 units.

14. A certain number contains 8 tens and c units. Express this number.

15. A certain number contains x tens and y units. Express this number.

16. Write a number with 3 hundreds, a tens, and b units.

17. Also one containing x hundreds, a tens, and 6 units.

18. Also one containing x hundreds, y tens, and z units.

19. How would you write a number whose digits in order from left to right are l, m, and n? Why may not such a number be expressed as lmn?

20. Express in symbols a number whose digits in order are a, b, c, and d. Whose digits are a, b, and c. a and b.

21. If 1 lb. of gold when immersed in water and weighed by a spring balance loses $\frac{1}{19}$ of its weight, 5 lb. will lose how much in weight? x pounds will lose how much?

22. If 1 lb. of silver when immersed in water loses $\frac{1}{10}$ of its weight, 7 lb. will lose how much? y pounds will lose how much?

23. Also an alloy of x pounds of gold and y pounds of silver when weighed in water will lose how much in weight?

24. If a given kind of milk contains 5 per cent of butter fat, x lb. of milk will contain how many pounds of butter fat?

25. If a given kind of cream contains 40 per cent of butter fat, y pounds of the cream will contain how many pounds of butter fat?

26. How many pounds of butter fat are there all together in a pounds of milk containing 4 per cent of butter fat and in b pounds of cream containing 45 per cent of butter fat?

EXERCISE 129

1. Separate .0015 into two parts such that one part equals three times the sum of .0001 and the other part.

2. The sum of two numbers is $1\frac{2}{3}$, and the larger is three times the sum of the smaller and $\frac{1}{3}$. Find the numbers.

3. Find two numbers such that the first exceeds the second by .6, while one quarter of the second exceeds one fifth of the first by .42.

4. A ton of fertilizer which contains 60 lb. of nitrogen, 100 lb. of potash, and 150 lb. of phosphate is worth $21.50; a ton containing 70, 80, and 90 lb. of these constituents in order is worth $19; and one containing 80, 120, 150 lb. of each in order is worth $25.50; what is the value of one pound of each of the constituents named?

5. The corn and wheat crops of the United States in the year 1909 were together 3,509,000,000 bu.; the corn and oat crops 3,779,000,000 bu.; and the wheat and oat crops 1,744,000,000 bu. How many bushels were in each crop?

6. One cubic foot of iron and one cubic foot of aluminum weigh 636 lb.; a cubic foot of iron and one of copper weigh 1030 lb.; a cubic foot of copper and one of aluminum weigh 706 lb. How much does one cubic foot of each of these materials weigh?

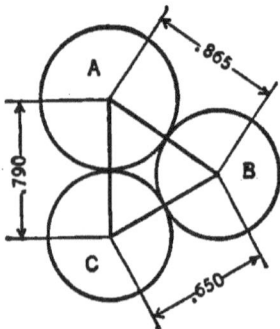

7. In boring holes in a metal plate, three circles touching each other are to be drawn, the distances between their centers being .865 in., .650 in., and .790 in., respectively. Find the radius of each of the three circles.

8. The Eiffel Tower is taller than the Metropolitan Life Building of New York, and the latter building is taller than the Washington Monument. If the difference between the heights of the first two is 284 ft.; between the first and last is 429 ft.; and the sum of the first and last is 1539 ft., find the height of each.

9. If two grades of coffee worth 15 cents and 30 cents a pound are to be mixed together to make 100 lb. which can be sold for 30 cents at a profit of 25 %, how many pounds of each must be used ?

10. If two grades of tea worth 50 cents and 75 cents a a pound are to be mixed together to make 100 lb. which can be sold for 72 cents at a profit of 20%, how many pounds of each must be used ?

11. The sum of two numbers is 22.8. When the larger is divided by the smaller, the quotient is 9 and the remainder 3. Find the numbers.

12. The sum of two numbers is 7. When the larger is divided by the smaller, the quotient is 3 and the remainder is 1.1. Find the numbers.

13. Find two numbers such that twice the greater exceeds three times the less by 1. But the sum of the numbers divided by the smaller number gives 2 for a quotient and 5 for a remainder. Find the numbers.

14. A party of boys purchased a boat and upon payment for the same discovered that if they had numbered 3 more, they would have paid a dollar apiece less; but if they had numbered 2 less, they would have paid a dollar apiece more. How many boys were there, and what did the boat cost?

15. After going a certain distance in an automobile, a driver found that if he had gone 3 mi. an hour faster, he would have traveled the distance in 1 hr. less time ; and that if he had gone 5 mi. faster, he would have gone the distance in 1½ hr. less. What was the distance ?

16. The sum of the numerator and denominator of a given fraction is 13. If the numerator of the fraction is increased by 1, the value of the fraction becomes $\frac{3}{4}$. Find the fraction.

17. If a baseball nine should play two games more and win both, it will have won ⅔ of the games played. If, however, it should play 7 more and win four of them, it will then have won ⅗ of the games played. How many games has it played so far and how many has it won?

18. If a physician should have 12 more cases of diphtheria and treat 10 of them successfully, he will have treated ¾ of his cases successfully. But if he should have 32 more cases and succeed with 30 of them, he will have succeeded with ⅘ of his cases. How many cases has he had so far and how many has he treated successfully?

19. A number consists of two digits whose sum is 13, and if 4 is subtracted from double the number, the order of the digits is reversed. Find the number.

20. The sum of the digits of a certain number of two figures is 5, and if 3 times the units' digit is added to the number, the order of the digits will be reversed. What is the number?

21. Twice the units' digit of a certain number is 2 greater than the tens' digit; and the number is 4 more than 6 times the sum of its digits. Find the number.

22. In a number of 3 figures, the first and last of which are alike, the tens' digit is one more than twice the sum of the other two, and if the number is divided by the sum of its digits, the quotient is 21 and the remainder 4. Find the number.

23. A tank can be filled by two pipes one of which runs 4 hr. and the other 5; or by the same two pipes if the first runs 3 hr. and the other 8. How long will it take each pipe running separately to fill the tank?

24. Two persons, A and B, can perform a piece of work in 16 days. They work together for 4 days, when B is left alone, and completes the task in 36 days. In what time could each do the work separately?

25. A and B can do a piece of work in 8 days; A and C can do the same in 10 days; and B and C can do it in 12 days. How long will it take each to do it alone?

26. A 21 lb. mass of gold and silver alloy when immersed in water weighed only 19 lb. If the gold lost $\frac{1}{19}$ of its weight when weighed under water, and the silver $\frac{1}{10}$ of its weight, how many pounds of each metal were in the alloy?

27. An alloy of aluminum and iron weighs 80 lb., but when immersed in water it weighs only 60 lb. If the specific gravity of aluminum is $2\frac{1}{2}$ while that of iron is $7\frac{1}{2}$, how many pounds of each metal are in the alloy?

28. A mass of copper and tin weighing 100 lb. when immersed in water weighed 87.5 lb. If the specific gravity of copper is 8.8 and that of tin is 7.3, how much of each metal was in the mass?

29. A train maintained a uniform rate for a certain distance. If this rate had been 8 mi. more each hour, the time occupied would have been 2 hr. less; but if the rate had been 10 mi. an hour less, the time would have been 4 hr. more. Find the distance.

30. If the greater of two numbers is divided by the less, the quotient is 3 and the remainder 3, but if 3 times the greater be divided by 4 times the less, the quotient is 2 and the remainder 20. Find the numbers.

31. In a number of two digits, the sum of the digits is 9. If the order of digits is reversed, the new number exceeds twice the first number by 18. Find the first number.

32. Two grades of spices worth 25 cents and 50 cents a pound are to be mixed together to make 200 lb. which can be sold at 52 cents per pound at a profit of 80%. How many pounds of each grade must be used?

33. A farmer wishes to combine milk containing 5 % of butter fat with cream containing 40 % of butter fat in order to produce 20 gal. of cream which shall contain 25 % of butter fat. How many gallons of milk and how many of cream must he use ?

34. A boy was engaged to work 50 days at 75 cents per day for the days he worked, and to forfeit 25 cents every day he was idle. On settlement he received $ 25.50 ; how many days did he work ?

35. If A gives B $ 10, A will have half as much as B; but if B gives A $ 30, B will have $\frac{4}{5}$ as much as A. How much has each ?

36. 2 lb. of tea and 5 lb. of coffee cost $ 2.50. If the price of tea should increase 10 % and that of coffee should diminish 10 %, the cost of the above amounts of each would be $ 2.45. Find the cost of a pound of each.

37. Two bins contain a mixture of corn and oats, the one twice as much corn as oats, and the other three times as much oats as corn. How much must be taken from each bin to fill a third bin holding 40 bu., to be half oats and half corn ?

38. Find two numbers whose sum is a and whose difference is b.

39. If a pounds of sugar and b pounds of coffee together cost c cents, while d pounds of sugar and e pounds of coffee together cost f cents, what is the price of one pound of each ?

40. If a bushel of oats is worth p cents, and a bushel of corn is worth q cents, how many bushels of each must be mixed to make r bushels worth s cents per bushel ?

41. Of c coins, some are worth a cents apiece and some are worth b cents apiece. If all the coins are together worth d cents, how many coins of each kind are there ?

CHAPTER XII

GRAPHS

PART I

109. Graphs of Equations. In an equation containing x and y, by letting x have certain values and finding the corresponding values of y, it is possible to construct a graph representing the equation. Such graphs are useful in a number of ways which will become evident as we proceed.

110. Framework of Reference. Axes are two straight lines perpendicular to each other used as an auxiliary framework in constructing graphs; as XX' and YY'.

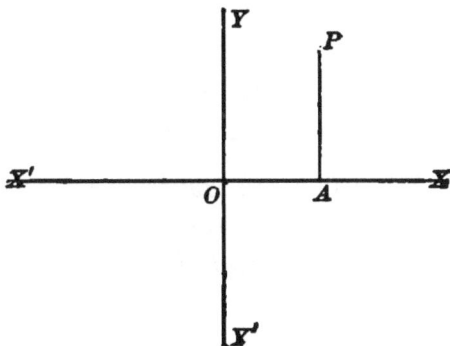

The **x-axis**, or **axis of abscissas**, is the horizontal axis; as XX'. The **y-axis**, or **axis of ordinates**, is the vertical axis; as YY'.

111. The **origin** is the point in which the axes intersect; as the point O.

112. The **ordinate of a point** is the line drawn from the point parallel to the y-axis and terminated by the x-axis. The **abscissa of a point** is the part of the x-axis intercepted

A FIRST BOOK IN ALGEBRA

between the origin and the foot of the ordinate. Thus, the
ordinate of the point P is AP, and the abscissa is OA.

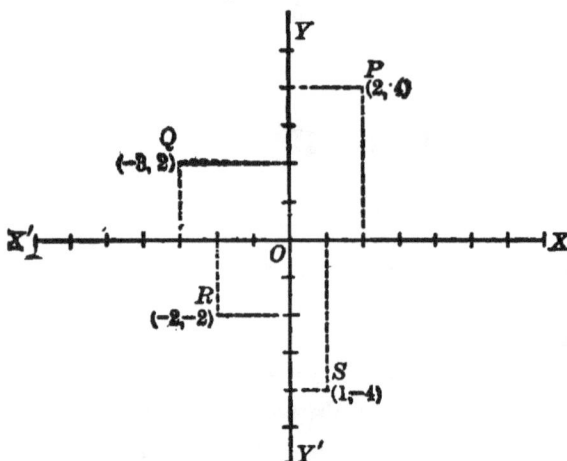

Ordinates
above the x-axis
are taken as
plus; those be-
low, as minus.
Abscissas to the
right of the
origin are plus;
those to the left
are minus.

The **coordi-
nates of a point**
are the abscissa and the ordinate taken together. They
are usually written together in parenthesis with the
abscissa first and a comma between.

Thus, the point (2, 4) is the point whose abscissa is 2 and
ordinate 4, or the point P of the figure. Similarly, the point
$(-3, 2)$ is Q; $(-2, -2)$ is R; and $(1, -4)$ is S.

113. The **quadrants** are the four parts into which the
axes divide a plane. Thus, the points P, Q, R, and S lie
in the *first*, *second*, *third* and *fourth* quadrants, respectively.

EXERCISE 130

Draw axes and locate each of the following points:

1. $(3, 2)$, $(-1, 3)$, $(-2, -4)$, $(4, -1)$.

2. $(2, \frac{1}{2})$, $(-3, -1\frac{1}{2})$, $(5, -\frac{5}{4})$, $(-2, \frac{1}{3})$.

3. $(2, 0)$, $(-3, 0)$, $(0, 4)$, $(0, -\frac{1}{2})$, $(0, 0)$.

4. Construct the triangle whose vertices are $(1, 1)$,
$(2, -2)$, $(3, 2)$.

5. Construct the quadrilateral whose vertices are
$(2, -1)$, $(-4, -3)$, $(-3, 5)$, $(3, 4)$.

6. Plot the points $(0, 0)$, $(1, 0)$, $(2, 0)$, $(5, 0)$, $(-1, 0)$.

7. All points on the x-axis have what ordinate?

8. All points on the y-axis have what abscissa?

9. Construct the rectangle whose vertices are $(1, 3)$, $(6, 3)$, $(1, -2)$, $(6, -2)$, and find its area.

10. Construct the rectangle whose vertices are $(-3, 4)$, $(4, 4)$, $(-3, -2)$, $(4, -2)$, and find its area.

11. Construct the triangle whose vertices are $(-3, -4)$, $(-1, 3)$, $(2, -4)$, and find its area.

GRAPHS OF EQUATIONS OF THE FIRST DEGREE

114. To Construct the Graph of an Equation of the First Degree Containing Two Unknown Quantities, as x, and y.

Let x have a series of convenient values, as $0, 1, 2, 3$, *etc.,* $-1, -2, -3$, *etc.;*

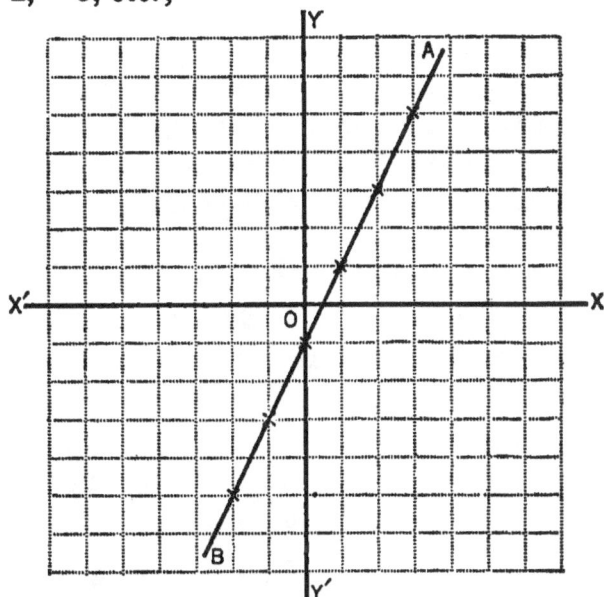

Find the corresponding value of y;

Locate the points thus determined, and draw a line through these points.

Ex. Construct the graph of the equation $y = 2x - 1$.

Construct the points $(0, -1)$, $(1, 1)$, $(2, 3)$, $(3, 5)$, $(-1, -3)$, $(-2, -5)$, etc., and draw a line through them. The straight line AB (p. 247) is thus found to be the graph of $y = 2x - 1$.

x	y
0	-1
1	1
2	3
3	5
etc.	etc.
-1	-3
-2	-5
etc.	etc.

115. An **Equation of the First Degree** is one which contains the first power of an unknown quantity and no higher power.

Thus, $3x = 5 - 2x$, is an equation of the first degree. So also is $3y = 2x - 5$.

But $x^2 + x = 7$, is an equation of the second degree.

116. Linear Equations. It will always be found that the graph of an equation of the first degree containing not more than two unknown quantities is a straight line. Hence,

A **linear equation** is an equation of the first degree.

Since a straight line is determined by two points, in order to construct the graph of an equation of the first degree it is sufficient to *construct any two points of the graph and draw a straight line through them.*

EXERCISE 131

Graph the following. (It is an advantage, if possible, to draw the graph line in red, the rest of the figure in black ink.)

1. $y = x + 1$.
2. $y = x + 2$.
3. $y = x - 3$.
4. $y = x$.
5. $y = 2x$.
6. $y = 3x$.
7. $y = 2x + 1$.
8. $y = 2x - 3$.
9. $2y = x$.
10. $2y = 3x + 6$.
11. $3y = 2x + 6$.
12. $2y = 3x + 3$.
13. $2y = x + 2$.
14. $3x - 2y = 6$.
15. $3x - 5y = 15$.
16. $4x - 5y = 1$.
17. $7y - 9x = 2$.
18. $x = 3(y - 1)$.
19. $3x - 5y = 15$.
20. $\dfrac{x-1}{2} = 3y$.

117. Graphic Solution of Simultaneous Linear Equations.
If we construct the graph of the equation $x - y = 3$ (the line AB) and the graph of $3x + 2y = 4$ (the line CD),

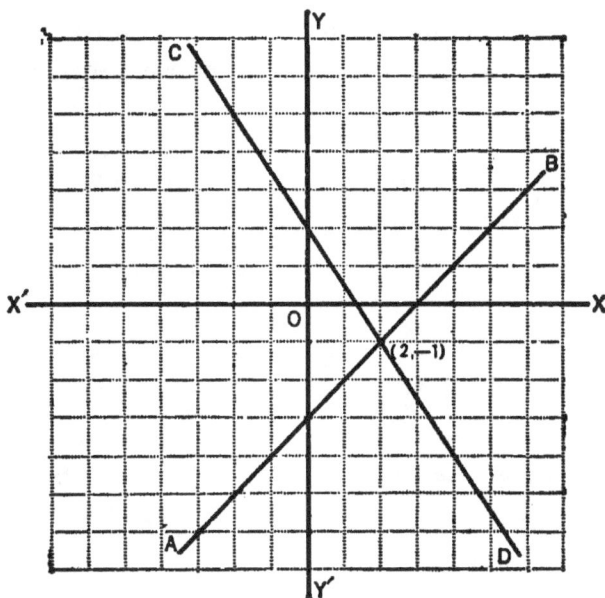

and measure the coördinates of their point of intersection, we find this point to be $(2, -1)$.

If we solve the pair of simultaneous equations

$$\begin{cases} x - y = 3, \\ 3x + 2y = 4, \end{cases}$$

by the ordinary algebraic method, we find that $x = 2$ and $y = -1$.

In general, *the roots of two simultaneous linear equations correspond to the coördinates of the point of intersection of their graphs;* for these coördinates are the only ones which satisfy both graphs, and their values are also the only values of x and y which satisfy both equations.

Hence, to obtain the graphic solution of two simultaneous equations,

Draw the graphs of the given equations, and measure the coordinates of the point (or points) of intersection.

Graphing two simultaneous equations forms a convenient method of testing or checking their algebraic solution.

EXERCISE 132

Solve each pair of the following equations graphically and algebraically, comparing the results in each example :

1. $\begin{cases} 2x + 3y = 7. \\ x - y = 1. \end{cases}$

2. $\begin{cases} y = 3x - 4. \\ y = -2x + 1. \end{cases}$

3. $\begin{cases} 3x - 2y = 0. \\ x - y + 1 = 0. \end{cases}$

4. $\begin{cases} y - x - 3 = 0. \\ y + \frac{1}{2}x = 0. \end{cases}$

5. $\begin{cases} y + x = 2. \\ y - x = 4. \end{cases}$

6. $\begin{cases} y = 2x + 3. \\ 2y = x + 3. \end{cases}$

7. $\begin{cases} y = x. \\ 2y = 3x + 3. \end{cases}$

8. $\begin{cases} x - 2y + 11 = 0. \\ 2x - 3y + 18 = 0. \end{cases}$

9. $\begin{cases} 2y = x. \\ x + y + 6 = 0. \end{cases}$

10. $\begin{cases} y = 2x. \\ x + y = 0. \end{cases}$

11. $\begin{cases} 4x - 7y = 13. \\ x - 8y = 22. \end{cases}$

12. $\begin{cases} 4x + y = 10. \\ 3x - 4y = 17. \end{cases}$

13. Construct the triangle whose sides are the graphs of the equations $5y - x - 7 = 0$, $3y - 5x + 9 = 0$, $y + 2x + 3 = 0$, and find the coördinates of the vertices of the triangle.

EXERCISE 133

1. In a certain manual training school the pupils make articles which they sell at a profit of 25 % above cost. If the cost be denoted by c and the selling price by s, show that $s = \dfrac{5\,c}{4}$.

2. Graph the formula $s = \dfrac{5\,c}{4}$.

Sug. By § 116 it is necessary only to locate two points on the graph and draw a straight line through them. Thus,

c	s
0	0
4	5

3. If the cost of a certain article is $\$\,6$, find from the formula of Ex. 2, what the selling price should be. Check your answer by means of the graph.

Sug. On the c-axis take $OA = 6$, and find the number of units in AB. AB should equal the answer obtained from the formula. This shows that the answer might have been obtained from the graph without the use of the formula.

4. If the cost of a certain article is $\$\,3$, find from the graph what the selling price should be. Check your answer by the use of the formula.

5. Also, by means of the graph, when the selling price is known, the cost (or limit of the cost) may be determined. Thus, if the selling price of an article is to be $\$\,10$, find the cost limit.

Sug. Find that point on the line OB whose ordinate is 10, viz., the point C, and then determine the abscissa of C. This is OD, or CF, which is 8. Hence, $\$\,8$. *Ans.*

6. If the selling price of an article is to be $\$\,6$, from the graph find the cost limit. (Check your answer by use of the formula of Ex. 2.)

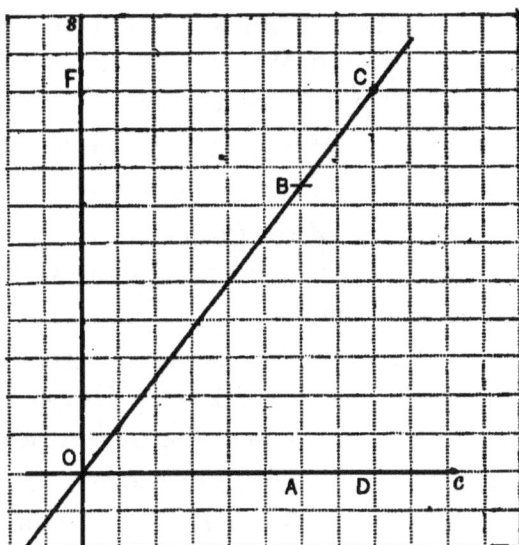

7. If the number of pounds in English money be denoted by p, and the corresponding number of dollars by d, $d = 4.87 \, p$. Make a graph of this formula.

8. By use of the formula of Ex. 7, find in dollars the value of £5. Check your answer by use of the graph.

9. By use of the graph of Ex. 7, find as closely as you can the number of pounds in the value of $8. Check by use of the formula.

10. If number of meters be denoted by m, and of feet by f, show that $f = 3.28 \, m$. Graph this formula. From this graph find the number of feet in 6 meters.

11. Make a formula and also a graph showing the relation between miles and kilometers and illustrate the use of the graph by a numerical example.

12. If one pound avoirdupois contains 7000 grains and a Troy pound contains 5760 grains, make a formula, and also a graph, showing the relation between the two kinds of pounds, and illustrate the use of the graph by a numerical example.

13. A certain house cost d dollars and rents for m dollars a month. The taxes and other expenses are e dollars per year. Find a formula for r, the rate per cent of income from the investment. Make up a numerical example and solve it by use of this formula.

EXERCISE 134
REVIEW

1. Factor: (1) $6y^2 - 6y - 12$.   (4) $x^2 - 11x - 26$.

     (2) $bx^7 - bx^3$.   (5) $5x^3 - 20x^2y + 20xy^2$.

     (3) $16a^2x + 8abx + b^2x$.   (6) $4x^2 + 1 + 4x$.

2. Solve $\frac{1}{3}(5x - 1) - \frac{1}{2}(7x + 2) + 5 = 0$.

3. Simplify $\dfrac{2x^2 + 3}{6x^2} - \dfrac{3x^3 + 1}{12x^3} - \dfrac{3x^4 - 2}{36x^4}$.

4. Separate 20 into two parts such that twice one part equals three times the other part.

5. Solve $.6\,y - .7\,x = 2$, $2.5\,y + 1.2\,x = 49.5$.

6. A mixture of alcohol and water contains 18 gallons of which 5 are alcohol. How many gallons of water must be added to the mixture to make it a 20 per cent solution?

7. If 1 liter $= .9$ dry quart, make a formula and also a graph showing the relations between liters and quarts. Make up a numerical example and solve it by use of the formula and check your answer by use of the graph.

8. Copy the following tabulation, filling each vacant place:

SUBJECT	RULE	FORMULA
Area of rectangle	width No. = area No. ÷ length No.	
Percentage		$r = \dfrac{p}{b}$
Interest	rate = (interest) ÷ (principal × time)	
Volume of solid		$w = \dfrac{v}{lh}$
Percentage		$b = \dfrac{p}{r}$
Circle	radius = circumference ÷ 2π	
Motion		$d = vt$
Motion		$v = \dfrac{d}{t}$

9. Solve $R = a\left(\dfrac{1}{c} + d\right)$ for c.

10. Find the value of $82.7^2 - 47.3^2$ in a short way by the aid of factoring.

---

11. Factor: (1) $6 - 96\,a^2b^2x^2$.   (4) $\tfrac{4}{9}\,a^2 - 16\,b^2$.
           (2) $y^2 + 5\,y - 126$.   (5) $\tfrac{4}{9}\,a^2 - 4\,ab + 9\,b^2$
           (3) $9\,a^2x^2y^2 + 6\,axy + 1$.   (6) $9\,a^2 + x^2 - 6\,ax$.

12. Solve $\dfrac{a+x}{a-x}=\dfrac{1}{a}$ for x.   **13.** Simplify $\dfrac{a^2-9}{a^2+4a}\div\dfrac{a^2-3a}{a^2-16}$.

14. Separate $\frac{7}{8}$ into two parts such that $\frac{3}{4}$ of one part equals the other part.

15. Solve
$$\left.\begin{array}{r} x-2y+3z=-4 \\ 3x+4y-z=18 \\ 5x-2y=18 \end{array}\right\}\text{ and check your answer.}$$

16. How many pounds each of 20 and 32 cent coffee must be combined to make a mixture of 72 lb. worth 28 cents a pound?

17. Using the fact that 1 centimeter $=\frac{2}{5}$ of an inch, make a formula and also a graph, showing the relation between centimeters and inches. Illustrate the use of the formula and graph by a numerical example.

18. The sum of the numbers is 36. The larger number divided by the smaller gives 4 for a quotient and 1 for a remainder. Find the numbers.

19. Solve $h=\dfrac{v^2}{2g}$ for g.

20. When $\pi=\frac{22}{7}$, $R=28$, and $r=18$, compute the value of $\pi R^2-\pi r^2$ in a short way by the aid of factoring.

21. Solve $\dfrac{1}{p}+\dfrac{1}{x}=\dfrac{1}{q}-\dfrac{1}{x}$ for x and check your answer.

22. Simplify $\dfrac{3a+2}{3a-2}-\dfrac{9a^2+4}{9a^2-4}$.

23. Compute the value of $12\times9\times17^2+8\times9\times17^2-10\times9\times17^2$ in a short way by the aid of factoring.

24. Solve
$$\left.\begin{array}{l} \dfrac{3x-6}{10}+3-\dfrac{5y-4}{2}=\dfrac{5y}{2} \\[2mm] \dfrac{y-x}{4}+\dfrac{x}{8}-\dfrac{7x-5y}{3}=y-2x \end{array}\right\}\begin{array}{l}\text{and check your}\\ \text{answer.}\end{array}$$

25. What number must be added to the numerator of $\frac{11}{32}$ to make the value of the fraction equal to $\frac{5}{8}$?

26. Express the following in symbols: the square of the sum of a and b equals the square of a plus twice the product of a and b plus the square of b.

27. Construct a graph of $c = \frac{4}{4}\,r$, letting each space on the vertical or c-axis represent four linear units. What does this graph represent?

By use of this graph, find c when $r = 3$. (Check by use of the formula.) Also, find c when $r = 4.5$.

28. Solve $H = \dfrac{SCAN}{12000}$ for N.

29. In $5.2\,x - 2(1.7\,x - 3.3) = .52$, find the value of x to the nearest hundredth.

30. Solve $\left.\begin{array}{l} ax + by = p \\ cx + dy = q \end{array}\right\}$ for x and y.

PART II

118. Bar Graphs. All the graphs that have been studied thus far represent the progressive changes in some one quantity, as in temperature, or in money obtained or used in some way, the changes covering a considerable period of time.

Another important kind of graph is that which shows the comparative size of several distinct objects (of a given kind) at one time.

Ex. At a certain time the amounts in a savings bank to the credit of four different boys were as follows: $47, $32, $9, $26. Represent the numbers graphically.

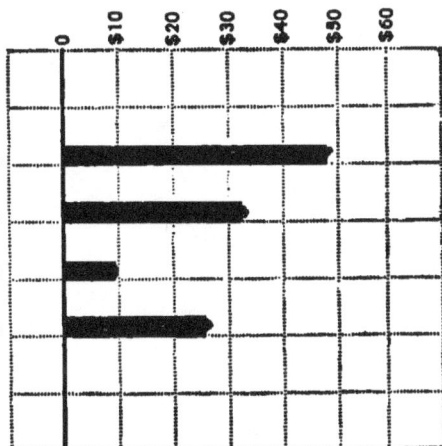

The graph is shown at the right.

A graph of this kind is called a **bar graph**, or a **bar pictogram**.

Make a bar graph for each of the following:

1. During a given summer the earnings of· five girls from canning the products of their tomato plots were as follows : $26, $37, $19, $12, $42.

2. The number of pupils in four different schools were 315, 260, 412, 192.

3. The average grades of five boys in a given half year were as follows : 56, 82, 65, 72, 91.

4. In a given year the principal grain crops of the world states were as follows in millions of bushels : oats, 3426; wheat, 3333; corn, 2766; rye, 1696; barley, 1225.

5. In a recent year the crops of tea in millions of pounds in the leading tea-producing countries of the world were as follows : India, 369; China, 206; Ceylon, 216; Java, 103; Japan, 84; Formosa, 34.

119. Circle Graphs. When a group of numbers are to be viewed as parts of a single whole, it is often suggestive to represent them as proportional parts of a single circle.

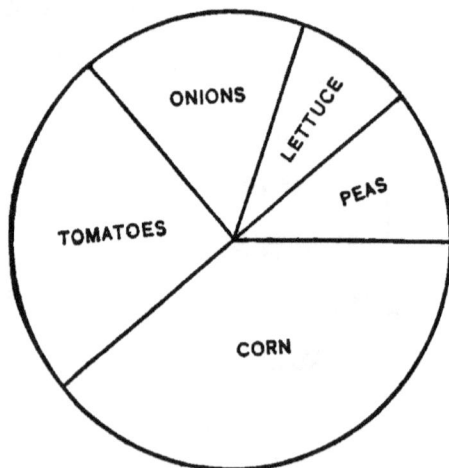

Ex. One summer a boy's profits from different crops raised in his garden were as follows: peas, $4; lettuce, $3; onions, $6; tomatoes, $9; corn, $14. Represent the relative size of these numbers by a circle graph.

Sug. The sum of all the items is $36. The profits from peas = $\frac{4}{36}$ or $\frac{1}{9}$ of the whole. $\frac{1}{9}$ of 360° = 40°. Draw a circle of convenient size. With a protractor at the center lay off an angle equal to 40°. Treat the other items in like manner.

Make a circle graph for each of the following :

1. During a certain year a girl earned money as follows : summer garden, $36 ; knitting garments, $20 . making and selling cakes, $16 ; confectionery, $8.

2. One year a man whose salary was $1200 used the money as follows : food, $480 ; rent, $216; clothing, $180; fuel and lights, $54 ; insurance, $18 ; miscellaneous expenses, $52 ; savings, $200.

3. During a given year the profits from five stores composing a chain were as follows : 1st store, $36,000 ; 2d, $38,000 ; 3d, $27,000 ; 4th, $16,000 ; 5th, $4200.

4. During a given year the sales in the four departments in a given store were as follows : $127,000, $216,000, $162,000, $295,000.

5. During a certain term the number of pupils in the four grades of a high school beginning with the highest were 212, 308, 416, 510.

REVIEW

1. From the product of $3x^2 - 2$ and $2x - 5$, subtract 7 times the product of x and $x - 2$.

2. Factor:
   (1) $\frac{1}{8}ah - \frac{1}{8}ak$.   (4) $228^3 - 228^2$.
   (2) $x^{16} - 1$.   (5) $.36x^2 + 1.2x + 1$.
   (3) $4(x + 2y)^2 - 9a^2$.   (6) $x^4 - (x - 6)^2$.

3. If five times a certain number is increased by 20.5, the result is equal to three times the number increased by 160. Find the number.

4. Simplify $\dfrac{1}{a} + \dfrac{a-b}{a^2+ab} - \dfrac{2}{a+b}$.

5. Solve $\dfrac{6x}{.5} + \dfrac{3y}{2} = 24, \ \dfrac{7x}{.2} - \dfrac{5y}{.4} = 25$.

6. Solve $px = q(y-2), \ y - x = \dfrac{p^2+q^2}{pq}$.

7. The sum of the digits of a certain number is 9. If the number be divided by the sum of the digits, the quotient equals the units' digit of the number. Find the number.

8. In a recent year the crops of coffee in the leading coffee-producing countries of the world were as follows in millions of pounds: Brazil, 2253; Central America, 337; Venezuela, 137; Columbia, 149; other countries, 511. Make a circle graph of these numbers.

9. In $r = \dfrac{b}{e^2}\left(\dfrac{1}{c} + d\right)$ find c if $r = 20, b = 30, e = 4,$ and $d = 8$.

10. Solve $h = \dfrac{s^2}{4(2r-h)}$ for r.

11. Multiply $\frac{2}{3}x^2 - ax + \frac{2}{3}a^2$ by $\frac{3}{4}x^2 + \frac{1}{2}ax + \frac{1}{3}a^2$.

12. Factor:

(1) $x^2 - a^2 - b^2 + 2ab$. (4) $x^4 + 6x^3 - 135x^2$.

(2) $x^2 + (c+d)x + cd$. (5) $5(a+b)^2x - 3(a+b)^2y$.

(3) $(3x+5y)^2 - 4(2x-3y)^2$. (6) $.16 - 25y^2$.

13. Divide $2 - x$ by $1 + x$ to four terms, expressing both quotient and remainder.

14. Simplify $\left(2 - \dfrac{x-4}{x-3}\right) \div \left(\dfrac{x^2-4}{x^2-9}\right)$. Check by letting $x = 1$.

15. Solve $\dfrac{3}{a}(x - 4b) - \dfrac{5}{4b}(x - 3a) - 4 = 0$ and check.

16. Solve $\left.\begin{array}{l} (x+7)(y-4) - (x+4)(y-3) = 0 \\ (x-2)(y+5) - (x-1)(y+2) = 0 \end{array}\right\}$ and check.

17. If the value of one franc is $\$.193$, make a formula and graph showing the relation between the dollar and the franc, and illustrate the use of the formula and graph by a numerical example.

18. In $\dfrac{F}{P}=\dfrac{h}{2\,\pi r}$, find P if $F=50$, $r=21$, $\pi=\frac{22}{7}$, and $h=\frac{3}{4}$.

19. Solve $\dfrac{P}{W}=\dfrac{dr}{2\,\pi l R}$ for R.

20. What number will be added to both numerator and denominator of $\frac{13}{18}$ to make the value of the fraction $\frac{3}{4}$?

---

21. Compute the value of the following in a short way by the aid of factoring: $689 \times \frac{1}{365} \times .05 + 724 \times \frac{1}{365} \times .05 - 573 \times \frac{1}{365} \times .05$.

22. Divide $2.4\,x^3 - .12\,x^2y + 4.32\,y^3$ by $1.5\,x + 1.8\,y$ and check by letting $x=1$ and $y=1$.

23. Solve $2.37\,x - (.46 - 5.2\,x) - 4.93 = 0$ to the nearest thousandth.

24. Solve $\dfrac{9}{5\,x} - \dfrac{4\,x-1}{4\,x^2-1} = \dfrac{8}{10\,x-5}$ and check.

25. Solve $\dfrac{3\,r+x}{2\,r+x} = \dfrac{x-c}{x+c}$.

26. Write a binomial the factors of which are a monomial and two binomials.

27. If the length of a certain rectangle be increased by 2 ft. and the width by 3 ft., the area would be increased by 44 sq. ft. But if the length be diminished by 2 ft. and the width by 1 ft., the area would be diminished by 16 sq. ft. Find the dimensions of the rectangle.

28. Solve for x and y and check:

$$(p-q)x + (p-q)y = p^2 - q^2, \quad px + qy = 2\,pq.$$

29. In a certain school the number of pupils in the different grades were as follows: 1st, 62; 2d, 83; 3d, 76; 4th, 93; 5th, 87; 6th, 82; 7th, 75; 8th, 61. Make a bar graph of these numbers.

30. Simplify $\dfrac{x^2y + xy^2}{x^3y - xy^3} \div \dfrac{5\,xy(x+y)^2}{x^4 - y^4}$ and check the work by letting $x=2$ and $y=1$.

CHAPTER XIII

SQUARE ROOT

PART I

120. Square Root of a Single Term. In arithmetic you learned that the square root of a number is that number which when multiplied by itself produces the given number. Thus, the square root of 9 is 3, since $3 \times 3 = 9$. What is the square root of 25? Of 49?

Similarly in algebra, the square root of a term is that expression which when multiplied by itself produces the given term. Thus, the square root of $9\,a^2$ is $3\,a$ since $(3\,a)(3\,a) = 9\,a^2$. In like manner the square root of $16\,x^4y^6$ is $4\,x^2y^3$.

121. Cube Root of a Term. In arithmetic you learned that the cube root of 8 is 2 since $2 \times 2 \times 2 = 8$. What is the cube root of 27? Of 64?

Similarly in algebra, the cube root of $8\,a^3$ is $2\,a$, since $2\,a \cdot 2\,a \cdot 2\,a$ equals $8\,a^3$. What is the fourth root of 16? Of 81?

EXERCISE 138

ORAL

At sight give the value of each of the following :

1. $\sqrt{4} + \sqrt{25}$.
2. $\sqrt{64} - \sqrt{9}$.
3. $2\sqrt{9}$.
4. $3\sqrt{4} + \sqrt{25}$.
5. $4\sqrt{9} - \sqrt{16}$.
6. $7\sqrt{16} - 2\sqrt{9}$.

7. $\frac{1}{2}\sqrt{16} - \sqrt{4}$.
8. $8\sqrt{9} - 3\sqrt{25}$.
9. $2 + \sqrt{9}$.
10. $10 - \sqrt{16}$.
11. $12 - 3\sqrt{4}$.
12. $\sqrt{25} + \sqrt{4}$.

13. $\sqrt{64} \div \sqrt{4}$.
14. $\sqrt{4 \times 25}$.
15. $\sqrt{9 \times 16}$.
16. $\sqrt[3]{8} \times \sqrt[3]{27}$.
17. $\sqrt[3]{64} - \sqrt[3]{27}$.
18. $2\sqrt[3]{8} - \sqrt[3]{64}$.

19. $\sqrt{9} \div \sqrt[3]{-8}$.

20. $\sqrt{25} + \sqrt{9} - \sqrt{4}$.

21. $\sqrt{100} - \sqrt{64} + \sqrt{9}$.

22. $\sqrt{16} + 9 + \sqrt{4}$.

23. $\sqrt[3]{64} - \sqrt{9}$.

24. $\sqrt[3]{125} + \sqrt[3]{8} - \sqrt[3]{-27}$.

25. $\sqrt{64} - \sqrt[3]{27} + \sqrt{9}$.

26. $\sqrt{25} + \sqrt{36} - \sqrt[3]{27}$.

27. $9 \times \sqrt[3]{27}$.

28. $9 \div \sqrt[3]{27}$.

29. $\sqrt{9 + 16} + \sqrt{36}$.

30. $\sqrt{36 + 64} + 2\sqrt[3]{27} + \sqrt[3]{-8}$.

31. $\dfrac{2 + 4\sqrt{25}}{\sqrt{4}}$.

32. $\dfrac{3\sqrt{64} - \sqrt{9}}{3}$.

State the value

33. Of $\sqrt{a} + \sqrt{b}$ when $a = 16$ and $b = 9$.

34. Of $\sqrt{a} - \sqrt{b}$ when $a = 25$ and $b = 9$.

35. Of $3\sqrt{a} - 2\sqrt{b}$ when $a = 9$ and $b = 4$.

36. Of $\sqrt{2\,gs}$ when $g = 32$ and $s = \frac{1}{4}$.

37. Does $\sqrt{9 + 16}$ equal $\sqrt{9} + \sqrt{16}$? What is the correct value of each of these expressions?

38. Does $\sqrt{25 - 16}$ equal $\sqrt{25} - \sqrt{16}$? Give a reason for your answer.

122. The **Square Root of an Algebraic Expression** is that expression which when multiplied by itself produces the given expression. Thus the square root of $a^2 + 4\,ax + 4\,x^2$ is $a + 2\,x$.

In order to determine a general method for finding the square root of an algebraic expression containing several terms, we first consider the relation between $a + b$ and its square. This relation stated in inverse form gives us the required method. The essence of the method consists in writing $a^2 + 2\,ab + b^2$ in the form $a^2 + b(2\,a + b)$.

Ex. Extract the square root of $16\,x^2 - 24\,xy + 9\,y^2$.

$$
\begin{array}{r|l}
16\,x^2 - 24\,xy + 9\,y^2 & \underline{4\,x - 3\,y}. \quad \textit{Root.} \\
16\,x^2 & \\
\hline
8\,x - 3\,y \;\big|\; -24\,xy + 9\,y^2 & \\
 -24\,xy + 9\,y^2 &
\end{array}
$$

Taking the square root of the first term, $16\,x^2$, we obtain $4\,x$, which is placed to the right of the given expression as the first term of the root. Subtract the square of $4\,x$ from the given polynomial.

Taking twice the first term of the root, $8\,x$, as a trial divisor, and dividing it into the first term of the remainder, we obtain the second term of the root, $-3\,y$. This is annexed to the first term of the root and also to the trial divisor to make the complete divisor, $8\,x - 3\,y$.

123. Square Root to Three or More Terms. In squaring $a + b + c$, we may regard $a + b$ as a single quantity, and denote it by a symbol, as p, and obtain the square in the form $p^2 + 2\,pc + c^2$.

Evidently we may reverse this process, and extract a square root to three terms by regarding two terms of the root, when found, as a single quantity. So a fourth term of a root, or any number of terms, may be obtained by regarding the root already found as a single quantity.

Ex.   Extract the square root of $x^4 - 6\,x^3 + 19\,x^2 - 30\,x + 25$.

$$
\begin{array}{l|l}
x^4 - 6\,x^3 + 19\,x^2 - 30\,x + 25 & \underline{x^2 - 3\,x + 5.}\quad Root. \\
x^4 & \\
\hline
2\,x^2 - 3\,x \quad | -6\,x^3 + 19\,x^2 & \\
\quad\quad | -6\,x^3 + 9\,x^2 & \\
\hline
2\,x^2 - 6\,x + 5 \quad | +10\,x^2 - 30\,x + 25 & \\
\quad\quad | +10\,x^2 - 30\,x + 25 & \\
\end{array}
$$

The first two terms of the root, $x^2 - 3\,x$, are found as in the example in § 122.

To continue the process, we consider the root already found, $x^2 - 3x$, as a single quantity, and multiply it by 2 to make it a trial divisor.

Dividing the first term of the remainder, $10\,x^2$, by the first term of the trial divisor, $+2\,x^2$, we obtain the next term of the root, $+5$.

The process is then continued as before.

The work may be checked by squaring the result obtained, or by numerical substitution.

EXERCISE 139

Extract the square root of the following and check:

1. $9\,a^2 + 6\,ab + b^2$.
4. $t^2 + 2\,tu + u^2$.

2. $4\,x^2 + 20\,xy + 25\,y^2$.
5. $x^4 - 4\,x^3 + 6\,x^2 - 4\,x + 1$.

3. $81\,a^2b^2 + 90\,abc + 25\,c^2$.
6. $1 - 2\,a - a^2 + 2\,a^3 + a^4$.

7. $9\,x^4 - 12\,x^3 + 10\,x^2 - 4\,x + 1$.

8. $25 + 30\,x + 19\,x^2 + 6\,x^3 + x^4$.

9. $n^6 - 4\,n^5 + 4\,n^4 + 6\,n^3 - 12\,n^2 + 9$.

10. $4\,x^6 + 12\,x^5 + x^4 - 24\,x^3 - 14\,x^2 + 12\,x + 9$.

11. $1 + 16\,m^6 - 40\,m^4 + 10\,m - 8\,m^3 + 25\,m^2$.

12. $46\,n^2 + 25\,n^4 + 4\,n^6 + 25 - 44\,n^3 - 40\,n - 12\,n^5$.

13. $9\,x^6 + 9\,y^6 + 24\,x^5y + 24\,xy^5 - 8\,x^4y^2 - 8\,x^2y^4 - 50\,x^3y^3$.

14. $1 + 5\,x^2 + 2\,x^4 + x^6 - 4\,x^5 + 2\,x^3 + 2\,x$.

15. $28\,x^2 - 47\,x^4 + 49\,x^6 - 42\,x^5 - 4\,x^3 + 16\,x + 4$.

124. Square Root of Arithmetical Numbers. The same general method as that used in § 123 may be used to extract the square root of arithmetical numbers.

The details of the method of extracting the square root of numbers are explained in arithmetic. As illustrations of the process, we give the following examples:

Ex. 1. Extract the square root of 1849

```
            1849 |40 + 3                    1849 |43  Root
   40² =    1600                            16
2 × 40 = 80 |249      or more briefly,   83 |249
         3  |249                            |249
           ———
            83
```

Let the pupil check the work.

Ex. 2. Extract the square root of 18.550249.

$$\overline{18.55}\overline{02}\overline{49}\,|4.307 \ \textit{Root.}$$

$$\begin{array}{r} 16 \\ \hline 83\,|2\,55 \\ 2\,49 \\ \hline 8607\,|\,60249 \\ 60249 \end{array}$$

Let the pupil check the work.

Ex. 3. Extract the square root of $\frac{2}{3}$ to the nearest thousandth.

$$\tfrac{2}{3} = .66666666^{+}$$

Extracting the square root of $.66666666^{+}$, we obtain $.816^{+}$ as the desired result.

EXERCISE 140

Find the square root and check where possible :

1. 7225.
2. 2601.
3. 105,625.
4. 182,329.
5. 337,561.
6. 567,009.
7. 36,144,144.
8. 8114.4064.
9. 199.204996.
10. 10.30731025.
11. .0291419041.
12. 1,513,689.763041.

Find to the nearest thousandth the square root of

13. 7.
14. 11.
15. 12.5.
16. 12.273.
17. $3\frac{1}{3}$.
18. $2\frac{1}{5}$.
19. .9.
20. 5.192.
21. $6\frac{2}{3}$.
22. $1\frac{5}{8}$.
23. $\frac{7}{60}$.
24. 482.
25. .049.
26. 1.0064.
27. $36\frac{1}{11}$.
28. 4.967.

EXERCISE 141

ORAL

1. What expression is obtained by dividing the product of s and t by the sum of u and v ?

2. How many tiles each a by b inches will be required to cover a floor m feet by n feet ?

3. A train travels 30 miles an hour. How far will it go in h hours ? In m minutes ?

4. Find the expression for x per cent of x.

5. The side of a square is $2\,a$. Find the perimeter. The area.

6. Express the sum of the squares of two consecutive numbers, the smaller of which is x.

7. If Arthur has m marbles, how many more must he buy in order to have x marbles ?

8. A man has a capital of c dollars. He invests $\frac{1}{4}$ of it at p per cent and the rest at 5 per cent. What is his income ?

9. Find the cost of a pecks, b quarts of potatoes at $\$2$ a bushel.

10. A picture is 6×8 in. The frame is i inches wide. Express the outside dimensions of the frame.

EXERCISE 142

1. The adjoining triangle is a right triangle. What is the special name of c, the side opposite the right angle ?

In arithmetic you learned the following rule concerning a right triangle :

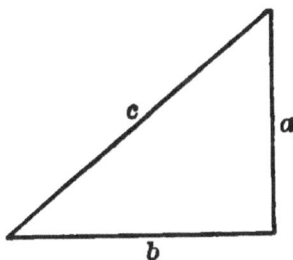

The square of the hypotenuse equals the sum of the squares of the other two sides.

Expressing this rule as a formula, we obtain, $c^2 = a^2 + b^2$, or $c = \sqrt{a^2 + b^2}$.

Use this formula in solving Exs. 2–6.

Find the hypotenuse of a right triangle in which the other two sides are :

2. 8 and 15. **4.** 112 and 15.

3. 28 in. and 45 in. **5.** 50 ft. and 60 ft.

6. If a city park is $\frac{1}{2}$ mi. long and 300 yd. wide, how much is saved by walking from a corner to the opposite corner along the diagonal instead of along the sides?

7. If $4 \pi R^2 = 10$, and $\pi = \frac{22}{7}$, find R.

8. On squared paper construct the triangle whose vertices are the points $(3, 0)$, $(1, 5)$, and $(-2, -3)$, and find the length of its sides to the nearest hundredth.

9. The area of Texas is 265,780 sq. mi. Find the side of a square having an equivalent area. Think of some distance familiar to you which is approximately equal to a side of this square, then picture to yourself the whole square and thus visualize the area of Texas.

10. By the method of Ex. 9, visualize the area of the state in which you live.

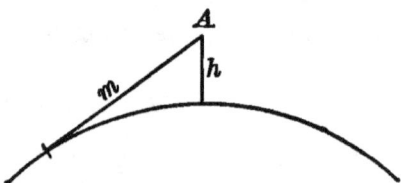

If a body is at A, h feet above the surface of the earth, the number of miles m, at which it can be seen, is limited by the curvature of the earth. This distance is determined by the formula $m = \sqrt{\dfrac{3\,h}{2}}$. Use this formula in solving Exs. 11–13.

11. A light on the mast of a certain lightship is 30 ft. above the water. What is the greatest distance in miles at which the light can be seen?

12. The light of a certain lighthouse is 180 ft. above the sea level. At a distance of how many miles can it be seen?

13. Taking the periscope of a submarine as on the sea level, what is the greatest distance at which it can be seen by an observer who is on a vessel and 18 ft. above the surface of the water?

PART II

125. Examples in Extracting Square Root.

Ex. 1. Extract the square root of $a^2 + 4ab + 4b^2 + 6a + 12b + 9$.

$$a^2 + 4ab + 4b^2 + 6a + 12b + 9 \underline{|\ a + 2b + 3}\ \ Ans.$$

$$\underline{a^2}$$

$$
\begin{array}{r|l}
2a+2b & 4ab+4b^2 \\
 & \underline{4ab+4b^2} \\
\end{array}
$$

$$
\begin{array}{r|l}
2a+4b+3 & 6a+12b+9 \\
 & \underline{6a+12b+9} \\
\end{array}
$$

Check the work by multiplying $a + 2b + 3$ by itself.

Ex. 2. Compute the value of $\sqrt{5 + 2\sqrt{3}}$ to the nearest thousandth.

The outline of the computation is as follows:

Compute the $\sqrt{3}$ to six decimal places and find it 1.732050^+.

Then $\sqrt{5 + 2\sqrt{3}} = \sqrt{5 + 2(1.732050^+)} = \sqrt{8.464100^+}$.

Computing the square root of 8.464100^+, we find 2.909^+. *Ans.*

EXERCISE 143

Extract the square root of the following and check :

1. $a^2 + 2ab + b^2 + 2a + 2b + 1$.

2. $p^2 + 6pq + 9q^2 + 6p + 18q + 9$.

3. $x^2 - 4xy + 4y^2 - 6x + 12y + 9$.

4. $4a^2 + 4ax + x^2 - 16a - 8x + 16$.

5. $m^2 + 9 + x^2 + 6m + 6x + 2mx$.

6. $x^2 + px + \dfrac{p^2}{4}$.

7. $x^2 - \frac{1}{2}xy + \frac{1}{16}y^2$.

8. $25 - 5x + \frac{1}{4}x^2$.

9. $a^4 + 2a^3 + 2a^2 + a + \frac{1}{4}$.

10. $x^4 - 2x^3 + 2x^2 - x + \frac{1}{4}$.

11. $x^4 + 2x^3 - x + \frac{1}{4}$.

12. $x^4 + 2x^3y - xy^3 + \dfrac{y^4}{4}$.

13. $b^4 + 4b^3 + \frac{5}{2}b^2 - 3b + \frac{9}{16}$.

14. $36 - 4a + \frac{55}{9}a^2 - \frac{1}{3}a^3 + \frac{1}{4}a^4$.

Compute to three decimal places the value of :

15. $\dfrac{7 + \sqrt{5}}{2}$.

17. $\dfrac{17 - \sqrt{7}}{4}$.

19. $\dfrac{-1 - \sqrt{7}}{2}$.

16. $\dfrac{7 - \sqrt{3}}{2}$.

18. $\dfrac{-3 + \sqrt{11}}{2}$.

20. $\sqrt{2 + \sqrt{3}}$.

21. $\sqrt{\sqrt{13} - \sqrt{3}}$.

22. $\sqrt{3\sqrt{3} + \sqrt{5}}$.

24. $\sqrt{\dfrac{\sqrt{5} - \sqrt{2}}{4}}$.

23. $\sqrt{3\sqrt{6} - 2\sqrt{7}}$.

25. $\sqrt{\dfrac{5(\sqrt{3} - \sqrt{2})}{2}}$.

26. By the aid of factoring compute in a short way the numerical value of $\sqrt{.62^2 - .57^2}$ to the nearest thousandth.

Sug. $\quad \sqrt{.62^2 - .57^2} = \sqrt{(.62 + .57)(.62 - .57)}$
$$= \sqrt{1.19 \times .05}$$
$$= \sqrt{.0595}$$
$$= .244^-. \quad \textit{Ans.}$$

In a short way compute to the nearest thousandth the numerical value of :

27. $\sqrt{.62^2 - .38^2}$.

30. $\sqrt{19.84^2 - 5.36^2}$.

28. $\sqrt{12.72^2 - 1.28^2}$.

31. $\sqrt{8.88^2 - 3.18^2}$.

29. $\sqrt{13.36^2 - 2.64^2}$.

32. $\sqrt{7.95^2 - 6.27^2}$.

33. Find the value of the expression $a\sqrt{x^2 - y^2}$ when $a = 7.2$, $x = 4.8$, and $y = 3$, carrying the result to the nearest tenth.

34. Find the value of $\sqrt{a^2b + c^3d}$ when $a = 7$, $b = 27$, $c = 25$, and $d = 12$.

EXERCISE 144
ORAL

At sight give the value of :

1. $\sqrt{9} + \sqrt[3]{27} + \sqrt[4]{81}$.

5. $\dfrac{\sqrt{16} + \sqrt[4]{16}}{2}$.

2. $\sqrt[4]{16} - 2\sqrt[3]{8} + \sqrt{16}$.

3. $\sqrt{36} + 2\sqrt[4]{16} - 2\sqrt[3]{8}$.

6. $\dfrac{2\sqrt{16} - \sqrt[4]{81}}{2}$.

4. $\frac{1}{2}\sqrt{64} - \frac{1}{4}\sqrt[3]{27} + 5$.

7. $\sqrt[4]{256} - \sqrt[3]{27}$.

Give the value of $\frac{1}{4} (\sqrt{a} + \sqrt[3]{b}.)$

8. When $a = 9$, $b = 27$. 　　**10.** When $a = 1$, $b = 64$.

9. When $a = 16$, $b = 8$. 　　**11.** When $a = 36$, $b = 1$.

12. Give the value of $\sqrt{2\,gs}$ when $g = 32$ and $s = 1$.

13. Of $\sqrt{2\,gs}$ when $g = 64$ and $s = \frac{1}{8}$.

EXERCISE 146

1. In a right triangle whose hypotenuse is $c \cdot$ and whose other sides are a and b, $a^2 + b^2 = c^2$.

From this formula, obtain a formula for a in terms of b and c. Express this formula as a rule. By use of this formula solve Exs. 2–8.

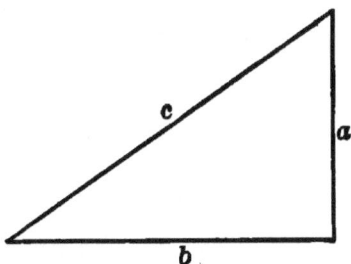

In a right triangle, to the nearest hundredth,

2. Find a when $c = 17$ and $b = 15$.

3. Find a when $c = 12.9$ and $b = 9.1$.

4. Find a when $c = 13.87$ and $b = 10.93$.

5. Find a when $c = 12.79$ and $b = 6.48$.

6. When the hypotenuse is 25 and one side is 24, find the other side.

7. When the hypotenuse is 12.68 and one side is 6.94, find the other side.

8. When one side of a rectangle is 50 ft. and the diagonal is 130 ft., find the other side.

When the three sides of a triangle are denoted by a, b, and c and $s = \frac{1}{2}(a + b + c)$, and the area of the triangle is denoted by K,

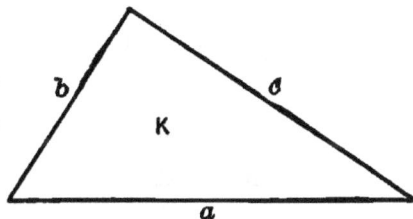

$$K = \sqrt{s(s - a)(s - b)(s - c)}.$$

Use this formula in solving Ex. 9–13.

Find to the nearest hundredth the area of the triangle whose sides are

9. 11, 12, and 13.   **10.** 25, 63, 74.

11. During the Great War, the sides of an army camp were 5, 4, and 7 miles. Find the area of the camp in square miles and also in acres.

12. The sides of another camp were 4.7, 3.8, 3.5 miles. Find its area in square miles.

13. Another camp was a quadrilateral, one of whose diagonals was 7 miles. If the two sides on one side of the diagonal were 5 and 4 miles, and the two sides on the other side of the diagonal were 3 and 6 miles, find the area of the camp in square miles.

The velocity (v) attained by a body falling through s feet, is given by the formula $v = \sqrt{2\,gs}$ where $g = 32.16$ ft. Use this formula in solving Exs. 14, 15.

14. A baseball dropped from the top of the Washington Monument has been caught by a player standing at the foot of the monument. The height of the monument is 555 ft. Find the velocity of the ball when caught.

15. During the war a steel dart was dropped from an airplane a mile above the earth's surface. Find the velocity with which it struck the earth.

16. In the formula $t = \pi\sqrt{\dfrac{l}{g}}$, find the value of t to the nearest hundredth when $l = 39$ and $g = 32.2$.

17. In $D = \sqrt{\dfrac{G}{.34\,LN}}$, when $G = 250$, $L = 17$, and $N = 8$, find the value of D to the nearest hundredth.

CHAPTER XIV

RADICALS

PART I

126. A **Root** of a quantity is a quantity which, taken as a factor a given number of times, will produce the given quantity.

127. A **Radical** is a root of a quantity indicated by the use of the radical sign; as $\sqrt{x}$, $\sqrt[3]{27}$.

The **radicand** is the quantity under the radical sign.

Since $2 \times 2 = 4$, and also $(-2)(-2) = 4$, $\sqrt{4}$ has two values, viz., $+2$ and -2.

128. The **Principal Root** of a number is that real root of the number which has the same sign as the number itself.

Thus, the principal root for $\sqrt{4}$ is 2; so the principal root for $\sqrt[3]{27}$ is 3; for $\sqrt[3]{-27}$ is -3.

In this chapter only principal roots of numbers are considered.

129. Simplification of a Quantity under the Radical Sign.

Ex. 1.   Simplify $\sqrt[3]{56}$.
$$\sqrt[3]{56} = \sqrt[3]{8 \times 7} = 2\sqrt[3]{7}. \quad Ans.$$

Ex. 2.   Simplify $5\sqrt{18\ a^3 b^2 c^5}$.
$$5\sqrt{18\ a^3 b^2 c^5} = 5\sqrt{9\ a^2 b^2 c^4 \times 2\ ac} = 15\ abc^2\sqrt{2\ ac}. \quad Ans.$$

Hence, in general,

Separate the quantity under the radical sign into two factors, one of which is the greatest perfect power of the same degree as the radical;

Extract the required root of this factor, and multiply the coefficient of the radical by the result:

The other factor remains under the radical sign.

271

EXERCISE 146

ORAL

At sight simplify:

1. $\sqrt{8}$.

2. $\sqrt{12}$.

3. $\sqrt{27}$.

4. $\sqrt{50}$.

5. $\sqrt{a^3}$.

6. $\sqrt{4\,a}$.

7. $\sqrt{9\,x}$.

8. $\sqrt{16\,b}$.

9. $\sqrt{x^5}$.

10. $2\sqrt{y^3}$.

11. $2\sqrt{8}$.

12. $\frac{1}{2}\sqrt{8}$.

13. $\frac{1}{3}\sqrt{27}$.

14. $\frac{2}{5}\sqrt{50}$.

15. $3\sqrt{4\,x}$.

16. $2\sqrt{a^5}$.

17. $\frac{2}{3}\sqrt{12}$.

18. $\sqrt{4\,r^3}$.

19. $\frac{1}{2}\sqrt{3\,r^2}$.

20. $\sqrt{\frac{5}{4}}$.

21. $\sqrt{\frac{r}{4}}$.

22. $\sqrt[3]{16}$.

23. $\frac{1}{2}\sqrt[3]{24}$.

24. $\sqrt[3]{54}$.

25. $\sqrt{32}$.

26. $\sqrt[3]{32}$.

27. $\frac{1}{3}\sqrt{9\,x}$.

28. $\frac{2}{3}\sqrt{\frac{7}{4}}$.

EXERCISE 147

Express in the simplest form :

1. $3\sqrt{24}$.

2. $\frac{1}{9}\sqrt{54}$.

3. $-\sqrt{40}$.

4. $-3\sqrt{28}$.

5. $\frac{2}{3}\sqrt{45}$.

6. $\frac{4}{5}\sqrt{50}$.

7. $\sqrt[3]{48}$.

8. $\sqrt[3]{24}$.

9. $\sqrt[3]{54}$.

10. $\frac{1}{2}\sqrt[3]{72}$.

11. $-\frac{5}{3}\sqrt{108}$.

12. $\sqrt[4]{48}$.

13. $\sqrt[4]{128\,a^5x^2}$.

14. $\sqrt[3]{250\,a^2b^6}$.

15. $\sqrt{99\,a}$.

16. $2\sqrt{4\,a^5x^2}$.

17. $a\sqrt{8\,a^3x^2}$.

18. $\sqrt{200\,a^7}$.

19. $\sqrt{147\,x^5y^6}$.

20. $-2\sqrt{63\,x^{15}y^{10}}$.

21. $\sqrt[3]{-81\,a^3x^5}$.

22. $\sqrt{a^2(x-y)^3}$.

23. $\sqrt{49x^3(a+1)^5}$.

24. $10\sqrt{\dfrac{12\,a^3c^4n}{25\,x^4}}$.

25. $\frac{3}{4}\sqrt{\dfrac{112\,x^5z^{11}}{9\,a^2}}$.

26. $\sqrt{9\,a^3bc^4d^5}$.

27. $\sqrt{8\,a^2x^3y^5}$.

28. $\sqrt{24\,a^3x^5y^2}$.

29. $\sqrt{(a+b)^2x}$.

30. $\sqrt{(a-2\,b)^2y}$.

31. $\sqrt{8(a+b)^2x}$.

32. $\sqrt{(x+y)^2(x-y)}$.

33. $\frac{1}{3}\sqrt{72(a+b)^2x}$.

34. $\sqrt{16(x+y)(x^2-y^2)}$.

35. $\sqrt{2\,x^2-4\,xy+2\,y^2}$.

36. Given $\sqrt{3}=1.73205^+$, find in the shortest way the value of $\sqrt{12}$. Of $\sqrt{27}$. Of $\sqrt{75}$. Of $\sqrt{243}$.

37. Using $\sqrt{5}=2.23607^+$, make up and work an example similar to Ex. 36.

130. Similar Radicals are those which have the same quantity under the radical sign and the same index, that is, number indicating the root to be extracted. (The coefficients and signs of the radicals may be unlike. Hence, similar radicals must be alike in two respects, and may be unlike in two other respects.)

Thus, $5\sqrt{3}$, $-4\sqrt{3}$ are similar radicals.

131. The Addition of Similar Radicals is performed like the addition of similar terms, by *taking the algebraic sum of the coefficients of the terms.*

The addition of **dissimilar radicals** *can only be indicated.*

Ex. 1.   Add $\sqrt{128} - 2\sqrt{50} + \sqrt{72} - \sqrt{18}$.

$$\sqrt{128} - 2\sqrt{50} + \sqrt{72} - \sqrt{18} = 8\sqrt{2} - 10\sqrt{2} + 6\sqrt{2} - 3\sqrt{2}$$
$$= \sqrt{2}. \quad Sum.$$

Ex. 2.   Simplify $5\sqrt[3]{8x} - 2\sqrt[3]{27x} + 8\sqrt[3]{x}$.

$$5\sqrt[3]{8x} - 2\sqrt[3]{27x} + 8\sqrt[3]{x}$$
$$= 10\sqrt[3]{x} - 6\sqrt[3]{x} + 8\sqrt[3]{x}.$$
$$= 12\sqrt[3]{x}. \quad Ans.$$

EXERCISE 148

ORAL

Simplify at sight:

1. $2\sqrt{3} + 5\sqrt{3}$.
2. $6\sqrt{5} - 2\sqrt{5}$.
3. $5\sqrt{3} + \sqrt{12}$.
4. $8\sqrt{7} - \sqrt{28}$.
5. $2\sqrt{12} - \sqrt{3}$.
6. $6\sqrt{48} - 7\sqrt{3}$.
7. $5\sqrt{7} - 3\sqrt{28}$.
8. $3\sqrt{27} - \sqrt{12}$.
9. $\sqrt{50} + 2\sqrt{8}$.
10. $\sqrt[3]{16} + \sqrt[3]{54}$.
11. $2\sqrt{9a} - \sqrt{16a}$.
12. $12\sqrt{4x} + 2\sqrt{25x}$.
13. $3\sqrt{2x^2} + 5x\sqrt{8}$.
14. $8\sqrt{2} - \sqrt{50} + \sqrt{8}$.
15. $\sqrt{3} - 2\sqrt{12} - \sqrt{27}$.
16. $4\sqrt{9x} - \sqrt{4x} - 2\sqrt{25x}$.
17. $\sqrt{a^2b} + 2a\sqrt{b} + \sqrt{9a^2b}$.
18. $\sqrt[3]{16} + 5\sqrt[3]{2} - 7\sqrt[3]{54}$.

Simplify:

1. $\sqrt{18} + 7\sqrt{8}.$

2. $\sqrt{50} - \sqrt{32}.$

3. $2\sqrt{27} + \sqrt{75}.$

4. $\sqrt{49\,x} - 3\sqrt{16\,x}.$

5. $\sqrt{8\,x^3} + \sqrt{50\,x^3}.$

6. $\sqrt{a^3b} + a\sqrt{ab^3}.$

7. $\sqrt{x^5y} - \sqrt{xy^5}.$

8. $\sqrt{50\,a^3} + a\sqrt{98}.$

9. $4\sqrt[3]{16} - 2\sqrt[3]{54}.$

10. $5\sqrt[3]{24} - 2\sqrt[3]{81}.$

11. $8\sqrt[3]{a} + 5\sqrt[3]{8\,a}.$

12. $\frac{2}{3}\sqrt{27} - \sqrt{12}.$

13. $3\sqrt{5} + \sqrt{20} + \sqrt{45}.$

14. $3\sqrt{18} - \sqrt{8} + 2\sqrt{32}.$

15. $\frac{1}{2}\sqrt{12} - \sqrt{243} + 2\sqrt{192}.$

16. $2\sqrt{20} - \frac{3}{4}\sqrt{80} + \sqrt{45}.$

17. $\sqrt{18\,x^2y} - 2\sqrt{2\,x^4y^3} + \sqrt{8\,y^5}.$

18. $3\sqrt{a^3bc} + 5\sqrt{ab^3c} - \sqrt{abc^3}.$

19. $\sqrt[3]{24} + \sqrt[3]{81} - \sqrt[3]{16}.$

20. $\sqrt{25\,x^3} - 7\,x\sqrt{64\,x} + 4\sqrt{x^3}.$

21. $2\sqrt{25\,b} + 3\sqrt{4\,b} - 2\sqrt{36\,b}.$

22. $3\sqrt[3]{2\,c} + 3\sqrt[3]{54\,c} - \sqrt[3]{16\,c}.$

23. $\sqrt{12\,ab^2} + b\sqrt{48\,a} - 6\sqrt{3\,ab^2}.$

24. $\sqrt{a^2bc^2} - 3\,a\sqrt{16\,bc^2} + 5\,c\sqrt{9\,a^2b}.$

25. $b\sqrt[3]{2\,a} + \sqrt[3]{250\,ab^3} - 2\,b\sqrt[3]{27\,a}.$

26. $\sqrt{2} + \sqrt{18} - \sqrt{50} + \sqrt{162}.$

27. $\sqrt{75} - \sqrt{243} + 2\sqrt{108}.$

28. $\frac{1}{3}\sqrt{27} - \sqrt{18} + \sqrt{300} - \sqrt{162} + \sqrt{8}.$

29. $3\sqrt[3]{16} - 3\sqrt{12} + 2\sqrt[3]{54} - 5\sqrt{27}.$

30. $7\,a\sqrt{12\,ab^2} - 9\,b\sqrt{27\,a^3} + 5\sqrt{300\,a^3b^2}.$

31. How many more symbols are used in Ex. 28 as it is given than in your answer?

32. Compute the value of $\sqrt{128} - 2\sqrt{50} + \sqrt{72}$ by extracting the square root of each radical and combining

results. Now obtain the value of the whole expression by first reducing each radical to its simplest form and then collecting. Compare the labor in the two processes.

33. Make up and work an example similar to Ex. 32.

132. Multiplication of Radicals. As a simple instance of the multiplication of radicals we have

$$\sqrt{9} \times \sqrt{4} = 3 \times 2 = 6.$$

Also $\sqrt{9 \times 4} = \sqrt{36} = 6.$

Hence $\sqrt{9} \times \sqrt{4} = \sqrt{9 \times 4}$, since each of these equals 6. So in general it can be shown that $\sqrt{a} \cdot \sqrt{b} = \sqrt{ab}.$

Hence, in general, to multiply two indicated square roots.

Multiply the coefficients for a new coefficient;
Multiply the quantities under the radical sign together for a new quantity under the radical sign;
Simplify the result.

Ex. 1. Multiply $5\sqrt{6}$ by $2\sqrt{3}$.
$$5\sqrt{6} \times 2\sqrt{3} = 10\sqrt{18} = 30\sqrt{2}. \quad Product.$$

Ex. 2. $(3\sqrt{8x})(5\sqrt{2x}) = 15\sqrt{16x^2}$
$$= 60x. \quad Ans.$$

EXERCISE 150

ORAL

At sight multiply and simplify:

1. $\sqrt{3} \cdot \sqrt{2}$.	7. $\sqrt{2x} \cdot \sqrt{3x}$.	13. $3\sqrt{2} \cdot \sqrt{2}$.
2. $\sqrt{5} \cdot \sqrt{3}$.	8. $\sqrt{2a^2} \cdot \sqrt{3}$.	14. $5\sqrt{3} \cdot \sqrt{3}$.
3. $\sqrt{2} \cdot \sqrt{2}$.	9. $2\sqrt{3} \times \sqrt{2}$.	15. $2\sqrt[3]{9} \cdot \sqrt[3]{3}$.
4. $\sqrt{a} \times \sqrt{2a}$.	10. $(\frac{1}{2}\sqrt{2})(\sqrt{8})$.	16. $(\frac{1}{3}\sqrt{3})(\frac{1}{2}\sqrt{3}.)$
5. $\sqrt{2} \times \sqrt{18}$.	11. $(\sqrt{3x})(\sqrt{x^3})$.	17. $\sqrt[3]{x} \cdot \sqrt[3]{x^2}$.
6. $\sqrt[3]{2} \cdot \sqrt[3]{4}$.	12. $(\sqrt{3})(2\sqrt{2})$.	18. $\sqrt{2} \cdot \sqrt{3} \cdot \sqrt{5}$.

EXERCISE 151

Multiply and simplify :

1. $3\sqrt{5} \cdot \sqrt{15}$. 4. $5\sqrt{12} \cdot \sqrt{30}$. 7. $\sqrt{2x} \cdot \sqrt{3x^5}$.
2. $\sqrt{27} \cdot \sqrt{12}$. 5. $4\sqrt{30} \cdot \sqrt{\tfrac{4}{5}}$. 8. $\sqrt{3y} \cdot \sqrt{12y}$.
3. $\sqrt{a^2 b} \cdot \sqrt{ab^3}$. 6. $6\sqrt{ab} \cdot \sqrt{a^3}$. 9. $\tfrac{1}{2}\sqrt[3]{x} \cdot \sqrt[3]{x^4}$.

10. $\tfrac{3}{8}\sqrt{ab} \cdot \tfrac{6}{5}\sqrt{6\,ab^2}$.   16. $\sqrt{12} \cdot \sqrt{6} \cdot \sqrt{20}$.
11. $(2\sqrt{15})(3\sqrt{35})$.   17. $\sqrt{\tfrac{2}{3}} \cdot \sqrt{\tfrac{9}{5}} \cdot \sqrt{\tfrac{10}{3}}$.
12. $(\tfrac{2}{3}\sqrt{28})(\tfrac{3}{4}\sqrt{35})$.   18. $\sqrt{ab} \cdot \sqrt{bc} \cdot \sqrt{ac}$.
13. $(\tfrac{1}{3}\sqrt{\tfrac{5}{6}})(\tfrac{2}{5}\sqrt{\tfrac{15}{32}})$.   19. $\sqrt{a^2 b} \cdot \sqrt{b^2 c} \cdot \sqrt{a^3 c}$.
14. $(\tfrac{1}{6}\sqrt[3]{6})(4\sqrt[3]{4})$.   20. $\sqrt{\tfrac{2}{3}} \cdot \sqrt{\tfrac{3}{5}} \cdot \sqrt{2\tfrac{1}{2}}$.
15. $3\sqrt{5} \times 2\sqrt{5} \times 2\sqrt{2}$.   21. $\sqrt{6\,a} \cdot \sqrt{\tfrac{4}{3}\,a} \cdot \sqrt{8}$.

22. Find the numerical value of $\sqrt{ab}$ when $a = 5$ and $b = 7$, carrying the result to the nearest hundredth.

EXERCISE 152

ORAL

At sight simplify the following miscellaneous examples :

1. $2\sqrt{8}$.   10. $3\sqrt{2} + 2\sqrt{8}$.   19. $\sqrt{3} \cdot \sqrt{6}$.
2. $\sqrt{6} \cdot \sqrt{2}$.   11. $4\sqrt{xy^3}$.   20. $\sqrt{3\,r^2} + r\sqrt{3}$.
3. $\tfrac{1}{3}\sqrt{27}$.   12. $5\sqrt{3} - \sqrt{27}$.   21. $3\sqrt{6} \cdot \sqrt{2}$.
4. $2\sqrt{a} \cdot \sqrt{a}$.   13. $\sqrt{a^2 x} + \sqrt{ax^2}$.   22. $\sqrt{2\,a^2} + a\sqrt{2}$.
5. $\sqrt{a^5}$.   14. $\sqrt{16} \cdot \sqrt{2}$.   23. $5\sqrt{8} - 2\sqrt{2}$.
6. $\sqrt{3} \cdot \sqrt{6}$.   15. $\sqrt{2\,a} \cdot \sqrt{8\,a}$.   24. $\sqrt{a^3} - a\sqrt{a}$.
7. $\sqrt{2} \cdot \sqrt{10}$.   16. $4 + 5\sqrt{9}$.   25. $\sqrt{8} \cdot \sqrt{\tfrac{1}{2}}$.
8. $\sqrt{12} - \sqrt{3}$.   17. $\sqrt{a-b} \cdot \sqrt{a-b}$.   26. $\tfrac{1}{4}\sqrt{4\,xy^3}$.
9. $\sqrt{a^2 b}$.   18. $\sqrt{12} + \sqrt{27}$.   27. $3\sqrt{2} \cdot 5\sqrt{2}$.

28. $\sqrt{4} + \sqrt{8} + \sqrt{9}$.   30. $\sqrt{6 \times 24} - \sqrt{45}$.
29. $\sqrt{16} - \sqrt[3]{8} + \sqrt{2}$.   31. $\sqrt{12} + \sqrt{27}$.

1. If one of the sides of a square is s, and the diagonal is d, obtain a formula for d in terms of s, and reduce it to the form $d = s\sqrt{2}$.

Use this formula in solving Exs. 2 and 3.

2. Find the diagonal of a square whose side is 5 in. Also of a square whose side is 12.5 in.

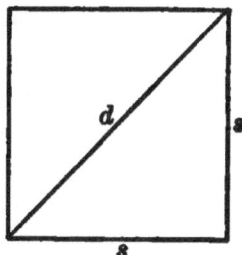

3. One side of a public square in a city is 320 yd. long. How much is saved by walking from a corner of it to the opposite corner along the diagonal, instead of along the sides?

4. Construct a graph of the equation $d = s\sqrt{2}$.

5. By means of this graph find the diagonal of a square whose side is 3 in. Then find the value of d by substituting $s = 3$ in the formula $d = s\sqrt{2}$. Compare the two results.

6. From the graph find the value of d when $s = 3.5$. Also from the formula find the value of d when $s = 3.5$. Compare the two results.

7. From the graph find s when $d = 4$. Find the same from the formula. Compare the two results.

8. From the graph find the side of a square whose diagonal is 6 in. Find the same from the formula and compare the two results.

9. From the graph find the side of a square whose diagonal is 3.25. Find the same from the formula. Compare the amount of work in the two processes. Which process gives the more accurate result?

10. On squared paper, by use of the compasses, con-struct a circle whose radius is four linear spaces, and whose center is a point where two lines cross. Obtain the area of this circle as accurately as you can by counting the number of square spaces inside the circle, regarding as a whole square each part of a square that is greater than a half square, and neglecting each part of a square that is less than a half.

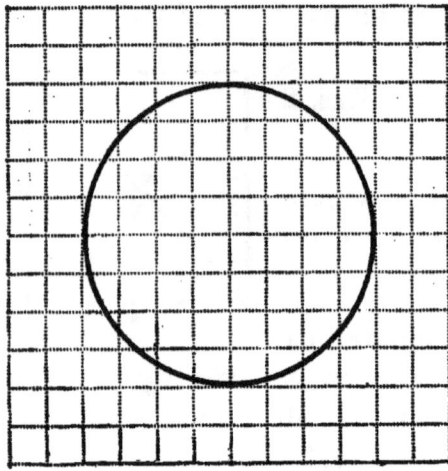

Then compute the area of the circle by the formula $a = \frac{22}{7} r^2$. By how much do the two results differ?

11. By use of the formula $a = \frac{22}{7} r^2$, find the area of a circle whose radius is 14 in.

12. Also of a circle whose diameter is 2.8 ft.

13. If a cow is tied to a stake by a tether, 20 yd. long, find by use of the formula over how many square yards she can graze. What fraction of an acre is this?

14. From the formula $a = \frac{22}{7} r^2$, obtain the formula $r = \sqrt{\frac{7a}{22}}$, that is $r = \sqrt{\frac{a}{\pi}}$.

Use this last formula in solving Exs. 15–17.

15. Find to the nearest hundredth of an inch, the radius of a circle whose area is 10 sq. in.

16. Find the radius of a circle whose area is 100 sq. yd. to two decimal places.

17. Find the length of the tether by which a cow must be tied in order that she may graze over exactly an acre.

PART II

133. Surds. An indicated root which may be exactly extracted is said to be *rational;* as $\sqrt[3]{27}$, since the cube root of 27 is 3.

A **surd** is an indicated root which cannot be exactly extracted; as $\sqrt{3}$, $\sqrt[3]{5}$.

134. The **Degree** of a radical is the number of the indicated root.

Thus, $\sqrt[3]{x}$ is a radical of the third degree.

135. Simplification of a Fraction under the Radical Sign. To simplify a radical when the quantity under the radical sign is a fraction,

Multiply both numerator and denominator of the fraction by such a quantity as will make the denominator a perfect power of the same degree as the radical;

Proceed as in § 129.

Ex. 1. Simplify $5\sqrt{\tfrac{2}{3}}$.

$$5\sqrt{\tfrac{2}{3}} = 5\sqrt{\tfrac{2}{3} \times \tfrac{3}{3}} = 5\sqrt{\tfrac{6}{9}}.$$

$$= 5\sqrt{\tfrac{1}{9} \times 6} = \tfrac{5}{3}\sqrt{6}. \quad Ans.$$

Ex. 2. Simplify $\sqrt[3]{\tfrac{40}{9}}$.

$$\sqrt[3]{\tfrac{40}{9}} = \sqrt[3]{\tfrac{40}{9} \times \tfrac{3}{3}} = \sqrt[3]{\tfrac{120}{27}} = \sqrt[3]{\tfrac{8}{27} \times 15} = \tfrac{2}{3}\sqrt[3]{15}. \quad Ans.$$

Ex. 3. Simplify $\sqrt{\dfrac{5\,ax^2}{18\,b}}$.

$$\sqrt{\frac{5\,ax^2}{18\,b}} = \sqrt{\frac{5\,ax^2}{18\,b} \times \frac{2\,b}{2\,b}} = \sqrt{\frac{10\,abx^2}{36\,b^2}}$$

$$= \sqrt{\frac{x^2}{36\,b^2} \times 10\,ab} = \frac{x}{6\,b}\sqrt{10\,ab}. \quad Ans.$$

EXERCISE 154

ORAL

Simplify at sight:

1. $\sqrt{\frac{1}{2}}$.   7. $\sqrt{\frac{4}{3}}$.   13. $\sqrt{\frac{a^2}{3}}$.   17. $\sqrt{\frac{2\,ab}{5}}$.

2. $\sqrt{\frac{1}{3}}$.   8. $\sqrt{\frac{4}{5}}$.   14. $\sqrt{\frac{r^2}{5}}$.   18. $\sqrt{\frac{9}{a}}$.

3. $\sqrt{\frac{1}{5}}$.   9. $\sqrt{\frac{4}{7}}$.

4. $\sqrt{\frac{1}{a}}$.   10. $\sqrt{\frac{9}{5}}$.   15. $5\sqrt{\frac{a^2}{5}}$.   19. $x\sqrt{\frac{4}{x}}$.

5. $\sqrt{\frac{2}{5}}$.   11. $3\sqrt{\frac{2}{3}}$.

6. $\sqrt{\frac{3}{5}}$.   12. $2\sqrt{\frac{2}{3}}$.   16. $\sqrt{\frac{xy}{3}}$.   20. $\sqrt{\frac{4\,r^2}{3}}$.

EXERCISE 155

Simplify:

1. $3\sqrt{\frac{5}{6}}$.   9. $\sqrt{\frac{5}{a^3}}$.   16. $\sqrt{\frac{xy}{12}}$.   22. $5\,a\sqrt[3]{\frac{8}{25}}$.

2. $\sqrt{\frac{25}{7}}$.   10. $\sqrt{\frac{5}{2\,b^3}}$.   17. $\sqrt{\frac{2\,ab}{5}}$.   23. $\frac{8}{3}\sqrt[4]{\frac{1}{2}}$.

3. $\sqrt{\frac{12}{5}}$.   11. $4\sqrt[3]{\frac{7}{4}}$.   18. $\sqrt{\frac{9}{2\,a}}$.   24. $3\sqrt[3]{\frac{3}{4}}$.

4. $\sqrt{\frac{24}{5}}$.   12. $5\sqrt[3]{\frac{8}{5}}$.   19. $x\sqrt{\frac{4}{x^3}}$.   25. $\sqrt{\frac{63\,a}{8\,x^2}}$.

5. $\sqrt{\frac{20}{3}}$.   13. $6\sqrt[3]{\frac{5}{6}}$.   20. $\sqrt{\frac{25}{24}}$.   26. $\frac{2\,a}{b}\sqrt{\frac{8\,b^2}{27\,a}}$.

6. $\sqrt{\frac{5}{12}}$.   14. $10\sqrt[3]{\frac{27}{2}}$.   21. $\frac{3}{2}\sqrt{\frac{2\,x}{3\,a}}$.   27. $\sqrt{\frac{5}{a+b}}$.

7. $\sqrt{\frac{5}{8}}$.   15. $\sqrt[3]{\frac{2\,b}{a^2}}$.

8. $\sqrt{\frac{7}{20}}$.

28. $\sqrt{\frac{3}{(x+1)^3}}$.   29. $\sqrt{4\frac{1}{6}}$.   30. $\sqrt{33\frac{1}{3}}$.

136. Examples in Simplifying and Adding Radicals.

Ex. 1. Simplify $3\sqrt{5}+4\sqrt{\frac{4}{5}}-3\sqrt{\frac{1}{5}}+3\sqrt{\frac{2}{3}}$.

$3\sqrt{5}+4\sqrt{\frac{4}{5}}-3\sqrt{\frac{1}{5}}+3\sqrt{\frac{2}{3}}=3\sqrt{5}+\frac{8}{5}\sqrt{5}-\frac{3}{5}\sqrt{5}+\sqrt{6}$
$$=4\sqrt{5}+\sqrt{6}.\ \ Ans.$$

Ex. 2. $(a - 2\,b)\sqrt{2} + \sqrt{2(2\,a + b)^2}.$
$$= (a - 2\,b)\sqrt{2} + (2\,a + b)\sqrt{2}$$
$$= (3\,a - b)\sqrt{2}. \quad \textit{Ans.}$$

EXERCISE 156

Simplify :

1. $\sqrt{\frac{2}{3}} + \sqrt{\frac{3}{2}}.$

2. $2\sqrt{\frac{3}{4}} + \sqrt{48}.$

3. $a\sqrt{\frac{x}{a}} + x\sqrt{\frac{a}{x}}.$

4. $6\sqrt{\frac{2}{3}} - 5\sqrt{24}.$

5. $3\sqrt{2} - (2\sqrt{8} - 5\sqrt{2},$

6. $\sqrt{27} - 3(\sqrt{3} - 5\sqrt{12}).$

7. $\sqrt{50} - 4(\sqrt{32} + \sqrt{12}).$

8. $\sqrt{20} - 2(\sqrt{45} - 6\sqrt{5}).$

9. $5\sqrt{\frac{3}{5}} - 12\sqrt{\frac{5}{3}} + 6\sqrt{60} - 30\sqrt{\frac{1}{15}}.$

10. $3\sqrt{5} - 10\sqrt{\frac{1}{5}} + 2\sqrt{45} - 5\sqrt{\frac{4.9}{5}}.$

11. $\sqrt{27} - \sqrt{18} + \sqrt{300} - \sqrt{162} + 6\sqrt{2} - 7\sqrt{3}.$

12. $2\sqrt{63} - 3\sqrt{\frac{1}{5}} - \sqrt{\frac{2}{7}} + \frac{1}{5}\sqrt{45} - \frac{4}{7}\sqrt{7}.$

13. $\sqrt{243} + \sqrt{48} - \sqrt{768} + 9\sqrt{\frac{1}{3}} + \sqrt{75} - 3\sqrt{33\frac{1}{3}}.$

14. $21\sqrt{\frac{2}{3}} - 5\sqrt{\frac{4}{5}} + 6\sqrt{4\frac{1}{6}} - 10\sqrt{3\frac{1}{5}} + \frac{40}{3}\sqrt{11\frac{1}{4}}.$

15. $5\,a\sqrt{12\,ab^2} - 3\,b\sqrt{27\,a^3} + 2\sqrt{300\,a^3b^2} - 40\,ab\sqrt{\frac{3}{4}\,a}.$

16. $3\sqrt[3]{16} - 3\sqrt{12} + 2\sqrt[3]{54} - 5\sqrt{27}.$

17. $3\sqrt[3]{54} - 2\sqrt{18} + 5\sqrt[3]{\frac{1}{4}} + 5\sqrt{\frac{1}{2}}.$

18. $2\sqrt{\frac{5}{3}} + \frac{1}{6}\sqrt{60} + \sqrt{15} + \sqrt{\frac{3}{5}}.$

19. $\sqrt[3]{128} + 2\sqrt[3]{\frac{1}{4}} - 3\sqrt[3]{81} + \sqrt[3]{54}.$

20. $3(2\,a + b)\sqrt{5} + \sqrt{5(2\,a + b)^2}.$

21. $3\sqrt{a^2x + 2\,abx + b^2x} + 2\sqrt{(a - b)^2x}.$

22. $8\sqrt{a(x + 2\,y)^2} - 3\sqrt{ax^2 + 4\,axy + 4\,ay^2}.$

23. $\sqrt{a^2b} + \sqrt{b^3} - \sqrt{(2\,a - 3\,b)^2b} - 2\,b^2\sqrt{\frac{1}{b}}.$

24. $\sqrt{5(a - b)^2} + \sqrt{20(a + b)^2} - a\sqrt{125}.$

25. Compute the numerical value of $5\sqrt{\frac{2}{3}} - 6\sqrt{\frac{3}{2}} + 7\sqrt{6}$ to the nearest thousandth before simplifying the expression. Then simplify first and compute afterward. Compare the amount of work in the two processes.

137. Examples in Multiplication of Radicals.

Ex. 1.   Multiply $\sqrt{a-b}$ by $\sqrt{a^2-b^2}$.

$$\sqrt{a-b}\cdot\sqrt{a^2-b^2}=\sqrt{a-b}\cdot\sqrt{(a-b)(a+b)}$$
$$=\sqrt{(a-b)^2(a+b)}$$
$$=(a-b)\sqrt{a+b}. \quad Ans.$$

Ex. 2.   Multiply $3\sqrt{2}+5\sqrt{3}$ by $3\sqrt{2}-\sqrt{3}$.

$$3\sqrt{2}+5\sqrt{3}$$
$$3\sqrt{2}-\sqrt{3}$$
$$\overline{18+15\sqrt{6}}$$
$$\underline{\quad -3\sqrt{6}-15}$$
$$3+12\sqrt{6}. \quad Product.$$

EXERCISE 157

Multiply and simplify:

1. $\sqrt{x-y}\cdot\sqrt{2(x-y)}$.
2. $\sqrt{5(a+b)}\cdot\sqrt{2(a+b)}$.
3. $4\sqrt{3(x-y)}\cdot3\sqrt{5(x-y)}$.
4. $\sqrt{(a+b)(a-b)}\cdot\sqrt{a-b}$.

5. $5\sqrt{x^2-y^2}\cdot3\sqrt{x-y}$.   8. $\sqrt{2a-b}\cdot\sqrt{4a^2-b^2}$.

6. $3\sqrt{2}\cdot\sqrt{2ab}$.   9. $\sqrt{a-b}\cdot\sqrt{(a-b)^3}$.

7. $3\sqrt{5}\cdot\sqrt{10(a-b)^2}$.   10. $\sqrt{2x-y}\cdot\sqrt{4(2x+y)}$.

11. $\sqrt{a^2-b^2}\cdot\sqrt{(a-b)(a-2b)}$.
12. $\sqrt{x^2-y^2}\cdot\sqrt{x^2-3xy+2y^2}$.
13. $2\sqrt{3}+\sqrt{6}-\sqrt{5}$ by $\sqrt{3}$.
14. $\sqrt{3}-\sqrt{6}+2\sqrt{10}$ by $2\sqrt{2}$.
15. $4\sqrt{6}-3\sqrt{3}+3\sqrt{2}$ by $2\sqrt{6}$.
16. $3+\sqrt{2}$ by $2-2\sqrt{2}$.
17. $2\sqrt{3}-3\sqrt{2}$ by $4\sqrt{3}+5\sqrt{2}$.

18. $4\sqrt{2} - 3\sqrt{3}$ by $3\sqrt{2} + 4\sqrt{3}$.

19. $\sqrt{2} + \sqrt{3} - \sqrt{5}$ by $\sqrt{2} - \sqrt{3} + \sqrt{5}$.

20. $2\sqrt{3} - 3\sqrt{6} - 4\sqrt{15}$ by $2\sqrt{3} + 3\sqrt{6} + 4\sqrt{15}$.

21. $3\sqrt{30} + 2\sqrt{5} - 3\sqrt{6}$ by $2\sqrt{5} + 3\sqrt{6}$.

22. $\sqrt{3x} + \sqrt{2x} - \sqrt{5x}$ by $\sqrt{x}$.

23. $\sqrt{a} + \sqrt{b} + \sqrt{c}$ by $\sqrt{abc}$.

24. $\sqrt{x+1} - \sqrt{3x} + 3\sqrt{x-1}$ by $\sqrt{x}$.

25. $\sqrt{a-b} - \sqrt{a+b} + \sqrt{a}$ by $\sqrt{a-b}$.

26. Make up and work an example where a binomial radical expression is multiplied by a monomial radical.

27. Make up and work an example where a binomial radical expression is multiplied by another binomial radical expression.

138. Rationalizing a Monomial Denominator. If the denominator of a fraction is a surd, in order to make the denominator rational,

Multiply both numerator and denominator by such a number as will make the denominator rational.

Ex.   Rationalize the denominator of $\dfrac{5}{3\sqrt{2}}$.

$$\frac{5}{3\sqrt{2}} = \frac{5}{3\sqrt{2}} \times \frac{\sqrt{2}}{\sqrt{2}} = \frac{5\sqrt{2}}{6}. \quad Ans.$$

One object in rationalizing the denominator of a fraction is to diminish the labor of finding the approximate value of the fraction.

Thus, if we find the approximate numerical value of $\dfrac{5}{\sqrt{2}}$ directly, we must find the square root of 2, and divide 5 by the decimal which we obtain. On the other hand, if we find the value of the equivalent expression, $\frac{5}{2}\sqrt{2}$, we extract the square root of 2, multiply it by 5, and divide by 2. In the latter process we therefore avoid the tedious long division, and diminish the labor of the process by nearly one half.

EXERCISE 158

Rationalize the denominator of :

1. $\dfrac{1}{\sqrt{2}}$.

6. $\dfrac{1}{2\sqrt{2}}$.

11. $\dfrac{\sqrt{3}}{5\sqrt{2}}$.

16. $\dfrac{\sqrt{7}}{\sqrt{3\,x}}$.

2. $\dfrac{2}{\sqrt{3}}$.

7. $\dfrac{3}{5\sqrt{7}}$.

12. $\dfrac{\sqrt{a}}{\sqrt{b}}$.

17. $\dfrac{8}{\sqrt{11}}$.

3. $\dfrac{5}{\sqrt{7}}$.

8. $\dfrac{a}{b\sqrt{x}}$.

13. $\dfrac{2\sqrt{x}}{3\sqrt{y}}$.

18. $\dfrac{9\sqrt{a}}{5\sqrt{3\,x}}$.

4. $\dfrac{2}{\sqrt{x}}$.

9. $\dfrac{ab}{\sqrt{y}}$.

14. $\dfrac{1}{\sqrt{5\,x}}$.

19. $\dfrac{\sqrt{7}}{2\sqrt{5}}$.

5. $\dfrac{a}{\sqrt{b}}$.

10. $\dfrac{\sqrt{2}}{\sqrt{5}}$.

15. $\dfrac{5}{3\sqrt{2\,x}}$.

20. $\dfrac{a\sqrt{x}}{b\sqrt{y}}$.

21. Compute to the nearest thousandth the value of $\dfrac{1}{\sqrt{7}}$ without rationalizing the denominator. Then rationalize the denominator of $\dfrac{1}{\sqrt{7}}$ and compute the numerical value of the result to the nearest thousandth. Compare the amount of work in the two processes.

22. What is an irrational number? What is a rational number? Why do we give the name "rationalizing the denominator" to the solution of examples in this exercise?

EXERCISE 159

Simplify the following miscellaneous examples concerning radicals:

1. $\sqrt{18\,a^3b^2c^5}$.

3. $\dfrac{2}{\sqrt{7}}$.

5. $\dfrac{3\sqrt{3}}{5\sqrt{5}}$.

7. $\sqrt{1\tfrac{6}{5}}$.

2. $a^2\sqrt{\dfrac{20}{a}}$.

4. $\sqrt[3]{2\tfrac{7}{k}}$.

6. $\sqrt{\tfrac{7}{18}}$.

8. $\sqrt{\dfrac{a^8}{3}}$.

9. $\sqrt{3(a^2-4b^2)(a+2b)}$.   **10.** $\sqrt{3(a+b)}\cdot\sqrt{6(a+b)}$.

11. $\sqrt{2}(3\sqrt{8}-2\sqrt{\tfrac12}+\sqrt{18})$.

12. $(7\sqrt{2}-\sqrt{5})(3\sqrt{2}+4\sqrt{5})$.

13. $(\sqrt{5}-\sqrt{2})(\sqrt{5}+\sqrt{2})$.

14. $\sqrt{a^5b^2}+\sqrt{a^2b^5}$.   **15.** $\sqrt{(x-y)^3}\cdot\sqrt{x^2-y^2}$.

16. $\sqrt{(a-b)^2x}-\sqrt{(2a-3b)^2x}$.

17. $(3\sqrt{a}-5\sqrt{b})(4\sqrt{a}-7\sqrt{b})$.

18. $3\sqrt{2}\times2\sqrt{5}\times2\sqrt{2}$.   **19.** $4\sqrt{a}+7\sqrt{a}+3\sqrt{a^3}$.

20. $5\sqrt{2}-(2\sqrt{2}-3\sqrt{3})$.

21. $5\sqrt{12}-3(\sqrt{27}-2\sqrt{8})$.

22. $3\sqrt{8}-5\sqrt{\tfrac12}+7\sqrt{18}+\dfrac{3}{\sqrt{2}}$.

23. $\sqrt{3}+7\sqrt{\tfrac13}-3\sqrt{5\tfrac13}+2\sqrt{1\tfrac13}$.

Given that $\sqrt{2}=1.414213^+$, $\sqrt{3}=1.732050^+$, and $\sqrt{6}=2.44948^+$, compute to the nearest hundredth in the shortest way the numerical value of

24. $3\sqrt{8}-\sqrt{18}+\sqrt{32}+4\sqrt{\tfrac12}$.

25. Also of $\sqrt{3}\times\sqrt{6}$.   **26.** Also of $5\sqrt{5}\cdot\sqrt{10}$.

27. Extract the square root of $8\times72\times27\times12$ by simplifying the radical $\sqrt{8\times72\times27\times12}$.

28. Similarly extract the square root of $48\times16\times18\times24$.

29. Also of $27\times12\times98\times50$.

EXERCISE 160

ORAL

At sight simplify:

1. $\sqrt{\tfrac52}$.   **4.** $5\sqrt{\tfrac23}$.   **7.** $\sqrt{\tfrac45}$.   **10.** $\sqrt{\tfrac{R}{2}}$.   **13.** $\sqrt{\tfrac{4r^2}{3}}$.

2. $\sqrt{\tfrac23}$.   **5.** $\sqrt{\tfrac35}$.   **8.** $2\sqrt{\tfrac52}$.   **11.** $\sqrt{\tfrac{r^2}{2}}$.   **14.** $\sqrt{\tfrac{3r^2}{4}}$.

3. $\sqrt{\tfrac{2}{a}}$.   **6.** $\dfrac{3}{\sqrt{5}}$.   **9.** $b\sqrt{\tfrac{a}{b}}$.   **12.** $\dfrac{\sqrt{2}}{\sqrt{3}}$.   **15.** $\sqrt[3]{16}$.

16. $\sqrt{9} + \sqrt[3]{27}.$ **18.** $3\sqrt{3} - \sqrt{18}.$ **20.** $\sqrt{3} \times 2\sqrt{3}.$

17. $\sqrt{8} + 3\sqrt{2}.$ **19.** $2\sqrt{5} + \sqrt{20}.$ **21.** $\sqrt{2a^2} + a\sqrt{2}.$

22. $\sqrt{ab} \cdot \sqrt{a-b}.$ **23.** $3\sqrt{x} \cdot \sqrt{3x}.$

24. $(2\sqrt{5})(3\sqrt{5}).$ **27.** $2\sqrt{\frac{1}{2}} + \sqrt{8}.$ **30.** $3\sqrt{\frac{2}{3}} - \sqrt{6}.$

25. $\sqrt{50} - \sqrt{20}.$ **28.** $3\sqrt{\frac{1}{3}} + \sqrt{27}.$ **31.** $\sqrt{\frac{1}{3}} + \frac{1}{\sqrt{3}}.$

26. $4 + \sqrt{9 \times 25}.$ **29.** $\sqrt{18} - \sqrt{8}.$

                           32. $2\sqrt{\frac{1}{2}} - \frac{1}{4}\sqrt{2}.$

EXERCISE 161

1. The formula for the volume, v, of a cylinder whose altitude is h, and the radius of whose base is r, is $v = \pi r^2 h$. Using this formula, find the volume of a cylinder whose altitude is 12 in. and the diameter of whose base is 10.5 in.

2. A cylindrical pail is 9 in. in diameter and 11 in. deep. How many gallons will it hold if one gallon is 231 cu. in. ? Solve by use of the formula of Ex. 1, and find the answer to the nearest hundredth.

3. From $v = \pi r^2 h$, find the value of r in terms of the other letters.

By use of the formula which you have obtained, solve Exs. 4 and 5.

4. A cylinder is to contain 4 cu. ft. If the height is to be 30 in., find the radius of the base.

5. A cylinder tank is to contain 200 gallons. If the height is to be 3 ft., find the diameter of the base.

6. The formula for the surface of the sphere is $s = 4\pi r^2$. Solve this formula for r.

By use of the formula which you have just obtained, solve Exs. 7 and 8.

7. The surface of a sphere is 100 sq. in. Find the radius of the sphere to the nearest hundredth of an inch.

8. A man has 1000 sq. yd. of silk out of which to make a spherical balloon. What is the largest possible diameter of the balloon?

9. The formula for the volume of a sphere is $v = \frac{4}{3}\pi r^3$. Solve this formula for r.

10. If s denotes the surface and e the edge of a cube, obtain a formula for s in terms of e.

SUG. Show that $s = 6 e^2$.

11. Solve the formula $s = 6 e^2$ for e. By use of the formula which you have obtained, find in inches the edge of a cube whose surface is a square yard.

12. In selling a certain article a merchant wishes to fix an asking price such that he can give a discount of 20 % and still make a profit of 10 % above the cost price. If c is the cost price, a the asking price, and s the selling price, find a formula for a in terms of c.

SUG. $s = c + \frac{1}{10}c$. Also $s = a - \frac{1}{5}a$. Put the two values of s equal to each other and hence find a in terms of c.

By use of this formula solve Exs. 13 and 14.

13. A merchant buys shoes at $4 a pair and in selling them wishes to grant a discount of 20 % yet make a profit of 10 % on the cost. What price must he ask for them?

14. A merchant bought an article for $16. What must be his asking price for it in order that he may give a discount of 20 % from the asking price and yet gain 10 % on the cost?

15. Solve the formula $a = \dfrac{11\,c}{8}$ (see Ex. 12) for c.

Use the result in solving Exs. 16 and 17

16. A merchant wishes to advertise suits of clothes at $20 each, be able to give a discount of 20 %, and yet make a profit of 10 %. What can he afford to pay for the suits?

17. What can I afford to pay for a property, if I wish to advertise a selling price of $3000, make a discount of 20 per cent, and a profit of 10 per cent.

18. In selling a certain article a merchant wishes to fix an asking price such that he can give a discount of 10% and the same time make a profit of 25% above the cost price. If c is the cost price, a the asking price, and s the selling price, find a formula for a in terms of c.

19. Make up and solve a numerical example by use of the formula obtained in Ex. 18.

20. In selling a certain article, a merchant wishes to fix an asking price such that he can give a discount of 5% and at the same time make a profit of 30% above the cost price. If c is the cost price, a the asking price, and s the selling price, find a formula for a in terms of c.

21. Make up and solve a numerical example by use of the formula obtained in Ex. 20.

22. Fixing your own per cent of discount and profit, make up and work an example similar to Ex. 20.

CHAPTER XV

QUADRATIC EQUATIONS

PART I

139. A **quadratic equation** of one unknown quantity is an equation containing the second power of the unknown quantity, but no higher power.

140. A **pure quadratic equation** is one in which the second power of the unknown quantity occurs, but not the first power.

Ex. $5 x^2 - 12 = 0$.

A pure quadratic equation is sometimes termed an incomplete quadratic equation.

141. An **affected** (or complete) **quadratic equation** is one in which both the first and second powers of the unknown quantity occur.

Ex. $3 x^2 - 7 x + 12 = 0$.

Pure Quadratic Equations

142. Solution of Pure Quadratics. Since only the second power, x^2, of the unknown quantity occurs in a pure quadratic equation, in solving such an equation, we

Reduce the given equation to the form $x^2 = c$;
Extract the square root of both members.

Ex. 1. Solve $\dfrac{x^2 - 12}{3} = \dfrac{x^2 - 4}{4}$.

Clearing of fractions, $4 x^2 - 48 = 3 x^2 - 12$.
Hence, $\qquad\qquad\qquad x^2 = 36$.

Extracting the square root of each member,

$$x = 6, \text{ or } -6.$$

That is, since the square of $+6$ is 36, and also the square of -6 is 36, x has two values, either of which satisfies the original equation. These two values of x are best written together. Thus,

$$x = \pm 6. \quad Roots.$$

CHECK. For $x = 6$.

$$\frac{x^2 - 12}{3} = \frac{36 - 12}{3} = \frac{24}{3} = 8.$$

$$\frac{x^2 - 4}{4} = \frac{36 - 4}{4} = \frac{32}{4} = 8.$$

CHECK. For $x = -6$.

$$\frac{x^2 - 12}{3} = \frac{36 - 12}{3} = 8.$$

$$\frac{x^2 - 4}{4} = \frac{36 - 4}{4} = 8.$$

Ex. 2.   Solve $\dfrac{a}{x^2 - b} = \dfrac{b}{x^2 - a}$.

$$ax^2 - a^2 = bx^2 - b^2.$$
$$ax^2 - bx^2 = a^2 - b^2.$$
$$x^2 = a + b.$$
$$x = \pm \sqrt{a + b}. \quad Roots.$$

Let the pupil check the work.

EXERCISE 162

1. $5x^2 = 80.$

2. $3x^2 - 5 = x^2 + 3.$

3. $\frac{7}{3}x^2 - 1 = \frac{1}{3} - 3x^2.$

4. $1 - \frac{2}{5}x^2 = x^2 - 4\frac{3}{5}.$

5. $\dfrac{5y^2}{3} = 1 - \dfrac{y^2}{3}.$

6. $\dfrac{3 - x^2}{11} + \dfrac{x^2 + 5}{6} = 3.$

7. $\dfrac{3}{4\,l^2} - \dfrac{1}{3\,l^2} = \dfrac{5}{12}.$

8. $\dfrac{1}{2\,m - 1} - \dfrac{1}{2\,m + 1} = 3\frac{5}{9}.$

9. $ax^2 + a^3 = 5\,a^3 - 3\,ax^2.$

10. $ax^2 + c = b.$

11. $\dfrac{2\,c}{x - c} + \dfrac{5\,x + 2\,c}{3\,x} + 1 = 0.$

12. $(x + a)(x - b) + (x - a)(x + b) = 2(a^2 + b^2 + ab).$

13. $3(2\,x - 5)(x + 1) - 2(3\,x + 2)(2\,x - 3) = x - 9.$

14. If $x^2 = \dfrac{28}{1 + 3\,v}$, find the value of x when $v = -\dfrac{2}{7}$.

15. $3\,x^2 = .27.$ **17.** Solve $a = \tfrac{4}{7}\,r^2$ for r.

16. $.2\,x^2 - .098 = 0.$ **18.** Solve $s = \tfrac{1}{2}\,gt^2$ for t.

19. The square of a certain number increased by 9 equals twice the square of the same number diminished by 27. Find the number.

20. A certain field contains 256 sq. rd. and the field is four times as long as it is wide. Find the dimensions of the field.

21. A certain field is four times as long as it is wide. If each of its sides is increased by one half, its area is increased 180 sq. rd. Find the dimensions of the field.

AFFECTED QUADRATIC EQUATIONS

143. Completing the Square. An affected quadratic equation may in every instance be reduced to the form

$$x^2 + px = q.$$

An equation in this form may then be solved by a process called *completing the square*. This process consists in adding to both members of the equation such a number as will make the left-hand member a perfect square. The use of familiar elementary processes then gives the value of x.

Thus, to solve $x^2 + 6\,x = 16,$

take half the coefficient of x (that is, 3), square it, and add the result (that is, 9) to both members of the original equation. We obtain

$$x^2 + 6\,x + 9 = 25,$$
$$\text{or,} \qquad (x + 3)^2 = 5^2.$$

Extract the square root of both members,

$$x + 3 = \pm 5$$

Hence, $x = -3 \pm 5.$

That is, $x = -3 + 5 = 2.$ $\left.\vphantom{\begin{matrix}a\\b\end{matrix}}\right\}$ *Roots.*

Also, $x = -3 - 5 = -8.$

Hence we have the general rule :

By clearing the given equation of fractions and parentheses, transposing terms, and dividing by the coefficient of x^2, reduce the given equation to the form $x^2 + px = q$;

Add the square of half the coefficient of x to each member of the equation;

Extract the square root of each member;

Solve the resulting simple equations.

Before clearing an equation of fractions it is important to *reduce each fraction to its simplest form.*

Ex. 1. Solve $6x^2 - 14x = 12$.

Dividing by 6, $\qquad\qquad\qquad x^2 - \tfrac{7}{3}x = 2$.

Completing the square, $x^2 - \tfrac{7}{3}x + (\tfrac{7}{6})^2 = 2 + \tfrac{49}{36} = \tfrac{121}{36}$.

Extracting the square root, $\qquad x - \tfrac{7}{6} = \pm\tfrac{11}{6}$.

$$x = \tfrac{7}{6} \pm \tfrac{11}{6}.$$

$$x = 3, \text{ or } -\tfrac{2}{3}. \quad \textit{Roots.}$$

CHECK. For $x = 3$ $\qquad\qquad$ CHECK. For $x = -\tfrac{2}{3}$

$6x^2 - 14x = 6 \times 9 - 14 \times 3 \qquad 6x^2 - 14x = 6 \times \tfrac{4}{9} + 14 \times \tfrac{2}{3}$

$\qquad = 54 - 42 = 12 \qquad\qquad\qquad = \tfrac{8}{3} + \tfrac{28}{3} = 12.$

EXERCISE 163

Convert each of the following into a complete square :

1. $x^2 + 10x$. $\qquad$ 5. $p^2 - 7p$. $\qquad$ 9. $x^2 - \tfrac{2}{5}x$.

2. $x^2 - 8x$. $\qquad$ 6. $x^2 + \tfrac{1}{3}x$. $\qquad$ 10. $y^2 + \tfrac{8}{7}y$.

3. $p^2 - 5p$. $\qquad$ 7. $x^2 - \tfrac{6}{5}x$. $\qquad$ 11. $x^2 - \tfrac{10}{13}x$.

4. $x^2 + 11x$. $\qquad$ 8. $x^2 - \tfrac{5}{2}x$. $\qquad$ 12. $x^2 + \tfrac{11}{15}x$.

Solve by completing the square, and check :

13. $x^2 + 6x = 16$. $\qquad\qquad$ 19. $p^2 + 11p + 24 = 0$.

14. $x^2 + 10x = 24$. $\qquad\qquad$ 20. $3x^2 + 4x = 7$.

15. $x^2 + 2x = 8$. $\qquad\qquad$ 21. $5x^2 - 6x = 8$.

16. $x^2 - 4x = 5$. $\qquad\qquad$ 22. $2x^2 - 5x = 7$.

17. $x^2 - 8x - 20 = 0$. $\qquad\qquad$ 23. $3x^2 + 7x = 26$.

18. $x^2 - 5x = 6$. $\qquad\qquad$ 24. $2x + 3\tfrac{3}{4}x^2 = 4$.

25. $3\,r^2 = \frac{1}{3}r + 2\frac{2}{3}$.

26. $x^2 - 5\,x = 4\,x - 18$.

27. $y - 3 - \dfrac{4\,y - 12}{y} = 0$.

28. $\dfrac{1}{2} = \dfrac{7\,x}{6} + x^2$.

29. $\dfrac{3\,x + 5}{x + 4} = 3 - \dfrac{2\,x - 5}{x - 2}$.

30. $x - \dfrac{1}{x} = 4 - \dfrac{x}{2} + \dfrac{1}{2\,x}$.

31. $\dfrac{x}{5 - x} - \dfrac{5 - x}{x} = \dfrac{15}{4}$.

32. $\dfrac{1}{x + 2} - \dfrac{x}{x - 2} = \dfrac{4}{3}$.

33. $\dfrac{2\,x + 1}{x - 1} + x = \dfrac{x - 1}{3}$.

34. $\dfrac{2}{3\,x - 1} + \dfrac{3\,x}{2\,x - 5} = 0$.

35. $\dfrac{3\,x + 1}{5\,x + 4} = \dfrac{x + 3}{4\,x + 5}$.

36. $\dfrac{1}{2}(x + 1) - \dfrac{x}{3}(2\,x - 1) = -12$.

37. $\dfrac{x - 3}{2\,x} + \dfrac{3\,x - 1}{3} = 1 - \dfrac{1 - x}{4\,x} + \dfrac{3 + 2\,x}{6}$.

38. Is 5 a root of $\dfrac{p + 3}{p} - \dfrac{p}{p + 2} = 8$?

39. Is -2 a root of $\dfrac{x - 1}{x} = \dfrac{13}{6} - \dfrac{x}{x - 1}$?

40. Is .3 a root of $x^2 - .5\,x + .06 = 0$?

144. Literal Quadratic Equations are solved by the methods employed in solving quadratic equations with numerical coefficients.

Ex.   Solve $x^2 - 3\,ax + 2\,a^2 = 0$.

$$x^2 - 3\,ax = -2\,a^2.$$

$$x^2 - 3\,ax + \left(\frac{3\,a}{2}\right)^2 = \frac{9\,a^2}{4} - 2\,a^2 = \frac{a^2}{4}.$$

$$x - \frac{3\,a}{2} = \pm\frac{a}{2}.$$

$$x = \frac{3\,a}{2} \pm \frac{a}{2}.$$

$$x = 2\,a,\ a.\quad \textit{Ans.}$$

Let the pupil check the work.

Solve and check :

1. $x^2 + 4\,ax = 12\,a^2.$

2. $x^2 + 4\,bx = 21\,b^2.$

3. $x^2 + 3\,cx - 10\,c^2 = 0.$

4. $x^2 - 5\,ax = -6\,a^2.$

5. $x^2 - 7\,bx + 12\,b^2 = 0.$

6. $x^2 = 6\,a^2b^2 - 5\,abx.$

7. $6\,x^2 = 12\,b^2 + bx.$

8. $3\,x^2 + 4\,cdx = 15\,c^2d^2.$

9. $x^2 + px + q = 0.$

10. $x^2 - 6\,mx = 3\,mx - 18\,m^2.$

11. $x + a = \dfrac{2\,ax + 6\,a^2}{x}.$

12. $x - 7\,b - \dfrac{15\,b^2 - 5\,bx}{x} = 0.$

13. $2\,a^2x^2 + ax = 3.$

14. $7\,c^2x^2 - 10\,acx + 3\,a^2 = 0.$

15. $x^2 + \dfrac{5\,x}{a} = \dfrac{6}{a^2}.$

16. $x^2 + \dfrac{bx}{a} = \dfrac{3\,b^2}{4\,a^2}.$

17. Solve $s = vx + \frac{1}{2}\,gx^2$ for x.

18. Solve $s = vt + \frac{1}{2}\,gt^2$ for t.

145. Use of Formula. Any quadratic equation can be reduced to the form

$$ax^2 + bx + c = 0.$$

This equation may be solved by completing the square (§ 143). Thus,

$$ax^2 + bx = -c.$$

$$x^2 + \frac{b}{a}\,x = -\frac{c}{a}.$$

$$x^2 + \frac{b}{a}\,x + \left(\frac{b}{2\,a}\right)^2 = \frac{b^2}{4\,a^2} - \frac{c}{a} = \frac{b^2 - 4\,ac}{4\,a^2}.$$

$$x + \frac{b}{2\,a} = \frac{\pm\sqrt{b^2 - 4\,ac}}{2\,a}.$$

$$x = \frac{-b \pm \sqrt{b^2 - 4\,ac}}{2\,a}.$$

This result may be regarded as a formula for the solution of any quadratic equation.

Ex. Solve $5x^2 + 3x - 2 = 0$ by use of the formula.

Here $\qquad a = 5,\ b = 3,\ c = -2.$

Substituting for a, b, c in the above formula,

$$x = \frac{-3 \pm \sqrt{9 + 40}}{10} = \frac{-3 \pm 7}{10} = \frac{2}{5},\ -1.\quad \textit{Roots.}$$

Let the pupil check the work.

EXERCISE 165

1. $x^2 + 5x = 6.$

2. $3x^2 - x = 2.$

3. $6x^2 + 5x = 4.$

4. $8x^2 - 2x = 3.$

5. $4x^2 + 4x = 35.$

6. $16x^2 - 40x + 9 = 0.$

7. $6x^2 - fx = 2f^2.$

8. $4a^2x^2 + 5ax = 21.$

9. $x(x + 2) = 15.$

10. $x^2 + 7x = 18.$

11. $\dfrac{12}{x^2 - 9} = 1 - \dfrac{2}{x + 3}.$

12. $\dfrac{x - 1}{2} + \dfrac{2}{x - 1} = \dfrac{5}{2}.$

13. $l^2 - 4l = 3l - 12.$

14. $2m^2 + 3m = 3 - 2m.$

15. Also solve the odd examples in Exercises 163 and 164 by the use of the formula.

EXERCISE 166

ORAL

1. The difference of two numbers is 10. The greater number is t. Find the quotient of the greater divided by the less.

2. If l denotes the length of your ruler in inches, what is the length of one which is 4 in. more than half as long?

3. What is the reciprocal of a? of 5? of $\frac{2}{3}$?

4. A man walks 2 yd. a second. How long will it take him to walk f feet?

5. A piece of cloth y yards long is cut into two pieces, one of which is 2 ft. more than one yard long. How long is the other piece?

6. What per cent is m of x?

7. A man sold a horse for h dollars and in so doing lost l dollars. Find the cost of the horse. What per cent of the cost was the loss?

8. If h denotes the number of inches in the height of a boy, what is the height of a boy $\frac{1}{8}$ taller?

9. How far will a boat go in m hours, at the rate of s miles in t minutes?

10. A building lot is 3 times as long as it is wide. If it is w ft. wide, find its area.

EXERCISE 167

1. The square of a certain number diminished by 4 times the number equals 45. Find the number.

2. What number plus its square equals 12?

3. Find two consecutive numbers whose product is 72.

4. If to 3 times the square of a certain number we add 4 times the number, the result equals 39. Find the number.

5. The depth of a certain lot equals three times the front, and the area of the lot is 7500 sq. ft. Find the dimensions of the lot.

6. The sum of 6 and the square of a certain number equals 5 times the number. Find the number.

7. Find a number which when multiplied by 7 less the number gives 10.

8. Find two consecutive numbers the sum of whose squares is 61.

9. Find three consecutive numbers the sum of whose squares is 110.

10. There are two consecutive numbers such that if the larger be added to the square of the less the sum will be 57. Find the numbers.

11. There are two numbers whose difference is 3, and if twice the square of the larger be added to 3 times the smaller, the sum is 56.   Find the numbers.

12. Seven times a certain number is one less than the square of the number next larger than the original number. Find the number.

13. What number increased by its reciprocal equals $\frac{65}{8}$?

14. Find three consecutive numbers such that their sum is 15 less than the square of the smallest.

15. If the length of a rectangle exceeds the width by 5 yd. and the width be denoted by x, express the length and the area in terms of x.

16. The area of a given rectangle is 36 sq. yd., and the length exceeds the width by 5 yd.   Find the dimensions of the rectangle.

17. The length of a certain rectangle is twice its width. The rectangle has the same area as another, $1\frac{1}{3}$ times as wide, and shorter by $4\frac{1}{2}$ ft.   Find the length of the first rectangle.

18. A rectangular garden contains one half an acre and the length of the rectangle exceeds its width by 2 rd. Find its dimensions.

19. A square garden contains 100 sq. rd.   By how much must one of its sides be lengthened in order that its area be doubled?

20. A rectangle is 30 × 40 ft.   By what per cent must the length and width be increased in order that the area be increased by 528 sq. ft.?

21. A rectangular park is 80 × 100 rd.   By adding the same amount to its length and width the area of the park is to be increased by 50 %.   What is the amount added to each dimension?

22. A rectangular lot is 8 rd. long and 6 rd. wide, and is surrounded by a drive of uniform width, which occupies $\frac{2}{3}$ as much area as the lot. Find the width of the drive.

23. A farmer has a wheat field 80 rd. long and 60 rd. wide. How wide a strip must be cut around the outside of the field in order to cut 15 A.?

24. The numerator of a given fraction exceeds its denominator by 2. Also the given fraction exceeds its reciprocal by $\frac{16}{15}$. Find the fraction.

25. Find two consecutive numbers, the difference of whose cubes is 217.

26. If the side of a square is 2 ft., how much must this be increased to increase the area of the square by 153 sq. in.?

27. A bin is to be constructed to hold 9 T. of coal. If the bin is to be 5 ft. deep and twice as long as it is wide, and if 40 cu. ft. are allowed for 1 T., what will the dimensions of the base of the bin be?

28. The walls and ceiling of a room together contain 104 sq. yd. The room is twice as long as it is wide, and its ceiling is 9 ft. high. Find the length and breadth of the room.

EXERCISE 168

REVIEW

1. Simplify $3 - [(2a - 5) - (7 - 3a)]$.

2. Factor: (1) $x^3 - 2x^2 - 143x$.   (4) $x^2 - ax - 6a^2$.
        (2) $5x^4 - 125x^2$.   (5) $81 - a^4$.
        (3) $1 - 8ax + 16a^2x^2$.   (6) $x^3 + x^2 - 3x$.

3. Simplify and check : $\dfrac{1}{x-1} + \dfrac{2}{x-2} - \dfrac{3}{x-3}$.

4. Extract the square root $4b^4 - 12b^3 + 11b^2 - 3b + \frac{1}{4}$.

5. Find three consecutive numbers such that the sum of the squares of the first and last shall be 44 more than the product of the last two.

6. If 1 kg. $= 2\frac{1}{5}$ lb., the relation between kilograms and pounds is expressed by the formula $p = \frac{11}{5} k$. Construct a graph of this formula. From the graph determine how many kilograms are in 10 lb.

7. Simplify $\sqrt[3]{64\,x} - \frac{1}{8}\sqrt[3]{8\,x^4} + 5\sqrt[3]{27\,x}$.

8. Solve $\left. \begin{array}{l} 2\,x - 3\,y + z = -2 \\ 4\,x - 4\,y - 3\,z = 2 \\ 6\,x + y - 4\,z = 6 \end{array} \right\}$ and check your answers.

9. Solve $t = \pi\sqrt{\dfrac{l}{g}}$ for l.

10. Solve $cx^2 + dx + f = 0$ for x by completing the square.

11. Compute 999.7^2 in a short way.

12. Simplify $\dfrac{3\,x^2 - 12\,y^2}{8\,a^2b^2} \times \dfrac{30}{7} \div \dfrac{5\,x + 10\,y}{28\,ab^2}$ and check the answer.

13. Extract the square root of 19.3 to the nearest thousandth.

14. The area of a given rectangle is 105 sq. ft.; the length exceeds the width by 8 ft. Find the dimensions.

15. Solve $\dfrac{4\,x - 6}{x^2 - 9} - \dfrac{1}{x - 3} = 1$.

16. Simplify $\sqrt{ax^3} \cdot \sqrt{a^3b^2x} \cdot \sqrt{axy}$.

17. Solve $\dfrac{x}{b} - \dfrac{1}{2} = \dfrac{2\,x}{3\,b}$ for x and check your answer.

18. Solve $\left. \begin{array}{l} rx + sy = 2\,rs \\ sx - ry = s^2 - r^2 \end{array} \right\}$ for x and y and check.

19. Solve $R = \dfrac{l^2}{6\,d} + \dfrac{d}{2}$ for l.

20. The following table gives the pressure for various velocities of the wind:

Velocity of wind in mi. per hr.	10	20	30	40	50	60	70	80	90	100
Pressure in lb. per sq. ft.	1.5	2	4.5	8	12.5	18	24.5	32	40.5	50

Graph the above table of facts. From this graph determine as exactly as you can the pressure when the velocity of the wind is 25 mi. per hour. 45 mi. 65 mi.

PART II

146. Examples of Solution of a Quadratic Equation.

Ex. 1. Solve $3x^2 = 2(1 + 2x)$ and find the value of x to the nearest hundredth.

$$3x^2 = 2 + 4x.$$

$$3x^2 - 4x - 2 = 0.$$

Using the formula of § 145, viz., $x = \dfrac{-b \pm \sqrt{b^2 - 4ac}}{2a}$,

$$a = 3, b = -4, c = -2.$$

Hence, $\quad x = \dfrac{4 \pm \sqrt{16 + 24}}{6} = \dfrac{4 \pm \sqrt{40}}{6}.$

$$= \dfrac{4 \pm 2\sqrt{10}}{6} = \dfrac{2 \pm \sqrt{10}}{3}.$$

But $\quad \sqrt{10} = 3.1622^+$ (see § 124).

$$\therefore x = \dfrac{2 \pm 3.1622^+}{3}$$

$$= 1.72^+, -.39^-. \quad Ans.$$

Ex. 2. Solve $.7p^2 + 1.8p - .3 = 0$ and find the value of p to the nearest thousandth.

Here $\quad a = .7, b = 1.8, c = -.3.$

Hence, $p = \dfrac{-1.8 \pm \sqrt{3.24 + .84}}{1.4} = \dfrac{-1.8 \pm 2.0199^+}{1.4}$

$$= .157^+, \; - 2.729^-. \quad Ans.$$

Note that in solving an example like $3x^2 + 2x + 5 = 0$, the square root of a negative number appears in the answer. The consideration of such examples lies beyond the scope of this book.

EXERCISE 169

Solve and find the roots to the nearest thousandth :

1. $4x^2 - 4x - 1 = 0.$

2. $4x^2 - 8x + 1 = 0.$

3. $9x^2 - 6x - 2 = 0.$

4. $x^2 - x - 1 = 0.$

5. $\dfrac{7x}{2} - \dfrac{23}{9} = x^2 - x.$

6. $2x^2 - 3x = 3x - 1.$

7. $\dfrac{3x^2 - 4x}{3} - \dfrac{12x - 5}{9} = 0$

Solve and check :

8. $x^2 - .5x + .06 = 0.$

9. $x^2 - .85x + .105 = 0.$

10. $x^2 - .6x - .72 = 0.$

11. $x^2 + .16x - .17 = 0.$

12. $x^2 + 1.09x + .204 = 0.$

13. $x^2 - .5x = .09x - .084.$

Solve and find the roots to the nearest tenth :

14. $1.4x^2 - 5.6x + 2.3 = 0.$

15. $3.4x^2 + 1.7 = 8.6x.$

16. $.3x^2 + 2.3x - 2.5 = 0.$

17. $.87x^2 = 3.4x + 6.9.$

18. $y^2 - 2y - \dfrac{.5y + .8}{.6} = 0.$

19. $\dfrac{5.6r - 1}{.2} = \dfrac{2.9r^2 + 1.7}{.4}.$

Solve to the nearest hundredth :

20. $.5x^2 - 9.1x + 2.7 = 0.$

21. $4.9p^2 + 16.3p - 1.8 = 0.$

22. $1.3y^2 + 5.6y - 1.72 = 0.$

23. $.52l^2 - 6.9l - 3.19 = 0.$

Solve to the nearest thousandth :

24. $x^2 - .9x + .13 = 0.$

25. $x^2 - 3.1x + .042 = 0.$

26. $x^2 + .16x - .15 = 0.$

27. $3x^2 - .47x - .28 = 0.$

28. $\frac{1}{2}x^2 + 6\frac{1}{4}x = 3\frac{1}{2}.$

29. $\frac{1}{4}x^2 + 5\frac{1}{2}x = 2\frac{1}{4}.$

147. Equations in the Quadratic Form. An equation containing only two powers of the unknown quantity, the index of one power being twice the index of the other power, is an equation of the quadratic form. It may be solved by the methods already given for affected quadratic equations.

Ex. 1.   Solve $x^4 - 5\,x^2 = -4$.

Adding $(\frac{5}{2})^2$ to both members will make the left-hand member a perfect square. Thus,

$$x^4 - 5\,x^2 + (\tfrac{5}{2})^2 = \tfrac{9}{4}.$$

Hence, $$x^2 - \tfrac{5}{2} = \pm\,\tfrac{3}{2}.$$

$$x^2 = 4 \text{ or } 1.$$

$$x = \pm\,2,\ \pm\,1.\quad \textit{Roots.}$$

Let the pupil check the work.

An equation of this kind may also be solved by use of the formula of § 145.

EXERCISE 170

Solve and check :

1. $x^4 - 10\,x^2 + 9 = 0$.

2. $x^4 - 17\,x^2 = -16$.

3. $x^4 - 4\,x^2 = 9\,x^2 - 36$.

4. $x^2 - 4 = \dfrac{x^4 - 4\,x^2}{16}$.

5. $4\,x^4 - 13\,x^2 + 9 = 0$.

6. $4\,x^4 - 29\,x^2 + 25 = 0$.

7. $x^6 - 9\,x^3 + 8 = 0$.

8. $x^4 - 11\,x^2 + 18 = 0$.

9. $\dfrac{x^4 - 5\,x^2}{6} = 5\,x^2 - 36$.

10. $x^2 - 3 = \dfrac{x^4 - 3\,x^2}{4}$.

11. $x^4 - 5\,x^2 + 6 = 0$.

Solve and find the roots to the nearest hundredth :

12. $x^4 - 3\,x^2 + 1 = 0$.

13. $2\,x^4 - 4\,x^2 + 1 = 0$.

14. $4\,x^4 - 12\,x^2 + 5 = 0$.

15. $4\,x^4 - 8\,x^2 + 1 = 0$.

16. Is $2 - \sqrt{5}$ a root of $x^2 - 6\,x + 4 = 0$?

17. Is $4 - \sqrt{5}$ a root of $x^2 - 8\,x + 11 = 0$?

18. Is $\frac{1}{2}(3 - \sqrt{2})$ a root of $x^2 - 3\,x + \frac{7}{4} = 0$?

Simultaneous Equations Involving Quadratics

148. Degree of a Term. The degree of a term is determined by the number of literal factors which the term contains. Hence, the degree of a term is equal to the sum of the exponents of the literal factors in the term.

Thus, $7\,a^3bc^2$ is a term of the 6th degree, since the sum of the exponents in it is $3 + 1 + 2$, or 6.

149. The Degree of an Equation containing One Unknown Quantity is equal to the highest exponent of the unknown quantity occurring in the equation when simplified.

Thus, the equation $5\,x^2 - 6\,x^3 + 7 = 0$ is of the third degree.

The Degree of an Equation containing Two Unknown Quantities is equal to the highest sum of the exponents of the unknowns in any single term.

Thus, $x^4 + 3\,x^3y^2 - 5\,y^3 = 8$ is an equation of the fifth degree.

150. Two Simultaneous Equations, one of the first degree, and the other of the second, *can always be solved by the method of substitution.*

Ex. Solve $\begin{cases} 2\,x - 3\,y = 2. & (1) \\ x^2 - 2\,xy = -7. & (2) \end{cases}$

From (1), we obtain, $x = \dfrac{3\,y + 2}{2}.$

Substituting for x in (2), $\left(\dfrac{3\,y + 2}{2}\right)^2 - 2\,y\left(\dfrac{3\,y + 2}{2}\right) = -7.$

Hence, $\dfrac{9\,y^2 + 12\,y + 4}{4} - 3\,y^2 - 2\,y = -7.$

$$9\,y^2 + 12\,y + 4 - 12\,y^2 - 8\,y = -28.$$

Whence, $\begin{aligned} y &= 4,\ -\tfrac{8}{3}. \\ x &= 7,\ -3. \end{aligned}\Bigg\}\ Ans.$

Check.

For $x = -3$ and $y = -\tfrac{8}{3}$,
$2\,x - 3\,y = -6 + 8 = 2.$
$x^2 - 2\,xy = 9 - 16 = -7.$

Check.

For $x = 7$ and $y = 4$,
$2\,x - 3\,y = 14 - 12 = 2.$
$x^2 - 2\,xy = 49 - 56 = -7.$

EXERCISE 171

Solve and check:

1. $3x^2 - 2y^2 = -5.$
$x + y - 3 = 0.$

2. $x - 2y = 3.$
$x^2 + 4y^2 = 17.$

3. $2x^2 + xy = 2.$
$3x + y = 3.$

4. $x^2 - 3y^2 - 1 = 0.$
$x + 2y - 4 = 0.$

5. $x - 3y = 1.$
$7xy - x^2 = 12.$

6. $2x + y + 3 = 0.$
$3x^2 - 7y^2 = 5.$

7. $2x + 5y = 1.$
$2x^2 + 3xy = 9.$

8. $\frac{1}{3}x - \frac{1}{2}y = \frac{1}{3}.$
$(x - y)^2 = y^2 - 7.$

9. $x^2 - 3xy + 2y^2 = 0.$
$2x + 3y = 7.$

10. $3x - 7y = -9.$
$2x^2 + 3xy - 2y^2 = 0.$

11. $3x - 5y - 1 = 0.$
$2x^2 + 3xy - 5y^2 - 6x + 7y = 4.$

12. $4x^2 - 4xy = y^2 + x + 3y - 1.$
$4x - 2 - 5y = 0.$

13. $xy = 48.$
$(x + 2)(y - 1) = 50.$

Sug. From the last of the two equations

$$xy + 2y - x - 2 = 50.$$

From this subtract $xy = 48,$ (1)
$$2y - x = 4. \qquad (2)$$

Solve (1) and (2) by substituting from (2) in (1).

14. $(x - 3)(y + 2) = 63.$ **15.** $xy = 60.$
$xy = 60.$ $(x - 10.5)(y + 1.5) = -3.75.$

Without solving the equations determine

16. Whether $x = 3$ and $y = 7$ satisfy the equations
$$x + y + 3xy = 83 \text{ and } 3x - y = 1.$$

17. Whether $x = 2$ and $y = 1$ satisfy the equations
$$3y - 1 = x \text{ and } 5y^2 - x^2 = 1.$$

18. Whether $x = -\frac{2}{7}$ and $y = -\frac{8}{7}$ satisfy

$$\frac{x}{y} + \frac{y}{x} = 1\frac{1}{4} \text{ and } x - 2y = 2.$$

19. Whether $x = \frac{4}{3}$ and $y = \frac{2}{3}$ satisfy

$$x + y = 2 \text{ and } \frac{2}{x} + \frac{6}{y} = 6.$$

20. The difference of two numbers is 1.3 and their product is 6.6. Find the numbers to the nearest hundredth.

EXERCISE 172

ORAL

1. I sold c cars for d dollars each and thus gained m dollars. What was the entire cost of the cars? How much did each car cost?

2. The left digit of a number of two digits is twice the right digit. If the right digit is m, what is the number?

3. In a given right triangle, the two sides forming the right angle are x and y. What denotes the hypotenuse of the right triangle? Its area? Its perimeter?

4. John has a times as many marbles as James, and b times as many as Will. If Will has c marbles, how many have the three boys together?

5. How many quarters must be taken from d dollars to leave $7x$ cents?

6. What number must be added to 7 to make the sum equal to twice the number added?

7. An alloy contains c parts of copper, and t parts of tin. How many pounds of each are in p pounds of the alloy?

8. If the interest on $1 for 1 year is c cents, what will be the interest on c dollars for 3 years at the same rate?

Express the following in the form of an equality:

9. The product of 9 and 7 exceeds 60 by 3.

10. The product of x and y exceeds 8 times their difference by 3.

11. 32 is separated into two parts, x and y.

12. Draw a rectangle and mark its length by x and its width by y. Inside the rectangle write 140 sq. yd. to denote the area of the rectangle. Form an equation concerning the numbers x, y, and 140. If the perimeter of the rectangle is 60 yd., form another equation concerning x, y, and 60.

13. If the length of a rectangle is x, and this length be increased by 20 %, what does the length become?

14. If the area of a rectangle is xy sq. yd., and this area be increased by 44 %, what does the area become?

15. If a train travels x miles an hour, how far will it go in y hours? If it travels $x + 6$ miles an hour, how far will it go in $y - 1$ hours?

16. If a given fraction is $\frac{x}{y}$, express the difference between the fraction and its reciprocal.

17. A team has won 5 games out of 17. What is its average of games won? If it should play x more games and win them all, what would be its average of games won? What if instead it should play x more games and lose them all? What if it should play x more games and win half of them?

EXERCISE 173

1. The sum of two numbers equals 10 and their product equals 21. Find the numbers.

2. The sum of two numbers is 5.7, and the product of the numbers is 7.7. Find the numbers.

3. The difference of two numbers is 1.3 and their product is 9.5. Find the numbers.

4. The difference of two numbers is 1.8 and their product is 6.6. Find the numbers to the nearest hundredth.

5. Separate 32 into two parts such that their product shall be 112.

6. Two numbers when added produce 5.7, and when multiplied produce 8. Find the numbers.

7. What are the two parts of 18 whose product exceeds 8 times their difference by 1?

8. The sum of two numbers increased by three times their product is 83; also three times the less number exceeds the larger number by 1. Find the numbers.

In working the following examples concerning rectangles, *draw a diagram* for each rectangle considered.

9. The area of a rectangle is 84 sq. ft. and the distance around it (perimeter) is 38 ft. Find the length and breadth (dimensions) of the rectangle. Verify your answer by making on squared paper a drawing to scale of your rectangle (1 space = 1 ft.).

10. The perimeter of a given rectangle is 13.8 ft. and the area is 11.34 sq. ft. Find the dimensions.

11. The area of a rectangular garden is 1200 sq. yd. If the width were increased by 5 yd. and the length by 10 yd., the area would be 1750 sq. yd. Find the dimensions of the rectangle. Verify your answer by making a drawing to scale on squared paper (1 space = 5 yd.).

12. The area of a (double) tennis court is 312 sq. yd., and the perimeter is 76 yd. Find the dimensions of the court in feet.

13. If the dimensions of a rectangular field were each increased by 3 rd., its area would be 140 sq. rd.; but if its width were increased by 8 rd. and its length diminished by 2 rd., its area would be 135 sq. rd. Find its actual dimensions.

14. The area of a given rectangle is 800 sq. ft. If the length of the rectangle were increased by 20% and the width by 4 ft., the area will be increased by 44%. Find the dimensions of the rectangle.

15. By going 6 mi. an hour faster, a train would have required 1 hr. less to run 180 mi. How fast did it travel.

16. A man divided $9 equally among some boys. If he had begun by giving each boy 5 cents more, 6 of then would have received nothing. How many boys were there?

17. A number of men agreed to buy a boat for $7200, but 3 of their number died, and each survivor was obliged to contribute $400 more than he otherwise would have done. How many men were there?

18. A certain club owes a debt of $400, but is informed by the treasurer that if 5 new members are admitted, the assessment to meet the debt will be $4 less per member. How many members has the club?

19. A given fraction when reduced to its lowest terms equals $\frac{3}{4}$. Also if 3 is subtracted from the numerator of the fraction, the fraction is the same as if 6 had been added to its denominator. Find the fraction.

20. The numerator of a given improper fraction exceeds its denominator by 1. Also the given fraction exceeds its reciprocal by $\frac{7}{12}$. Find the fraction.

21. The sum of the numerator and denominator of a fraction is 8. If $2\frac{1}{2}$ be added to each term of the fraction, its value will be increased by $\frac{2}{15}$. Find the fraction.

22. One basket ball team has won 5 games out of 17 played, and another team has won 6 games out of 12 played. How many straight (*i.e.* consecutive) games must the first team win from the second in order that their averages of games won may be equal?

23. A baseball nine has won $\frac{2}{3}$ of the games played. If it should play 16 more games and win half of them, its average of games won would be $\frac{4}{5}$ of what it would be if it should play 8 more games and win all of them. How many games has it played, and how many has it won?

24. Divide the number 12 into two parts such that the sum of the fractions obtained by dividing 12 by the parts shall be $6\frac{4}{15}$.

25. Find two numbers whose product is 42, such that if the larger be divided by the less, the quotient is 4 and the remainder 2.

26. In placing telephone poles between two places, it was found that if the poles were placed 10 ft. farther apart than was first planned, 4 poles less per mile were needed. How far apart were the poles placed at first?

27. A girl has 12,000 words to write. Using a typewriter, she can write 25 words more per minute than with the pen, and it will take $8\frac{1}{3}$ hours less to write the 12,000 words. What is her rate per minute with the pen?

EXERCISE 174

1. If a body is dropped from an elevated place, as the top of a tower or an airplane, and the distance it falls (resistance of the air being neglected) in t seconds is denoted by s, then $s = \frac{1}{2} gt^2$, g being some number determined by the force of gravitation at the place of the falling body.

If $g = 32.2$ ft., by use of the above formula find how far a body will fall in 4 seconds.

2. Using the formula $s = \frac{1}{2} gt^2$ (where $g = 32.2$ ft.), find the distance a body will fall from the end of 5.2 sec. to the end of 7 sec.

3. A stone dropped from the top of a precipice was observed to strike the ground at the foot of the precipice in $11\frac{1}{2}$ seconds. How high was the precipice?

4. Solve $s = \frac{1}{2} gt^2$ for t.

By use of the formula obtained, solve Exs. 5 and 6.

5. How many seconds will it take a body to fall from rest a distance of 1000 ft. (resistance of air neglected)?

6. If an airplane is 5000 ft. high, how long will it take an object dropped from it to reach the ground ?

7. If the number of lumber feet in a log be denoted by L, the length of the log in feet by l, the diameter of the small end in inches by d, then $L = l\left(\dfrac{d-4}{4}\right)^2$. State this formula as a rule. Using this formula, find the number of lumber feet in a log 12 ft. long and 10 in. in diameter.

8. Solve the formula of Ex. 8 for d. By use of the formula obtained, determine what must be the diameter of a log 15 ft. long in order to contain 50 lumber feet. To contain 100 lumber feet.

9. In selling an article which cost c dollars, what must be the asking price a, in order to grant a discount of d per cent and yet make a gain of g per cent on the cost price ? (d and g represent per cents expressed decimally.)

Sug. $s = c + gc$. Also $s = a - da$. Put the two values of s equal to each other and find the value of a in terms of the other letters.

By use of this formula, solve Exs. 10–12.

10. If an article cost $6, in selling it what must be the asking price if a gain of 20 % is to be made after allowing a discount of 40 % ?

11. A merchant can buy a certain kind of hat for $2. Owing to liability of change of style, he wishes to make a profit of 80%, and yet wishes to give a discount of 10% on cash sales. What will be his asking price for the hat ?

12. A dealer can buy automobiles of a certain make for $2400. At what price should he advertise them if he wishes to be able to give a discount of 20 % and yet make 20 %?

13. From the formula found in Ex. 9, derive a formula for c in terms of d, a, and g. State this as a rule.

Solve Ex. 14 by the use of this formula.

14. A 10-cent store wishes to offer an article for 10 ¢, to be able to give a discount of 5 % when selling in large quantities, and yet make a profit of 20 per cent on the cost. What can the store afford to pay for the manufacture of the article? Find to the nearest hundredth of a cent.

15. Make up and work an example similar to Ex. 14, but using different per cents.

16. Solve the formula $a = \pi r^2$ for r. State your result as a rule.

17. Make up and solve a numerical example which is an application of the formula $r = \sqrt{\dfrac{a}{\pi}}$.

151. Graph of a Quadratic Equation of Two Unknown Quantities.

Ex. 1.   Construct the graph of $y = x^2 - 3x + 2$.

x	y	x	y
0	2	−1	− 6
1	0	−2	12
2	0	etc.	
3	2		
$\frac{3}{2}$	$-\frac{1}{4}$		
etc.			

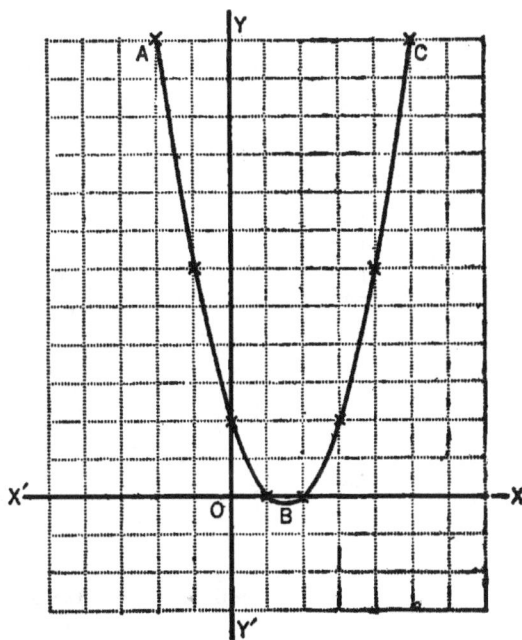

The graph obtained is the curve *ABC.* A curve of this kind is called a *parabola.* The path of a projectile, for instance that of a baseball when thrown or batted (resistance of the air being neglected), is an arc of a parabola.

It will be noted that the above method of graphing is the same as that given in § 114 (p. 247), but that here it is sometimes advantageous to let x have fractional values, as $\frac{1}{2}$, $\frac{1}{4}$, $\frac{1}{10}$, $\frac{3}{2}$, etc. The observant pupil will also find methods of abbreviating the work in certain cases.

In general, it will be found that the graph of a quadratic equation of two unknown quantities is a curved line, and, in particular, either a circle, parabola, ellipse, or hyperbola.

Ex. 2. Construct the graph of $y^2 = 2x$.

Solving the given equation for y, we obtain $y = \pm\sqrt{2x}$.

x	y
0	0
1	± 1.4
2	± 2
3	± 2.4
4	± 2.8
8	± 4
etc.	

x	y
-1	$\pm\sqrt{-2}$
-2	$\pm\sqrt{-4}$
etc.	

When x is negative no value of y can be constructed on the graph, for there is no positive or negative number which when multiplied by itself produces -2, -4, or any other negative number. Hence no part of the curve exists to the left of the origin.

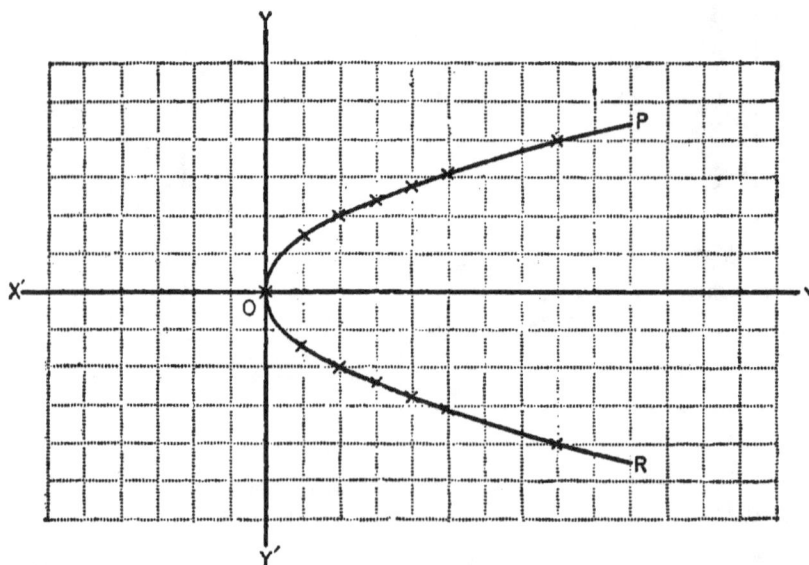

The graph obtained is the curve POR, which is also a parabola.

Ex. 3.   Graph $y=15\,x^2$.

x	y	x	y
0	0	-1	15
1	15	-2	60
2	60	-3	135
3	135	-4	240
4	240	etc.	
etc.			

Since most of the values of y are very large in comparison with the values of x, it is convenient to let each space on the y-axis represent twenty times as large a number as each space on the x-axis.

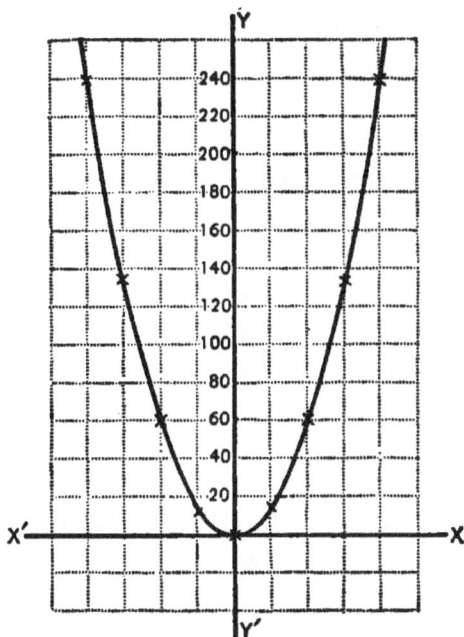

Graph the following:

1. $y = x^2 - 1$.
2. $y = x^2 - 2\,x - 3$.
3. $y = x^2 - 4\,x + 4$.
4. $y = x^2 + 3\,x - 4$.
5. $y = \frac{1}{4}\,x^2$.
6. $y = x^2 + 1$.
7. $y = 1.3\,x^2$.

8. $y^2 = 4\,x$.
9. $y^2 = x$.
10. $y^2 = x - 1$.
11. $y^2 = 2\,x - 3$.
12. $y^2 = 3\,x + 1$.
13. $2\,y^2 = 3\,x$.
14. $7\,y = 22\,x^2$.

15. The formula for the area of a circle is $a = \pi r^2$. Graph this formula, letting each space on the a (or vertical) axis represent 10 times as large a number as each space on the horizontal or r axis. If the paper is available, much more accurate results can be obtained by constructing the graph on a large scale on squared paper ruled in centimeters (see p. 38).

16. By use of the formula $a = \pi r^2$, find the area of the circle whose radius is 2.5 in. Check your result roughly by use of the graph obtained in Ex. 15.

17. By use of this graph find as nearly as you can the radius of a circle whose area is 30 sq. in.

18. If the only sizes of pipe available are those whose diameters are an exact number of inches, find from the graph the diameters of the pipe to be used when the area of the cross section of the pipe is to be at least 40 sq. in. Is to be 60 sq. in. To be 100 sq. in.

19. If $g = 32$, the formula $s = \frac{1}{2} gt^2$ reduces to $s = 16\, t^2$. Graph $s = 16\, t^2$, letting each space on the vertical or s-axis represent 30 times as large a number as each space on the horizontal or t-axis.

20. From this graph determine the number of feet a body will fall in $3\frac{1}{2}$ seconds. Check your result by use of the formula $s = 16\, t^2$.

21. From the graph obtained in Ex. 19, determine the number of seconds it will take a body to fall 100 ft. Check by use of the formula.

22. Graph the formula $s = \sqrt{\dfrac{3\,h}{2}}$ (see Ex. 10, p. 266). By use of the formula, determine at a distance of how many miles the light of a lighthouse may be seen, if the light is 85 ft. above sea level. Can you check your answer by use of the graph?

23. By use of the graph, determine how high above sea level a light must be in order to be visible at a distance of 10 miles.

GENERAL REVIEW

1. Draw a line 8 in. long and on it mark the whole, half, and quarter inches. Denote the middle point of the line by the letter O. Mark with another letter the point on the line corresponding to $+3$ in. To -4 in. To $+3\frac{1}{4}$ in. To $-2\frac{1}{4}$ in.

2. Find the results of the following:

(1) $-8.2 + 16.3$.   (5) $-66 - (-1.7)$.   (9) $45 + -3$.

(2) $-4\frac{1}{2} + (-8\frac{3}{4})$.   (6) -2.7×3.4.   (10) $-58 + (-.2)$.

(3) $7.3 - (-3.6)$.   (7) $-5\frac{1}{2} \times (-3\frac{1}{2})$.   (11) $(4\,x^2y)(-1.5xy)$.

(4) $47 - 3.7$.   (8) $-4.8 \div 3$.   (12) $p^4r^2s^3 \div p^3r^2s$.

3. Add $-5\,a^2$, $9\,a^2$, $3\,ab - 4\,b^2$, $16\,ab + 9\,b^2$, $-3\,a^2 - 2\,b^2$.

4. Add ay^3, by^3, $-3\,cy^3$.

5. From $15\,x^3 - x^2y + 5\,xy^2 - y^3$ subtract $-2\,x^3 - 4\,x^2y + 2\,xy^2$.

6. Subtract $rx^2 - py^2$ from $ax^2 + 2\,bxy - 3\,cy^2$.

7. Simplify $3.7\,p^2 - [8.4\,p^2 - (p^2 - 4) + 9.2]$.

8. Multiply $5\,x^2 - y^2$ by $x^2 - 4\,y^2$.

9. Multiply $a^2 + 3\,ab + b^2$ by $a^2 - 3\,ab + b^2$.

10. Divide $3\,x^4 - 7\,x^3 - x^2 + 7\,x - 2$ by $x^2 - 3\,x + 2$.

11. Divide $16\,a^4 + 4\,a^2x^2 + x^4$ by $4\,a^2 + 2\,ax + x^2$ and check by letting $a = 1$ and $x = 1$.

12. If $t = 10$ and $u = 1$, perform the arithmetical operation expressed in algebraic form by $(3\,t + u)(4\,t + 3\,u) = 12t^2 + 13\,tu + 3\,u^2$.

13. The formula for the area of a trapezoid is $a = \frac{1}{2}\,h(b + b')$. Find a when $h = 8$, $b = 6$, and $b' = 7\frac{1}{2}$.

14. Factor:

(1) $16\,p^2x^2 - x^2$.   (5) $x^2 - 4\,x - 77$.

(2) $16\,a^2b^2 - 8\,ab + 1$.   (6) $(3\,a + b)^2 - 4\,x^2$.

(3) $25\,a^2b^2 - 169\,x^2$.   (7) $(3\,r^2 + 1)^2 - (3\,s^2 - 1)^2$.

(4) $4\,x^3 - 12\,x^2y^3 + 9\,xy^4$.   (8) $x^3 - x^2y - 6\,xy^2$.

15. Arrange $15.8^2 - 9.7^2$ in a form more convenient for numerical computation and find the numerical value of the result.

16. Reduce $\dfrac{9\,x^3 - 25\,a^2x}{9\,x^4 - 30\,ax^3 + 25\,a^2x^2}$ to its lowest terms.

17. Change $\dfrac{15\,x^2 - 7}{x^2 + 5}$ to a fraction with $x^3 - 3\,x^2 + 5\,x - 15$ for its denominator.

18. Change $\dfrac{9\,m^2 + 3\,p^2}{3\,m - p}$ to a mixed number.

19. Change $3\,a^2 - 5\,a + 2 - \dfrac{4\,a + 5}{2\,a - 1}$ to a fraction.

20. Simplify $\dfrac{x^2 - 7}{9\,x^2 - 4} + \dfrac{x + 3}{3\,x + 2} - \dfrac{3\,x - 5}{3\,x - 2}$ and check your answer by letting $x = 2$.

21. Simplify $\dfrac{r^2 - 9\,s^2}{r^2 - 16\,s^2} \div \dfrac{r - 3\,s}{r + 4\,s}$ and check by letting $r = 1$ and $s = 1$.

22. By substituting numerical values for b, determine whether $\dfrac{b}{2} - \dfrac{3\,b}{5} = -\dfrac{b}{10}$. Is this an equation or an identity?

Solve to the nearest hundredth:

23. $5.72\,x - 2.83 = 3.19x + 5.16$.

24. $8\,x + 5\,y = 11,\ \ 15\,x - 5.3\,y = 8$.

25. Solve $\dfrac{b}{x} - 3\,c = \dfrac{a}{3\,x} + 7$, for x.

26. Solve $\dfrac{x}{p} + \dfrac{y}{q} = r,\ \dfrac{5\,x}{q} - \dfrac{3\,y}{p} = 7\,r$, for x and y.

27. Solve the formula $R = \dfrac{pq}{p + q}$, for q.

28. In $R = \dfrac{ae}{c}(1 + cd)$, find d in terms of the other letters.

29. Is 2.3 a root of the equation $5\,x - 7 = 3.5$?

30. Abbreviating the work as much as possible find the value of $p\sqrt{4\,a^2 - b^2}$ when $p = 9.5$, $a = 2.4$, and $b = 2.2$.

31. Find to the nearest tenth the value of $\sqrt{a^2p - b^2q}$ when $a = 5$, $p = 16$, $b = 4$, $q = 3$.

32. Simplify $5\sqrt{x^3y} - 3\sqrt{xy^3} - 4xy\sqrt{\dfrac{1}{xy}}$.

33. Simplify $5\sqrt{(x - y)^3} - 3\sqrt{(x^2 - y^2)(x + y)}$.

34. Multiply $\sqrt{a^2 - 5ab + 4b^2}$ by $\sqrt{a^2 - b^2}$ and simplify.

35. Extract the square root of $x^4 - 6x^3 + 12x^2 - 9x + \frac{9}{4}$.

36. Extract the square root of 19.876 to the nearest thousandth.

37. Solve $5 + p - \dfrac{p^2 - 7p}{5} = 12$.

38. Solve $2.3x^2 + 3.8x - 2 = 0$ to the nearest hundredth.

39. Solve $x^4 - 5x^2 + 2 = 0$ to the nearest tenth.

40. Determine whether 2.5 and -9 are the roots of $2x^2 + 13x - 45 = 0$.

41. Solve $11 = x + 2(y - 1)$, $6 = \dfrac{y}{2}(x + 1)$ and check.

42. Solve $T = 2\pi r(r + h)$, for r.

43. The population of a village in the year 1900 was 1210; in 1905 was 1308; in 1910 was 1452; in 1915 was 1723; in 1920 was 2163. Make a graph showing the growth in population.

44. Express algebraically the following statement: a divided by b gives c as a quotient and d as a remainder.

45. What is the dividend when the quotient is $x^3 + 2x^2 + 7x + 20$, the remainder $62x + 59$, and the divisor $x^2 - 2x - 3$?

46. What is the divisor if the quotient is $x^3 + 3x$, the dividend $x^5 - 8$, and the remainder $9x - 8$?

47. If $x = -\frac{2}{3}$ and $y = -\frac{3}{2}$, find the value of
$(3x - 2y)^2(9x^2 + 4y^2) - 6(y - x)\sqrt{6xy(x + 2y^2 + \frac{1}{8})}$.

48. Add a to b. Also add $3a - 5b$ to $4c + 7d$.

49. Subtract $3x - 2y + z$ from -7. From $-a$. From b.

50. Subtract $2a - 3b$ from 0. Also 5 from 0.

51. Can $3 + 2ab$ be united in a single term? Give a reason.

52. The product of an even number of negative factors has what sign? Of an odd number of negative factors? Give an example using not less than five factors.

53. Express the following in a simpler form: $5\,aaa(x-y)$ $(x-y)(x-y)(x \div y)$.

54. If a boy's mark on each of three recitations is 0, what is his average for the three recitations? Give the value of $0+0+0$. Of 3×0. Of $\frac{0}{3}$.

55. Find the value of $8 \times 0 - 5 + \frac{0}{7}$. Also of $\dfrac{a-a}{b}$. Of $\dfrac{a}{b} \cdot \dfrac{b}{a}$.

56. Prove that if half of the sum of any two numbers (as of a and b) is added to half their difference, the result will equal the greater of the numbers. Prove by two numerical examples.

57. Prove that if half the difference of any two numbers is subtracted from half their sum, the result will be the smaller of the two numbers.

58. Write the power whose base is a and exponent 4. Whose base is $3\,a$ and exponent 5.

59. Simplify $\sqrt{x^2 - 4\,xy + 3\,y^2} \cdot \sqrt{x^2 - 9\,y^2}$.

60. Why is it allowable to change both minus signs to plus in $-x = -3$, and not in $-x-3$?

61. Solve and check: $\dfrac{3}{x} + \dfrac{4}{y} = -6$, $\quad \dfrac{5}{x} - \dfrac{2}{y} = 16$.

62. In an election for two candidates 32,544 votes were cast. The successful candidate had a majority of 2416 votes. How many votes did each candidate receive?

63. Factor (1) $16\,x^2 + 2.4\,x + .09$. (2) $x^4 - .0081$.

64. In a certain kind of concrete, twice as much sand is used as cement, and twice as much gravel as sand. How many pounds of each are used in making 2800 lb. of concrete?

65. Letting the horizontal axis represent the days and the vertical axis the attendance, represent by bars of proper length the class attendance, given the statement that on Monday it was 30; on Tuesday, 32; on Wednesday, 28; on Thursday, 30; on Friday, 26.

66. Solve $\dfrac{x}{p} - \dfrac{2x}{q} = \dfrac{q^2 - 4p^2}{pq}$.

67. Represent algebraically the equality existing between the difference of the squares of two numbers and the product of the sum and difference of the numbers. Is this equality an identity or an equation?

68. Solve $\dfrac{2x}{3} + y = z$, $z - 2y = 2$, $x + z = 4$, and check.

69. A man has $20,000, of which part is invested at 5% and the rest at 7%. His annual income is $1160. How much has he invested at each rate?

70. Copy the following tabulation and fill in the vacant place with the needed rule or formula:

SUBJECT	BRIEF RULE	FORMULA
Area of floor	No. sq. yd. in floor = (no. ft. in length) × (no. ft. in width) ÷ 9	
Percentage		$b = \dfrac{p}{r}$
Area of circle	Area of circle equals π times the radius squared	
Circumference		$c = 2\pi r$
Division	Dividend diminished by remainder = product of divisor and quotient	
Multiplication		$M = \dfrac{p}{m}$
Number bricks in wall		$B = \dfrac{6\,lwh}{11}$.
Radius of circle		$r = \sqrt{\dfrac{a}{\pi}}$

71. Arrange and extract the square root of:
$$x^3 - \tfrac{3}{2}x + \tfrac{9}{4} + x^4 - \tfrac{11}{4}x^2.$$

72. During a certain year the rainfall at a certain place per month in inches was as follows :

Month	Jan.	Feb.	Mar.	Apr.	May	June	July	Aug.	Sept.	Oct.	Nov.	Dec.
Rainfall	3.2	2.2	2.8	2.7	1.8	1.2	0.8	2.3	2.7	1.7	4.6	2.1

Graph these facts by a broken line.

73. A man is c times as old as his younger son and is d years older than his elder son. If the younger son is y years old, find the sum of the ages of the three persons.

74. Write the number whose digits in order are p, q, r. Why may not such a number be written as pqr ?

75. Write a binomial the factors of which are three different binomials.

76. Write a trinomial the factors of which are two different binomials and a monomial.

77. Solve $\frac{1}{2}(x - y) = x - 4$, $xy = 2x + y + 2$, and check.

78. What factor exactly divides both $x^2 - 4$ and $x^2 - 4x + 4$?

79. What is the smallest expression into which both $x^2 - 4$ and $x^2 - 4x + 4$ can be divided exactly ?

80. If the diagonal of a rectangle is 13.73 and one side is 5.27, compute the other side to the nearest hundredth.

81. A boy goes to the store with $5 and buys x pounds of sugar at c cents each. How much change will he receive? Express the result in cents. Also in dollars.

82. In a certain high school the pupils in different grades numbered as follows : first year, 237 ; second, 202 ; third, 172 ; fourth, 142. Make a circle graph showing the relative size of the four grades.

83. Solve $\dfrac{8}{x} - \dfrac{x+1}{x-1} + \dfrac{x-3}{x+3} = 0$.

84. If post cards are bought at a cents per dozen and sold 2 for b cents, how much is gained on 12 dozen ?

85. Graph on the same diagram (1) the growth of railroads in the United States; (2) the growth of population of the United States. (See Ex. 35, p. 216, and Ex. 7, p. 150).

86. Solve $lx - br - p(a-y) = 0,$ $\dfrac{lx}{a} + r - p\left(1+\dfrac{y}{b}\right) = 0,$ for x and y.

87. In the following cancel the 3's where it is allowable:
$$\frac{x+3}{y+3}, \frac{3+x}{3y}, \frac{3(a+b)}{3(a-b)}, \frac{3(a+b)}{3x}, \frac{3(a+b)}{3a}.$$

88. A man starts on a journey of m miles and travels a miles an hour for b hours. What fractional part of the journey does he complete?

89. From $4\sqrt{4(a-2b)^2(a-b)} + 4\sqrt{(a^2-b^2)(a+b)}$ subtract $5\sqrt{(a-b)^3}$ and simplify. Check your work.

90. On the same diagram construct graphs of $c = 2\pi r$ and $a = \pi r^2$. (Let one space on the vertical axis represent 10, and one space on the horizontal axis represent 1.)

91. In the formula $e = 1 + [.000017\,(t-20)]$, find e when $t = 85$.

92. Is it allowable to divide each term of $16x = 96$ by 16? Is it allowable to divide each term of $16x - 96$ by 16? Give reasons.

93. Does $\sqrt{a^2+b^2}$ equal $\sqrt{a^2}+\sqrt{b^2}$? Does $\sqrt{a^2b^2}$ equal $\sqrt{a^2}\cdot\sqrt{b^2}$? Justify your answer by letting $a=3$ and $b=4$.

94. In $5x^2 - 3x - 3 = 0$ find the value of x to the nearest hundredth.

95. Show that the sum of two numbers (as a and b), divided by the sum of their reciprocals, equals the product of the two given numbers. Is this relation an equation or an identity?

96. Extract the square root of $x^2 - \dfrac{2xy}{3} + \dfrac{13y^2}{9} - \dfrac{4y^3}{9x} + \dfrac{4y^4}{9x^2}$.

97. When we change $x - 3 = 5$ to $x = 5 + 3$, what is the change called? What right have we to make this change? What is the advantage of making this change?

98. Simplify $\dfrac{a^2-4\,ab^2+4b^4}{3\,a^2+3\,a} \times \dfrac{12}{a^4-16\,b^8} \div \left(1-\dfrac{2\,b^2}{a}\right)$ and check.

99. If r per cent of p is x, find r per cent of y.

100. The United States 5 cent piece (or nickel) is 75 % copper and 25 % nickel. If a mass of nickel and copper weighing 80 pounds is 90 % copper, how many pounds of nickel must be added to it to make it ready for coinage into 5 cent pieces ?

101. Separate 200 into three such parts that the first divided by the second gives 2 for a quotient and 2 for a remainder; and the second divided by the third gives 4 for a quotient and 1 for a remainder.

102. Solve $\dfrac{s-1}{2}=\dfrac{5}{2}-\dfrac{2}{s-1}$ and check.

103. Extract the square root of

$$x^2 - 4\,ax + 4\,a^2 + \frac{2\,bx}{3} - \frac{4\,ab}{3} + \frac{b^2}{9}$$

104. A baseball player has been to the bat 150 times in a given season and made an average of .280 hits. How many more times will he need to bat to bring his average up to .375, provided that the number of base hits he makes in the future equals half the number of times he bats ?

105. Solve $3\,x - 2\,y = 9$, $2\,x^2 + 10 - 3\,y^2 = 5\,xy$ and check.

106. In the year 1913, the wheat crops in the leading wheat-producing countries of the world in millions of bushels were as follows :

Russia, 813	Canada, 211	Spain, 110
United States, 764	Italy, 209	Prussia, 94
India, 358	Argentina, 198	Australia, 95
France, 320	Hungary, 150	Roumania, 88

Make a bar graph of these numbers.

107. A given alloy of gold and silver weighs 106 lb.; when immersed in water it loses 7 lb. of its weight. If gold loses $\frac{1}{19}$ and silver $\frac{1}{10}$ of its weight when weighed in water, how many pounds of each of the metals are in the alloy ?

CHAPTER XVI

ADDITIONAL TOPICS

LITERAL AND FRACTIONAL EXPONENTS

152. Multiplication of Expressions Containing Literal Exponents.

Ex. 1. Multiply x^{n+1} by x^3.

Since $n + 1$ added to 3 gives $n + 4$, the product is x^{n+4}.

Ex. 2. Multiply $3\,a^{4x}b^{3v}c^{2z}$ by $5\,a^{2x}b^{5v}c^{z}$.

$$3\,a^{4x}b^{3v}c^{2z}$$
$$5\,a^{2x}b^{5v}c^{z}$$
$$\overline{15\,a^{6x}b^{8v}c^{3z}}. \quad Ans.$$

Ex. 3. Multiply $5\,a^{2n+3} - 3\,a^{2n+2}$ by $3\,a^{n+1} - 4\,a^{n}$.

$$5\,a^{2n+3} - 3\,a^{2n+2}$$
$$3\,a^{n+1} - 4\,a^{n}$$
$$\overline{15\,a^{3n+4} - 9\,a^{3n+3}}$$
$$\qquad\quad - 20\,a^{3n+3} + 12\,a^{3n+2}$$
$$\overline{15\,a^{3n+4} - 29\,a^{3n+3} + 12\,a^{3n+2}}. \quad Ans.$$

153. Division of Expressions Containing Literal Exponents.

Ex. 1. Divide $6\,x^{5a}y^{4b}z^{2c}$ by $2\,x^{3a}y^{2c}$.

$$\frac{6\,x^{5a}y^{4b}z^{2c}}{2\,x^{3a}z^{2c}} = 3\,x^{2a}y^{4b}. \quad Ans.$$

Ex. 2.   Divide $15\,a^{6x} + a^{3x}b^{2y} - 2\,b^{4y}$ by $3\,a^{3x} - b^{2y}$.

$$
\begin{array}{l}
15\,a^{6x} + \;\; a^{3x}b^{2y} - 2\,b^{4y} \;\big|\; 3\,a^{3x} - b^{2y} \\
\underline{15\,a^{6x} - 5\,a^{3x}b^{2y}} \;\;\big|\; \overline{5\,a^{3x} + 2\,b^{2y}.} \quad \textit{Ans.} \\
\qquad\quad 6\,a^{3x}b^{2y} - 2\,b^{4y} \\
\qquad\quad \underline{6\,a^{3x}b^{2y} - 2\,b^{4y}}
\end{array}
$$

EXERCISE 177

Multiply:

1. x^{n+3}
   x^{2}

2. $5\,x^{n-1}$
   $-3\,x^{4}$

3. $8\,a^{x-3}$
   $4\,a^{x-2}$

4. $3\,a^{n-1}b^{n}$
   $-2\,a^{n+3}b^{2n}$

5. $5\,a^{4x}b^{3x}c^{2x}$
   $6\,a^{3x}b^{x}c^{3x}$

6. $7\,x^{3a}y^{2b}z^{3c}$
   $\cdot\,4\,x^{2a}z^{5c}$

7. $8\,a^{2x}b^{n-1}$ by $3\,a^{x-2}b^{3}$.

8. Find the product of $3\,x^{n-2}$, $5\,x^{n+3}$, and x^{2n}.

9. Also of $5\,a^{3x}b^{2y}$, $4\,a^{2x}b^{3y}$, $5\,a^{4x}b^{7y}$.

Multiply:

10. $3\,x^{n+2} - 5\,x^{n+1} + 4\,x^{n}$ by x^{3}.
11. $3\,x^{n+2} - 5\,x^{n+1} + 4\,x^{n}$ by x^{3n}.
12. $4\,a^{3x}b^{2y} - 3\,a^{2x}b^{y}$ by $-5\,a^{x}b^{y}$.
13. $5\,x^{n+1} - 3\,x^{n}$ by $3\,x + 5$.
14. $x^{3n} + 4\,y^{2n}$ by $x^{3n} - 4\,y^{2n}$.
15. $8\,a^{x+4} - 5\,a^{x+3}$ by $3\,a^{x+1} - 2\,a^{x}$.
16. $a^{n+3} + 2\,a^{n+2} - 3\,a^{n+1}$ by $2\,a^{n+1} - 3\,a^{n}$.
17. $x^{3a} - 3\,x^{2a} + 3\,x^{a} - 1$ by $x^{2a} + 2\,x^{a} + 1$.

Divide:

18. $6\,a^{n+5}$ by $3\,a^{4}$.
19. $8\,a^{5x}$ by $4\,a^{2x}$.
20. $-27\,a^{3n+5}b^{4n}$ by $-9\,a^{n+2}b^{3n}$.
21. $a^{4x}b^{2y}c^{5x}$ by $a^{2x}b^{y}c^{5x}$.
22. $24\,x^{5n} - 15\,x^{4n} - 3\,x^{3n}$ by x^{2n}.
23. $28\,x^{4n} + x^{3n} - 15\,x^{2n}$ by $7\,x^{2n} - 5\,x^{n}$.
24. $4\,x^{2n+3} - 4\,x^{2n+2} + 7\,x^{2n+1} + 5\,x^{2n}$ by $2\,x^{n+1} + x^{n}$.
25. $6\,a^{5x} + 13\,a^{4x} + 8\,a^{3x} + 3\,a^{2x}$ by $2\,a^{3x} + 3\,a^{2x}$.

Factor:

26. $6x^{5n} - 3x^{3n}$.

27. $12a^{3x}b^{2y} - 8a^{2x}b^{y}$.

28. $x^{2n} + 6x^{n} + 9$.

29. $4x^{2n} - 12x^{n} + 9$.

30. $x^{2n} - 4x^{n} + 3$.

31. $x^{4n} - 6x^{2n} + 9$.

32. $x^{2n} - 4x^{n}y^{n} + 4y^{2n}$.

33. $x^{2n} - 1$.

34. $x^{6y} - 1$.

35. $9x^{2n} - 4y^{2}$.

CASE V IN FACTORING

154. Product of Two Binomials whose Corresponding Terms are Similar.

Ex. Multiply $2a - 3b$ by $4a + 5b$.

By actual multiplication,

$$\begin{array}{r} 2a - 3b \\ 4a + 5b \\ \hline 8a^2 - 12ab \\ + 10ab - 15b^2 \\ \hline 8a^2 - 2ab - 15b^2. \quad Product. \end{array}$$

We see that the middle term of this product may be obtained directly from the two binomials by taking the algebraic sum of the cross products of their terms. Thus,

$$(+2a)(+5b)+(-3b)(+4a)=10ab - 12ab = -2ab.$$

Hence, in general,

The product of any two binomials of the given form consists of three terms:

The first term is the product of the first terms of the binomials;

The third term is the product of the second terms of the binomials;

The middle term is formed by taking the algebraic sum of the cross products of the terms of the binomials.

Ex. Multiply $10\,x + 7\,y$ by $8\,x - 11\,y$.

To show the method of obtaining the middle term of the product, we write the given expression in the form

$$(\overbrace{10\,x + 7\,y)(8\,x - 11}\,y)$$

Hence,

$$(10\,x)(-11\,y) + (7\,y)(8\,x) = -110\,xy + 56\,xy = -54\,xy.$$

$$\therefore\ (10\,x + 7\,y)(8\,x - 11\,y) = 80\,x^2 - 54\,xy - 77\,y^2.\quad \textit{Product.}$$

EXERCISE 178

Write at sight the product of each of the following and check each result:

1. $(2\,x + 3)(x + 4)$.
2. $(2\,x - 3)(x - 4)$.
3. $(2\,x + 3)(x - 4)$.
4. $(2\,x - 3)(x + 4)$.
5. $(3\,a + 5)(2\,a + 3)$.
6. $(3\,a - 5)(2a + 3)$.
7. $(5\,x - 1)(x + 7)$.
8. $(x + 3\,y)(3\,x - 8\,y)$.
9. $(3\,a^2 + b)(4\,a^2 - 5\,b)$.
10. $(x + \frac{1}{2})(\frac{3}{4}x + \frac{1}{2})$.
11. $(a + .2\,b)(2\,a - .3\,b)$.
12. $(\frac{1}{2}x + \frac{3}{2}a)(\frac{2}{3}x - \frac{1}{3}a)$.

155. Factoring a Trinomial of the Form $ax^2 + bx + c$. — From § 154 it is evident that the essential part of the process of factoring a trinomial of the form $ax^2 + bx + c$ lies in determining two factors of the first term and two factors of the last term, such that the algebraic sum of the cross products of these factors equals the middle term of the trinomial.

Ex. Factor $10\,x^2 + 13\,x - 3$.

The possible factors of the first term are $10\,x$ and x, $5\,x$ and $2\,x$. The possible factors of the third term are -3 and 1, 3 and -1. In order to determine which of these pairs will give $+13x$ as the sum of their cross products, it is convenient to arrange the pairs thus:

$$10x, -3 \qquad 5x, -1$$
$$\diagdown \diagup \qquad \diagdown \diagup$$
$$\diagup \diagdown \qquad \diagup \diagdown$$
$$x, \quad 1 \qquad 2x, \quad 3$$

Variations may be made mentally by transferring the minus sign from 3 to 1; and also by interchanging the 3 and the 1.

It is found that the sum of the cross products of

$$\begin{matrix} 5x, & -1 \\ 2x, & 3 \end{matrix} \quad \text{is} +13x.$$

Hence, $10x^2 + 13x - 3 = (5x - 1)(2x + 3)$. *Factors.*
Let the pupil check the work.

Hence, in general, to factor a trinomial of the form

$$ax^2 + bx + c,$$

Separate the first term into two such factors, and the third term into two such factors, that the sum of their cross products equals the middle term of the trinomial;

As arranged for cross multiplication, the upper pair taken together and the lower pair taken together form the two factors.

EXERCISE 179

Factor and check:

1. $2x^2 + 3x + 1$.
2. $3x^2 - 14x + 8$.
3. $2x^2 + 5x + 2$.
4. $3x^2 + 10x + 3$.
5. $6x^2 + 7x - 5$.
6. $2x^2 + 5x - 3$.
7. $6x^3 + 20x^2 - 16x$.
8. $3x^4 - 4x^3 - 4x^2$.

9. $8a^2 + 2a - 15$.
10. $2x^2 + x - 10$.
11. $12x^2 - 5x - 2$.
12. $4x^2 + 11x - 3$.
13. $5x^2 + 24x - 5$.
14. $9x^3 - 15x^2 - 6x$.
15. $6x^2y - 2xy - 4y$.
16. $.06a^2 + 2.3a + 20$.

17. $12\,x^2 + xy - 63\,y^2.$ **23.** $25\,a^4 + 9\,a^2b^2 - 16\,b^4.$

18. $32\,a^2 + 4\,ab - 45\,b^2.$ **24.** $16\,x^4 - 10\,x^2y^2 - 9\,y^4.$

19. $4\,x^4 - 13\,x^2 + 9.$ **25.** $4.4\,a^2 - 2.33\,ab + .15\,b^2.$

20. $9\,x^4 - 148\,x^2 + 64.$ **26.** $25\,a^4 - 41\,a^2b^2 + 16\,b^4.$

21. $.15\,p^2 - .07\,p - .04.$ **27.** $20 - 9\,x - 20\,x^2.$

22. $24\,x^3 + 104\,x^2y^2 - 18\,xy^4.$ **28.** $5 + 32\,xy - 21\,x^2y^2.$

29. $(a + b)^2 + 5(a + b) - 24.$

30. $3(x - y)^2 + 7(x - y)z - 6\,z^2.$

31. Make up and work an example in which are comprised Cases I and V in factoring.

32. Make up and work an example in which are comprised Cases III and V.

33. Make up and work an example in which are comprised Cases I, III, and V.

EXERCISE 180

Reduce to the simplest form :

1. $\dfrac{x^2 - x - 6}{2\,x^2 - 11\,x + 15}.$

2. $\dfrac{2\,x^2 - 8\,y^2}{4\,x^2 - 2\,xy - 12\,y^2}.$

3. $\dfrac{6\,x^2 - xy - 2\,y^2}{6\,x^2 - 7\,xy + 2\,y^2}.$

4. $\dfrac{12\,x^3 - 2\,ax^2 - 24\,a^2x}{4\,x^4 - 2\,ax^3 - 6\,a^2x^2}.$

Simplify :

5. $\dfrac{5}{2\,a^2 - 7\,ab + 6\,b^2} + \dfrac{3}{8\,a^2 - 10\,ab - 3\,b^2}.$

6. $\dfrac{x + 1}{6\,x^2 + 7\,x - 20} - \dfrac{2\,x}{6\,x^2 + 13\,x - 5}.$

7. $\dfrac{x + 2}{2\,x^2 + x - 1} - \dfrac{x - 3}{4\,x^2 - 1} + \dfrac{2\,x + 5}{2\,x^2 + 3\,x + 1}.$

8. $\dfrac{7}{2\,p^2+5\,pq-3\,q^2} - \dfrac{3}{8\,p^2+10\,pq-7\,q^2}$
$$+ \dfrac{4}{4\,p^2+19\,pq+21\,q^2}.$$

9. $\dfrac{12\,x^2-27\,y^2}{15\,x^2-7\,xy-2\,y^2} \cdot \dfrac{15\,x+3\,y}{8\,x+12\,y}.$

10. $\dfrac{4\,x^2-2\,x-2}{2\,x^2+x-1} \div \dfrac{4\,x^2-4}{4\,x^2-1}.$

11. $\dfrac{2\,x-3}{x+1} \cdot \dfrac{3\,x^2+x-2}{4\,x^2-4\,x-3} \div \dfrac{9\,x^2-4}{4\,x^2-1}.$

12. $\dfrac{6\,x^2-5\,x-4}{2\,x^2+7\,x-4} \cdot \dfrac{6\,x^2+x-2}{4\,x^2-4\,x-3} \div \dfrac{9\,x^2-6\,x-8}{2\,x^2+5\,x-12}.$

Solve and check : ❦

13. $\dfrac{3}{2\,x-1} - \dfrac{2}{x+3} - \dfrac{5}{2\,x^2+5\,x-3} = 0.$

14. $\dfrac{3}{3\,x^2-4\,x+1} = \dfrac{9}{4\,x^2-4}.$

FACTORIAL METHOD OF SOLVING QUADRATIC EQUATIONS

156. Illustrative Examples.

Ex. 1. Solve $x^2 + 5\,x - 24 = 0$ by the factorial method.
Factoring the left-hand member, we obtain

$$(x+8)(x-3)=0.$$

If any factor of a product equals zero, the entire product equals zero. Hence to obtain the roots for the above equation, we may let each factor in the left-hand member equal zero and obtain the value of x from the two resulting simple equations.

Hence we have for the above equation

$$x + 8 = 0$$
$$x = -8. \quad Root.$$
Check for $x = -8$
$$x^2 + 5x - 24$$
$$= 64 - 40 - 24$$
$$= 0$$

Also $x - 3 = 0$
$$x = 3. \quad Root.$$
Check for $x = 3$
$$x^2 + 5x - 24$$
$$= 9 + 15 - 24$$
$$= 24 - 24$$
$$= 0.$$

Ex. 2. Solve $3x^2 - 4x - 4 = 0$ by the factorial method.

Factoring, $(x - 2)(3x + 2) = 0.$

Hence, $x = 2, -\frac{2}{3}. \quad Ans.$

Let the pupil check the work.

EXERCISE 181

Solve by the factorial method and check :

1. $x^2 - 5x + 6 = 0.$

2. $x^2 - 4 = 0.$

3. $x^2 - x - 2 = 0.$

4. $x^2 + 8x + 7 = 0.$

5. $2x^2 - 3x + 1 = 0.$

6. $2x^2 - 5x + 2 = 0.$

7. $3x^2 + 4x - 4 = 0.$

8. $x^2 - x = 6.$

9. $3x^2 + 3 = 10x.$

10. $8r^2 + 2r = 15.$

11. $24x^2 = 2x + 15.$

12. $x^2 + \dfrac{11x}{4} = \dfrac{3}{4}.$

13. $15 = x^2 + \dfrac{7x}{6}.$

14. $x^2 - 7ax + 12a^2 = 0.$

15. $3x^2 - 4ax - 7a^2 = 6.$

16. $5x^2 - 6bx = 8b^2.$

17. $5a^2x^2 - 6ax = 8.$

18. $2x^2 + \dfrac{x}{a} = \dfrac{3}{a^2}.$

19. $3a^2x^2 + 10ax = 8.$

20. $x^4 - 10x^2 + 9 = 0.$

21. $4x^4 - 13x^2 + 9 = 0.$

22. $x^4 + 2 = 3x^2.$

23. $9x^4 - 10x^2 + 1 = 0.$

24. $9p^4 + 25 = 34p^2.$

25. $\dfrac{2x^2}{3} + 3\tfrac{1}{2} = \dfrac{x}{2} + 8.$ **26.** $\dfrac{2x-1}{x+3} - \dfrac{x+1}{2x-3} = -\dfrac{1}{2}.$

27. $\dfrac{2}{3x-12} - \dfrac{5}{4x-8} - \dfrac{1}{4} = 0.$

28. $\dfrac{3x+5}{x+4} + \dfrac{2x-5}{x-2} = 3.$

29. $\dfrac{x-2}{6} - \dfrac{8-x}{2} - \dfrac{2x-11}{x-3} = 0.$

30. The square of a certain number diminished by 4 times the number equals 45. Find the number.

31. The square of a certain number increased by 6 times the number equals 40. Find the number.

32. What number plus its square equals 12?

33. The square of a certain number diminished by 9 times the number equals zero. Find the number.

34. If to 3 times the square of a certain number we add 4 times the number, the result is 4. Find the number.

35. State Ex. 1 as a problem concerning the finding of an unknown number.

FRACTIONAL EXPONENTS

157. Meaning of a Fractional Exponent.

By addition of exponents, $a^2 \times a^2 \times a^2 = a^{2+2+2} = a^6.$

If we apply the same method to fractional exponents, we have, for instance, $a^{\frac{2}{3}} \times a^{\frac{2}{3}} \times a^{\frac{2}{3}} = a^{\frac{2}{3}+\frac{2}{3}+\frac{2}{3}} = a^2.$

Hence, $a^{\frac{2}{3}}$ may be regarded as one of three equal factors composing a^2; that is, $a^{\frac{2}{3}}$ is the cube root of $a^2.$

$$\therefore a^{\frac{2}{3}} = \sqrt[3]{a^2}.$$

Hence, in general, *in a fractional exponent the numerator denotes the power of the base that is to be taken, and the denominator denotes the root that is to be extracted.*

Ex. 1. $8^{\frac{2}{3}} = \sqrt[3]{8^2} = \sqrt[3]{64} = 4.$ *Ans.*

Ex. 2. $a^{\frac{2}{3}} \times a^{\frac{1}{2}} \times a^{\frac{1}{3}} = a^{\frac{2}{3}+\frac{1}{2}+\frac{1}{3}} = a^{1\frac{1}{2}}.$ *Ans.*

Ex. 3. $32^{\frac{6}{5}} = \sqrt[5]{32^6} = 2^6 = 64.$ *Ans.*

EXERCISE 182

Express with radical signs :

1. $a^{\frac{2}{3}}.$ 3. $a^{\frac{1}{4}}.$ 5. $c^{\frac{3}{4}}.$ 7. $3\,x^{\frac{5}{6}}.$

2. $x^{\frac{5}{6}}.$ 4. $y^{\frac{4}{5}}.$ 6. $2\,a^{\frac{2}{3}}.$ 8. $27^{\frac{2}{3}}.$

Express with fractional exponents :

9. $\sqrt[3]{x^4}.$ 11. $\sqrt{p^5}.$ 13. $a\sqrt[3]{y^2}.$ 15. $4\sqrt{x}.$

10. $\sqrt[5]{a^3}.$ 12. $2\sqrt[7]{a^2}.$ 14. $b\sqrt{x^5}.$ 16. $\sqrt{x}\sqrt[3]{y^2}.$

Find the value of

17. $4^{\frac{1}{2}}.$ 19. $27^{\frac{2}{3}}.$ 21. $\sqrt[4]{16^3}.$ 23. $5\sqrt[3]{8^2}.$

18. $8^{\frac{2}{3}}.$ 20. $64^{\frac{2}{3}}.$ 22. $8^{\frac{5}{3}}.$ 24. $\frac{5}{8}\sqrt[4]{16^2}.$

Simplify:

25. $\sqrt{18} - 8^{\frac{1}{2}}.$ 27. $5\sqrt[3]{a} + 6\,a^{\frac{1}{3}} - \sqrt[3]{8\,a}.$

26. $3\,x^{\frac{1}{2}} + 5\sqrt{x}.$ 28. $3\sqrt{2\,a} - 5\sqrt{2\,a} + 2(2\,a)^{\frac{1}{2}}.$

     29. $8\sqrt{x-y} - 4(x-y)^{\frac{1}{2}} + \sqrt{9(x-y)}.$

30. $x^{\frac{1}{2}}\sqrt{x}.$ 31. $x^{\frac{1}{2}}\sqrt{x^3}.$ 32. $a^{\frac{2}{3}}\sqrt{a}.$

33. $5(p-q)^{\frac{1}{2}}\sqrt{p-q}.$ 37. $8(a-b)^{\frac{1}{2}}\sqrt{4(a-b)}.$

34. $3^{\frac{1}{2}}\sqrt{2}.$ 38. $16^{\frac{3}{4}} - \sqrt[3]{8} + 8^{\frac{1}{3}}.$

35. $5\,y^{\frac{1}{3}}\sqrt[3]{y}.$ 39. $2^{\frac{1}{2}}\sqrt{8\,x}\sqrt{32}.$

36. $5^{\frac{1}{2}}\sqrt{20}\sqrt{8}.$ 40. $\sqrt{8(a-b)^2} - 3(a-b)2^{\frac{1}{2}}.$

41. Multiply $\sqrt{x^2 - y^2}$ by $(x-y)^{\frac{1}{2}}.$

42. Multiply $4(a-b)$ by $(a-b)^{\frac{1}{2}}.$

Rationalize the denominator of:

43. $\dfrac{1}{x^{\frac{1}{3}}}$. **45.** $\dfrac{7}{(5a)^{\frac{1}{3}}}$. **47.** $\dfrac{\sqrt{x}}{y^{\frac{1}{3}}}$. **49.** $\dfrac{7}{2(5)^{\frac{1}{3}}}$.

44. $\dfrac{5}{3\,a^{\frac{1}{3}}}$. **46.** $\dfrac{a^{\frac{1}{3}}}{\sqrt{b}}$. **48.** $\dfrac{b}{a^{\frac{1}{3}}\sqrt{x}}$. **50.** $\dfrac{a^{\frac{1}{3}}}{b^{\frac{1}{3}}\sqrt{c}}$.

RATIO AND PROPORTION

158. The **ratio** of one number to another is the quotient obtained by dividing the first of the two numbers by the second. A ratio is in certain respects the same as a fraction.

A ratio may be indicated in two different ways.

Thus, the ratio of 5 to 17 may be written as $\frac{5}{17}$, or as $5:17$. (The sign : may be regarded as an abbreviation of the division mark ÷.)

159. A **proportion** is an equality between two ratios. Hence a proportion is in certain respects the same as an equation.

A proportion may be expressed in different ways.

Thus, if the two ratios $\dfrac{x}{x+2}$ and $\dfrac{5}{6}$ are equal the proportion may be written either in the form $\dfrac{x}{x+2}=\dfrac{5}{6}$, or thus, $x:x+2=5:6$.

Ex. 1. Find the ratio of 1 ft. 3 in. to 1 yd.

$$\frac{1\text{ ft. 3 in.}}{1\text{ yd.}}=\frac{15\text{ in.}}{36\text{ in.}}=\frac{5}{12}. \quad Ans.$$

Ex. 2. Find the value of x which satisfies the proportion $x:8=3:2$.

The given proportion may be written in the form $\dfrac{x}{8}=\dfrac{3}{2}$.

Solving this equation, $x=12$. *Ans.*

1. Write the following ratios by use of the sign :

$$\frac{3}{7}, \frac{8}{15}, \frac{3}{x}, \frac{x-3}{x}, \frac{a}{b}.$$

2. Write the following ratios in the form of fractions :

$5:8$	$8:b+2$	$a+b:a-b$
$a:7$	$x:x-3$	$3x-2y:7x+12$

3. Write the following proportions in the form of equations :

$$5:8=7:y, \quad 7:x=5:x-2, \quad a:b=c:d.$$

4. Write the following equations as proportions in which the mark : is used.

$$\frac{3}{7}=\frac{5}{y}, \quad \frac{x-2}{x-5}=\frac{7}{12}.$$

5. Simplify the following ratios :

(1) 4 in. : 1 ft. 8 in.   (3) 1 qt. 1 pt. : 1 gal.

(2) 1 ft. 3 in. : 2 yd.   (4) $12\frac{1}{2}\% : 37\frac{1}{2}\%$.

Find the value of x in

6. $x:5=7:15$.   **8.** $2x-3:4x-5=4:7$.

7. $x:a=7:5$.   **9.** $x:x-3=x+1:x-4$.

10. Find y if $a:b=c:y$.

11. Find p if $p+3:2=12-p:3$.

Solve and check :

12. $2x+3:3x-1=3x+1:2x+1$.

13. $z+5:3-z=10+3z:z-10$.

14. What number must be added to each of the terms of $\frac{2}{5}$ to make the value of the fraction $\frac{2}{3}$?

15. What number added to each of the numbers 3, 7, 15, 25 will give results that are in proportion ?

16. Separate 1200 into two parts which shall be in the ratio 2 to 3.

Sug. Denote the two parts by 2 x and 3 x.

17. Separate 1200 into three parts which shall be proportional to 3, 4, and 5.

Sug. Denote the parts by 3 x, 4 x, and 5 x.

18. In a certain year the profits of a certain business were $39,260. Divide this number into three parts which shall be proportional to 2, 3, 8.

19. Divide $140 into parts proportional to 5, 21, 1, 1.

20. A and B are in business and their respective shares of the profits are in the ratio of 2 to 3. If the profits for a certain year are $16,000, and during the year A takes out $1200 and B $1000, at the end of the year how much of the profits does each receive?

21. A, B, and C are in business and their respective shares of the profits are proportional to 2, 3, 4. If for a given year the total profits are $28,440 and during the year A takes out $1250, B, $1735, and C, $2674, at the end of the year how much of the profits does each receive?

INDEX

NOTE. Italic numbers refer to pages in Part II, other numbers to Part I.

PAGE

Abbreviated multiplica-
tion 101, *112*
Abscissa of point . . . 245
Absolute value *33*
Addition . . 44, *53*, 167, *273*
Affected quadratic equa-
tion 289
Aggregation, signs of . . 49
Algebra defined *16*
Algebraic expression . . 11
numbers *33*
sum 43, *53*
Analysis of problems . . 4
Axes 245

Bar graphs *255*
Binomial expression . . *17*

Check 43, 47, 48, 61, 81, 86,
178, 218, 225
Circle graphs *256*
Coefficient 11, *17*, 83
literal *18*
numerical *17*
Common factor 140
Completing the square . 291
Coördinates of point . . 246

Degree of an equation . 248, *303*
radical *279*
term *303*
Division *83*, *95*, 171
Double use of + and −
signs *57*

Elimination 217
by addition and sub-
traction 218
by substitution . . . 220

PAGE

Equation . . . 61, 118, *131*
aids in solving *129*
degree of *303*
fractional *188*
linear 248
literal . 191, 223, *236*, 293
numerical 191
root of 61, *70*
solution of . . 61, 119, *205*
Exponent *18*, 83
fractional 331
law of 76, 83
literal 323

Factoring 140, 325
Factors 140
Formulas . . . 8, 201, 294
interest 51
percentage 10
rectangle 8
uniform rate 52
volume of box 51
Fractional exponents . . 331
Fractions . 160, 165, *176*, *279*
addition of *167*, *181*
division of *171*, *183*
equations containing . 188
lowest terms 160
multiplication of . *169*, *183*
reduction of mixed ex-
pressions to 164
subtraction of 167

Graphs 28, *36*, 91, *98*, 148, *157*,
246, 248, 250, *256*, *257*, *313*
bar *255*
broken-line 90
circle *256*
curved-line 27

337

Graphs — (*continued*)
of equations 245
linear 247
quadratic *311*
simultaneous . . . 249

Identity *131*
Improper fraction . . . 163
Independent equations . 217
Integer
consecutive 15
even *39*
odd *39*
Integral expression . . . *177*

Laws for + and − signs . *177*
Linear and quadratic systems *303*
Linear equations . . . 248
graph of 247
Literal coefficient . . . *18*
exponents 323
Lowest common denominator 165
Lowest terms 160

Members of equation . . 118
Mixed expression . . 164, *177*
Monomial *17*
Multiplication 76, *92*, 101, 169

Negative number . . . 24
utility of *35*
Number
consecutive 15
positive and negative . 24
prime *150*
square root of . . . 263

Oral Exercises *21*, *39*, 42, 46, 63,
70, 78, 84, 87, 110, 119, 121,
132, 161, 167, 170, 173, *185*,
189, 192, 194, *207*, 226, *238*,
264, 272, 273, 275, 276, *285*,
295, *305*

Order of operations . . *19*
Ordinate of point . . . 245
Origin 245

Parenthesis 49
insertion of *58*
removal of 49
Polynomial *17*
addition of 45
division of 86
multiplication of . . . 81
subtraction of 48
Positive numbers . . . 24
Power *18*
Prime number *150*
Principal root 271
Problems
analysis of 4
solution of 6
Proportion 333

Quadrant 246
Quadratic equations . . 289
affected 289
graph of *311*
literal 293
pure 289
solution of, by completing square . . . 291
by formula 294
by factorial method . 329
Quadratic and linear equations, systems of . . *303*

Radicals 271
addition of 273
degree of *279*
fractions under radical
sign *279*
multiplication of . . . 275
rationalizing monomial
denominator . . . *283*
similar 273
Radicand 271
Ratio 333

PAGE

Reciprocal *183*
Reviews 66, *73*, 126, *137*, 203, *214*, 252, *257*, 298, *315*
Root . . . *70*, 260, 271
of equation. . . . 61, *70*
principal 271

Sign of aggregation. . . 49
Sign of fraction *177*
changes of *177*
Similar radicals 273
terms 43
Simplification of radical 271, *279*
Simultaneous equations 217, *234*
graphic solution of . . 249
Solution of an equation 61, *205*, 329
Square of binomial . . 101, 102
Square root 260
of algebraic expression . 261
of arithmetical numbers 263
of three or more terms . 262
Squares, difference of . . 103

PAGE

Substitution, elimination by 220
Subtraction 47, *55*
Sum, algebraic . . . 43, *53*
Surd. *279*
Systems of equations . . 217
in three unknowns . . 225
quadratic *303*

Term 11, *17*
degree of *303*
dissimilar 44
similar 43
Transposition 118
Triangles *269*
Trinomial *17, 154*

Utility of algebra . . . *16*
of addition in algebra . *53*
of negative numbers . *35*

Zero
as factor 79
as numerator 85
division by *129*
multiplication by. . . *129*

A FIRST BOOK IN ALGEBRA

ANSWERS *

EXERCISE 1, Page 2

35. 5.	40. 2.	44. 5.	48. 1¼.	52. 27.
36. 3.	41. 1½.	45. 8.	49. 2.	53. ½.
37. 3.	42. 6.	46. 14.	50. 6.	54. 720.
38. 12.	43. 8.	47. 7.	51. 34.	55. 18, 24.
39. 2.				

EXERCISE 2, Page 5

1. 12, 24.
2. 8, 24.
3. 56, 28.
4. $8000, $4000.
5. $96.60, $32.20.
6. 24 qt., 12 qt.
7. $12.40, $6.20.
9. 11,250,000 bales.
10. $4000, $2000.
11. 49,200 sq. mi., 8200 sq. mi.
12. 200, 40.
13. 4.84, 2.42.
14. 4⅔, ⅔.
15. 12 ft., 24 ft., 96 ft.
16. $1000, $2000, $3000.

EXERCISE 3, Page 7

1. $11, $22, $33.
2. $12,000, $24,000, $48,000.
3. 3375, 6750, 16,875 cu. ft.
5. 500, 500, 1000 lb.
6. $190\frac{10}{21}, 952\frac{8}{21}, 2857\frac{1}{7}$ lb.
7. $9.90, $19.80.
8. 20, 40, 60.
9. 20, 40, 60.
10. 200, 400, 1200.
11. 80, 80, 160, 160.

EXERCISE 4, Page 9

1. 40 sq. in.
2. 480.
3. 480 sq. in.
4. 20.28.
5. 20.28 sq. in.
6. 1176 sq. in.
7. $8p$ sq. in.; rs sq. in.
8. 555 sq. yd.
9. 282 sq. ft.
10. 26⅔ sq. yd.
11. 7 yd.
12. 24 ft.
13. 54.45 ft.
15. 21.
16. $21.
17. 198.
18. 48; 80; 42⅔; 112.
19. $80.
20. $19,200.
21. 3%.
22. 3%.
23. .80.
24. .771.

EXERCISE 5, Page 11

16. 9.	20. 45.	24. 5.	28. 44.	32. $2\frac{3}{4}$.
17. 6.	21. 1.	25. 23.	29. 26.	33. $2\frac{1}{4}$.
18. 64.	22. 108.	26. 2.	30. 2.	34. $3\frac{1}{2}$.
19. 12.	23. 48.	27. 40.	31. 40.	35. 0.

EXERCISE 6, Page 14

1. 37, 49.	7. 25, 35.	13. 10, 11, 12.
2. 480, 730.	8. 35.	14. $1300, $1600, $2100.
3. 9, 14, 17.	9. 7258.	15. $1200, $1500, $2300.
4. 14, 26.	10. 9 hr. 14 min.	16. 112, 112, 124, 139.
5. $800, $1000, $1300.	11. $15\frac{2}{3}$, $12\frac{7}{8}$.	17. 7, 6, 9, 8.
6. $10.10, $14.70.	12. 7, 8.	18. 6, 8, 12.

EXERCISE 7, Page 18

8. 11.	14. 100.	20. 5.	26. .804.	32. 29 94.
9. $11\frac{2}{3}$.	15. $88\frac{1}{3}$.	21. 30.	27. .8.	33. .139.
10. 34.	16. 10.	22. 50.	28. 239.7.	34. 550.
11. 33.	17. 0.	23. 11.9.	29. 15.9996.	35. 150.
12. $8\frac{1}{3}$.	18. $15\frac{1}{4}$.	24. 1.63.	30. 5.	
13. 119.	19. 31.	25. .06.	31. 1.51.	

EXERCISE 8, Page 20

1. 2.	5. $10\frac{1}{2}$.	9. 16.	13. 37.	17. 63.	21. $\frac{2}{3}$.
2. 9.	6. 17.	10. 12.	14. 108.	18. 2.	
3. 11.	7. 18.	11. 5.	15. 21.	19. 15.	
4. 32.	8. 12.	12. 9.	16. 21.	20. 26.	

EXERCISE 10, Page 22

1. 1080.

2. $N = \dfrac{9\,lw}{2}$

3. 720 ; 591.

4. 618 ; 520.

5. $N = \dfrac{144\,lw}{ab}$.

6. 756.

7. $14\frac{4}{5}$.

8. $S = \dfrac{lw}{50}$.

9. 14.4.

10. 6.56.

11. 8.0025.

12. 9.9.

13. $262\frac{2}{3}$.

14. $Y = \dfrac{2\,el + w(p+e)}{9}$.

15. 210 sq. yd.

16. $258\frac{2}{3}$ sq. yd.

17. $D = \dfrac{c(2\,el + pw + ew)}{900}$.

18. $31.63.

EXERCISE 11, Page 24

2. 25°.   3. 7°.   4. 3° N. ; 11° S.   5. $9000.   6. − 2¾°.
11. (1), (2), (4), (5), (7) positive; (3), (6) negative.
15. 7°.   16. 15°.   17. 77.   18. 33.   19. $300 ; $1700.

EXERCISE 13, Page 31

1. 1200 sq. ft.
2. $24, $14.
3. $3600, $7200, $7200.
4. 155, 151 ft.
5. 500, 500, 1000 lb.
6. 240, 80.
7. $1880, $1340.
8. $8.20, $16.40.
9. $3.84, $8.84.
10. 121,391 sq. mi.
11. 6290 ft.
12. $4000, $8000, $12,000.
13. $3000, $5000, $6000.
14. $7600, $7600, $15,200, $9600.
15. 11, 12, 13.
16. 25, 26 ; 27, 28.
18. 716,555 sq. mi.
19. 34, 24, 48.
20. $12,000, $8000, $6000, $9000.
21. $12,000, $4000, $12,000, $8000.

EXERCISE 14, Page 34

4. 62°.   5. + 5 yd.   7. $50 ; − $25 ; − $50 ; $25.

EXERCISE 17, Page 40

1. ½, $\frac{1}{11}$.
2. .0036, .0009.
3. $90, $30.
4. .0062, .0124, .0186.
5. 30, 30, 60, 120.
6. $200, $600, $3000.
7. 35⅔ lb.
8. 750, 125, 125 lb.
9. 81, 19.
10. 1.07, 3.33.
11. 6, 8, 10.
12. 13, 15, 17 ;
   5, 7, 9, 11, 13.
13. 18, 20, 22 ;
   8, 10, 12, 14, 16.
14. 6,405,000 sq. in.
15. 150 lb.
16. 9⅔ sec.
17. $1410, $1675.
18. $12,000, $16,000,
   $36,000.

EXERCISE 19, Page 44

1. $43\,ab$.
2. $-36\,a^2x$.
3. $-13\,ax^2$.
4. $-10\,abc^2$.
5. $9.4\,x^2$.
6. $\frac{1}{4}\,xy$.
7. 0.
8. $.43\,ab^3$.
9. $2\,x$.
10. $-x^2-2$.
11. $3\,a^2-2\,x^2$.
12. $a+b$.
13. $2\,x^2-11\,y^2$.
14. $2\,by^3$.
15. $7\,x^2+2\,y^2$.
16. $m^2-mn-2\,n^2$.
17. $3\,x^2$.
18. 0.
19. $-2\,x-4\,y+2\,z$.
20. $-xy+2\,ax+y^2$
   $-3\,x^2$.
21. $-4\,x-y-2\,z$.
22. $2\,x+5\,y-3$.

EXERCISE 20, Page 45

1. $5a + b + c.$
2. $3a + b - 5.$
3. $3a + b + c + d.$
4. $5x - y - 5.$
5. 5 yd. 2 ft. 7 in.
6. 16 gal. 2 qt. 1 pt.
7. $3ab + 10 + x.$
8. $6 + 3a - 3b.$
9. $7a^2 + 13ab + 5b^2.$
10. $2y^2 - 6x^2 - 2xy.$
11. $8a - b + 5c.$
12. $n + 1.$
13. $4n - 2.$
14. $3a + b + c.$
15. $n.$

EXERCISE 22, Page 47

1. 37 tens.
2. $-47t.$
3. $-9x^2y^3.$
4. -57 in.
5. $-25x.$
6. 18 $ab.$
7. $-5abc.$
8. $-23ab^3.$
9. $-2x^2y^2.$
10. $-10.8x.$
11. $-8.8ab.$
12. $-4\frac{1}{4}m.$
13. 4 ft. 2 in.
14. $4f + 2i.$
15. $x^2 - 5x.$
16. $8x - 8.$
17. 2 gal. 2 qt.
18. $x^3 - 7.$
19. $4x^2 + 7x - 8.$
20. $-a^2 + 3ab - 6b^2.$
21. $-2x^3 + x + 6.$
22. $a + 4b - 4c + d.$
23. $-8 + x + 7x^2.$
24. $-x^2 - 2y^2 + z^2 - 2.$
25. $-1 - 2x + 2x^2 + x^3 + 3x^4.$
26. $12xy^2 - x^2y^2 - 9x^2y.$
27. $2 - 4ac + 12cd.$
28. $4x^2 - 6x.$
29. 2 yd. 1 ft. 5 in.
31. $2a + 2b + 2c.$
32. $n - 3.$
33. $5 - n.$
34. $x^3 - 2x^2 + 2x + 4.$
35. $9x + 2y - z.$
36. $7xy - 7xz + 5yz + 2x^2.$
37. $1 - 2x - 2x^3 + x^4 + x^5.$
38. $m - 3d - x + 3c.$
39. $2x^5 + 2x^4 - x^3 - 2x.$

EXERCISE 23, Page 50

1. $a + b + c.$
2. $a - b - c.$
3. $a + x - y.$
4. $a - x + y.$
5. $3a + 2x - 3y.$
6. $3a - 2x + 3y.$
7. $a^2 + 3a - 1.$
8. $10x^2 - 1.$
9. $x + 1.$
10. $1 - x.$
11. $-1.$
12. $2x + 1.$
13. $3p^3 - 2 + 5p^2.$
14. $2a^3 + a^2 - 1.$
15. $5a - b.$
16. $-x + 3y.$
17. $1 - 2x.$
18. $9x - 1.$
19. 4.
20. $4x - 1.$
21. 0.
22. $a - 1.$
23. $7a - 6b - (3a + 2b + 5c).$

EXERCISE 24, Page 51

2. 70 cu. yd.
3. 64 cu. in.
4. 5 ft.
7. $84.
8. .05.
9. .05.
10. 5 yr.
11. 5 yr.
12. 200 mi.; $40t$ mi.
13. vt miles.
15. 155 mi.
16. 12 mi. per hour.
19. 176.
20. 14 in.
21. 66 in.

ANSWERS

V

EXERCISE 25, Page 53

1. 5.03.
2. $6\frac{4}{5}$.
3. $-1.7x^2 + 12x - .2$.
4. $1\frac{11}{12}a - \frac{5}{8}b + 1\frac{5}{8}c$.
5. 7.11.
6. $-14\frac{1}{4}$.
7. $\frac{1}{4}$.
8. 4.22.
9. 1.3 rise.
10. $4x^4 - 8x^3 + 25x^2 + 9$.
11. $-\frac{71}{72}a^3 + 4b^2c + \frac{11}{12}abc + 6\frac{1}{4}bc$.
12. $3(a+b)$.
13. $-7(x-y)$.
14. $\sqrt{a+x}$.
15. $\frac{5}{8}\pi r^2$.
16. $(4a+3b)x$.
17. $(3a+c-2)x^2$.
18. $(2a+2b+2c)xy$.
19. 5×25.
20. $2a(x-y)$.
21. $(3a-4b+5)(a^2-b^2)$.
22. $(a-b+c)x + (a+b-3c)x^3$.
23. $(a+y+z)x + (b-2+c)x^2$.
24. $9x + 12xz$.
25. $2x^2y + 4xy^2$.
26. $-a^2b + a^2c$.
27. 2.

EXERCISE 26, Page 55

1. 7.08.
2. 8.52.
3. -12.2.
4. $4.1x$.
5. $3\frac{3}{4}ab$.
6. -5.27.
7. 5.28.
8. 17.25.
9. -3.46.
10. $-12.5x^3$.
11. $-3.44a$.
12. $(a-b)x$.
13. $(a+b)y$.
14. $(-a^2-b^2)x$.
15. $(-p+3c)y^2$.
16. $(-a+b)y^3$.
17. $-x^3 - 2.9$.
18. $6x^2 + 7x - 8$.
19. $10.2a^2 + 8.3a - .66$.
20. $2(a+b)$.
21. $10(x+y)$.
22. $5\sqrt{a+x}$.
23. $-6\sqrt{b-y}$.
24. $\frac{1}{2}x^2 + \frac{1}{3}y^2$.
25. $(2a+b)x$.
26. $(a-b)(x+y)$.
27. $12(x+y)$.
28. $(-a+b+c)x$.
29. $4a^2 - 2ab + 7b^2 + 10c^2$.
30. $12y^2 + 7xy + 5bc - bc^2$.
31. $-3ab - 4bc + 7ac$.
32. $-2a^2 + 8ac - c^2$.
33. $a + 4b - 4c + d$.
34. $7b + x + 7x^2 - 15a$.
35. $2p^2 - x^2 - 2y^2 - z^2 - 2$.
36. -7.2.
37. $a + b - c$.
41. $3x$; $-x+y$; $-3a^2 + 2ab - b^2$.
42. $4x^3 - 2x - 2$.
43. $2x^3 + 6x^2 - 2x - 4$.
44. $-2x^3 - 2x^2 - 8x - 2$.
45. $4x^3 - 4x^2 + 8x + 4$.

EXERCISE 27, Page 57

1. 17.2.
2. $-15\frac{1}{4}$.
3. -36.
4. -5.7.
5. 14.13.
6. -4.56.
7. 5.94.
8. 34.77.
9. 29.23.
10. $.9 - 2a$.
11. $3.1 - x$.
12. $1 + b$.
13. $1.4 + 2x$.
14. $9.5 + x$.
15. $8.8 - 3x$.
16. $3x + 6.5$.
17. -2.5.
18. $2.7 + x$.
19. -6.12.
20. $5a - 2$.
21. x.
22. $2z^2$.
23. $-.8m^2 - 4$.
24. $2.4a^2 - a + 5.2$.

EXERCISE 28, Page 58

1. $x^3 - (3x^2 - 3x + 1)$.
3. $1 - (-2a + a^2 + 1)$.

EXERCISE 29, Page 58

1. $T = \dfrac{lwh}{38}$.

2. $7\frac{7}{19}$.

3. $2\frac{7}{14}$.

4. 6.944.

5. 4 ft.

6. $409\frac{1}{11}$.

7. $B = \dfrac{5\,lwh}{11}$.

8. 360.

9. 2 ft.

10. 28.

11. $L = \dfrac{lwt}{12}$.

12. 36.

13. $13\frac{1}{3}$.

15. 80 bu.

16. $6.15.

17. $7.03.

18. $6.93.

EXERCISE 30, Page 62

1. 5 in.
2. 5 in.
3. 7 ft.
4. 5.
5. 5.
6. 13.
7. 28.
8. 17.

9. $4.
10. 3 in.
11. 2.
12. 7.
13. 3.
14. 7.
15. 6.
16. 4.

17. 5.
18. 4.
19. $1\frac{1}{2}$.
20. 5.
21. 3.
22. 3.
23. 4.
24. 3.

25. 4.
26. 2.74.
27. 2.2.
28. 4.
29. 1.
30. $\frac{1}{3}$.
31. $\frac{24}{11}$.
36. 3.

37. 7.
38. 3.
39. -2.
40. $6\frac{1}{4}$.
41. $1\frac{1}{4}$.

EXERCISE 32, Page 64

1. 56.
2. 18.
3. 5.8.
4. 40.
5. 56.
6. 35.

7. 60.
8. 169.9.
9. 69.8.
10. 50, 40.
11. $13, $8.
12. 42, 30.

13. 15.
14. 6315, 6105.
15. 16, 24.
16. $1.50.
17. 5, 44.
18. 30, 53.

19. 21, 55.
20. $2500, $6000.
21. 30 oz.
22. $2\frac{1}{4}$ lb.

EXERCISE 33, Page 66

1. 8.

2. 297.28.

3. 36, 144.

4. 11.

5. $7\,a^2 - 7\,ab + 6\,b^2 - 6\,c^2$.

6. $2\,x^4 + 2\,x^3 - 3\,x^2 - 7\,x - 2$.

7. $5\,x^3 + x^2 - 5\,x + 2$;
$-2\,x^3 + 3\,x^2y + xy^2 - 2\,y^3$.

8. $4.32, $17.28.

9. 21, 21, 42.

10. $9 - 2\,x$.

11. 7.

12. -4.

13. 25, 26, 27, 28.

16. 1260 sq. ft.

17. $.06\frac{1}{4}$.

18. 4.25^+ %.

20. $34, $46, $34, $68.

21. 45 %.

22. $\frac{1}{2}$.

23. $-\frac{1}{3}$.

24. $-8\,x^2 + 2\,ax^2 + 3\,ax + 2\,a^2 - a^4$.

25. 79.

26. $x^2 + 3\,ax + 5\,a^2$.

27. $6\,x^2 - 4\,ax - 6\,a^2$.

28. $56.25.

29. $73\frac{1}{2}$ cu. ft.

32. 866, 400.

33. $3600, $5600, $10,800.

34. 2162 mi.

EXERCISE 34, Page 70

1. .7.
2. -1.8.
3. .5.
4. a.
5. 1.2.

6. 1.23.
7. $3b$.
8. $2+a$.
9. -4.
10. $-.5$.

11. 1.875.
12. $-.1$.
13. 5.
14. 4.
15. .5.

16. .4775.
17. 1.78.
18. $b-a$.
19. $2a$.

20. $-\dfrac{3a}{2}$.
21. .29.
22. $2p$.

EXERCISE 36, Page 71

1. 6.42.
2. 11.88.
3. 5.03.
4. 3.07.
5. 139.5.
6. .856, 3.544
7. $\frac{2}{3}$, $\frac{11}{9}$.
8. 22, 11, 17.

9. 21, 31, 59.
10. 26, 37, 35.
11. $9\frac{3}{8}$ sec.
12. $12\frac{1}{4}$ mi.
13. 617°.
14. 120 ft. per second.
15. $27\frac{1}{3}$ sec.

16. 5240 mi.
18. 13, 14, 15, 16, 17.
19. 24 ft. $7\frac{1}{4}$ in.
20. 37, 24, 12.
21. $27\frac{1}{3}$ sec.
22. 234 mi.
23. 555 ft.

EXERCISE 37, Page 73

1. -1.3405.
2. $1\frac{1}{2}x^2 - 1\frac{1}{4}x$.
3. $\frac{2}{3}x^2 - \frac{1}{4}x - \frac{1}{2}$.
4. $.95a^2 + .45a + .3$.
5. $1.23a^2 + 2.12a + .6$.
6. $3x + 4y + z$.
7. $3a^3 - 10ab + 3a^2b^2$.
8. $x^3 - x^2 + x - 1$;
   $-x^2 + 4x - 1$;
   $-x^2 + x + 14$;
   $-x^2 + x - 1$.
9. $-x^3 + 2x^2 - 3x + 1$;
   $x^2 - 3x - 9$;
   $2x^2 - 3x - 6$;
   $2x^2 - 2x - a$.
10. 6.6.

11. 180, 540, 280.
12. $4x - 3.81$.
14. -1.2.
15. .11.
16. 14,147 ft.
17. $1093\frac{11}{18}$ yd.
18. $T = \dfrac{lwh}{40}$.
19. 54.4+.
21. $2.5x^3 + 4.5x^2y + .6y^3 - 11xy^2$
    $\qquad - 10y^2 + 5z^2 + .06x^2$.
24. $2x^3 - 2x^2y + 6xy^2 + 5y^3$.
25. $4x^3 - 2x^2y + 2xy^2 + 7y^3$.
26. $-2x^3 + 2xy^2 + 3y^3$.
27. $6x^3 - 2x^2y - 3y^3$.
28. $\frac{11}{6}lwh$.
29. 35,640.

EXERCISE 39, Page 78

1. -210.
2. $-260a$.

3. $-115ab$.
4. $-360x^2y^2$.

5. $-.8x^2$.
6. $1.5x^2$.

7. $-52a^2x^4$.
8. $42x^3y^4$.

9. $21\,a^3xy^2$.
10. $20\,a^2bcd^2$.
11. $-18\,c^3d^3$.
12. $16\,x^3y^3z^4$.
13. $60\,a^2bc$.
14. $-345\,acx^2$.
15. abx.

16. $-6\,abc^2d$.
17. $abxy$.
18. $65\,ax^2y^2$.
19. $.08\,x^5$.
20. $\frac{1}{4}\,a^2x^6$.
21. $.015\,x^2$.
22. $1.05\,y^5$.

23. $1\frac{1}{7}\,x^4$.
24. $\frac{2}{5}\,x^3$.
25. a^3bxy.
26. $-30\,a^2x^2y^2$.
27. $-18\,a^2b^6c$.
28. $-2.4\,a^2b^3x^2y^2$.
38. 0.

39. 0.
40. 0.
41. 16.
42. 4.
43. 1.

EXERCISE 40, Page 80

1. $6\,a^2x + 9\,ax^2$.
2. $-15\,x^2y + 10\,xy^2$.
3. $8\,x^3y^2 - 2\,x^2y^3$.
4. $-21\,a^2bx^2y - 12\,ab^2xy^2$.
5. $-12\,x - 20\,y$.
6. $-35\,a + 15\,b$.
7. $-12\,mx - 10\,nx$.
8. $-35\,ab^3y + 15\,abx^2y$.
9. $x^3 - 1.48\,x^2 + .204\,x$.
10. $\frac{1}{4}x^2 - \frac{1}{4}x^3 - \frac{3}{4}x$.
11. $-\frac{3}{15}\,a^2x^3 + \frac{7}{15}\,a^2x^2 + \frac{1}{8}\,a^2x$.
12. $-.48\,l + .6\,m + 2.4\,n$.
13. $40\,a^2c^2n - 15\,am^2n^2$.
14. $-7\,m^3n + 7\,m^4n + 21\,m^5n$.
15. $24\,x^3y^2 - 15\,x^2y^3 - 3\,xy^4$.
16. $.2\,a^2b - .6\,a^2b^2 - .15\,ab^3$.
19. $6\,a^3b - 9\,ab^3$.
20. $a^3b^2 + a^2b^3 + a^2b^2c$.
21. $9\,a^3x - 15\,a^2x^2 + 6\,ax^3$.
22. $3\,x - 3\,a$.
23. $3\,x^2y - 3\,xy^2$.
24. $2\,x + \frac{10}{3}$.

EXERCISE 41, Page 81

4. $x^2 - 6\,x + 8$.
5. $y^2 - 5\,y + 6$.
6. $3\,x^2 - 17\,x + 20$.
7. $16\,a^2 - 66\,a + 35$.
8. $2\,a^2 - 3\,ab + b^2$.
9. $12\,x^2 - 29\,xy + 15\,y^2$.
10. $x^2 + 7\,x + 10$.
11. $6\,x^2 + 19\,x + 15$.
12. $3\,x^2 + 10\,xy + 3\,y^2$.
13. $3\,a^4 + 11\,a^2b^2 + 6\,b^4$.
14. $x^2 - 2\,x - 35$.
15. $2\,x^2 - 3\,x - 20$.
16. $21\,x^2 - 23\,xy - 20\,y^2$.
17. $x^2 - 4\,x - 45$.
18. $6\,x^2 - 7\,x - 5$.
19. $12\,x^4 - 11\,x^2y^2 - 15\,y^4$.
20. $6\,x^2 - 19\,xy + 10\,y^2$.
21. $10\,x^4 + 57\,x^2y^2 + 54\,y^4$.
22. $12\,m^2 - 37\,mp^2 + 28\,p^4$.
23. $40\,a^2 + 43\,abc - 99\,b^2c^2$.
24. $30\,x^2y^2 + xy - 42$.
25. $6\,x^3 + 19\,x^2 + 18\,x + 20$.
26. $35\,x^3 + 24\,x^2 + 25\,x + 6$.
27. $9\,a^3 + 21\,a^2b + 13\,ab^2 + 5\,b^3$.
28. $8\,x^3 - 2\,x^2 + 7\,x + 5$.
29. $6\,x^3 - x^2 - 33\,x - 5$.
30. $21\,x^3 + 5\,x^2y - 3\,xy^2 + 2\,y^3$.
31. $x^3 - y^3$.
32. $6\,x^3 - 25\,x^2 + 33\,x - 20$.
33. $12\,a^3 - 35\,a^2b + ab^2 + 30\,b^3$.
34. $\frac{1}{4}a^2 - \frac{1}{4}b^2$.
35. $\frac{1}{2}x^3 - 17\frac{3}{4}x + 2$.
36. $.1\,a^2 - .23\,ab + .12\,b^2$.
37. $4.5\,x^3 - 7.1\,x^2 - .4\,x + .24$

38. $x^4 + 4x^3 + 5x^2 + 5x + 6.$

39. $3a^4 + 11a^3b + 21a^2b^2 + 22ab^3$
$+ 8b^4.$

40. $8x^4 - 2x^3 + x^2 - 1.$

41. $15m^4 - 32m^3p + 10m^2p^2$
$- 10mp^3 + 24p^4.$

42. $8r^4 - 44r^3s + 50r^2s^2 + 37rs^3$
$- 56s^4.$

43. $x^4 - y^4.$

44. $a^4 - b^4.$

45. $6x^3 - 19x^2 - 90x + 175.$

46. $6x^9 + 19x^6 - 22x^3 - 80.$

47. $4a^6 - 9a^4b^2 - 10a^2b^4 + 3b^6.$

48. $6a^3 - a^2 - 19a - 6.$

49. $18x^3 - 111x^2 + 200x - 112.$

EXERCISE 43, Page 84

1. $-3.$
2. $-3x^2.$
3. $-2a.$
4. $5xy.$
5. $-x.$
6. $-7xy.$
7. $-2bd^2.$
8. $-3x^4y^3z^2.$

9. $-2x.$
10. $n.$
11. $3x^2.$
12. $-16a.$
13. $-24y.$
14. $40m.$
15. $.5a.$
16. $.08.$

17. $\frac{10}{3}x.$
18. $-\frac{1}{2}x.$
19. $2r^2; 4\pi; 4r.$
20. $\frac{1}{2}t; 2t^2; gt.$
21. $\frac{3v^2}{4}; m; 2mv.$
28. $0.$
29. $0.$

30. $0.$
31. $0.$
32. $\frac{1}{4}$
33. $\frac{3}{4}$

EXERCISE 44, Page 85

1. $-x^2 + 3x.$
2. $5x - 2y.$
3. $-2b + 3ac.$
4. $6x^2 - 7x + 1.$
5. $3x^2 - 2xy - y^2.$
6. $1 + m - m^2 + m^3.$
7. $-14x^2y + 21yz^2 - 1.$
8. $3x^2 + 2x - 5.$
9. $-2x^2 + .4x - 30.$
10. $.04a^2 - .08ab - 1.6b^2.$

11. $-\frac{3}{4}x^2 + x + \frac{15}{4}.$
12. $\frac{4}{5}a^3b - \frac{1}{3}a^2b^2 - \frac{2}{3}ab^3.$
13. $2x^2 + 4x + 1.$
14. $-6a + 7b.$
15. $.2a^2b^3 + .4ab^2 + .9b.$
16. $3x - 2y - 4.$
17. $3a - 2b - 4.$
18. $-2x^3y + 3x^2y^2 + 4x^2.$
19. $1 - 2xy^4 - 3x^3.$
20. $-2p^2 + .4pq + q^2.$

EXERCISE 45, Page 87

1. $x + 5.$
2. $x + 4.$
3. $x + 4y.$

4. $a + 4b.$
5. $2x + 1.$
6. $x + 2y.$

7. $p + 3q.$
8. $x - 5.$
9. $x - 2y.$

10. $a + 3.$
11. $4x + 1.$
12. $x - 2y.$

13. $a + 2b.$
14. $8m - 9n.$
15. $2x^2 + 1.$

16. $x^3 - 3.$
17. $a - 12b.$
18. $7x^4 - 4.$
19. $3x^2 + 2x + 1.$

20. $4x^2 + 5x + 2.$
21. $5a^2 + 4ab + 3b^2.$
22. $7p^2 + 4p - 3.$

23. $3x^2 - 4x - 5.$
24. $5m^2 - mn + 4n^3.$
25. $2x^3 - 3x^2 + 4x - 5.$

EXERCISE 47, Page 88

1. $36\frac{1}{4}$ sq. yd.
2. $y = \dfrac{lw}{9}$.
3. 600 sq. yd.
4. 50.22 sq. yd.
5. 18 ft.
6. 20.
8. 23.
9. 23.
10. 20.
11. $93\frac{1}{3}$ cu. yd.
12. $Y = \dfrac{lwh}{27}$.
13. 96 cu. yd.
14. 50.
15. $7\frac{1}{4}$ ft.
16. $G = \dfrac{15\,abc}{2}$.

EXERCISE 49, Page 93

1. $-4.8\,x^3$.
2. $-.23\,ab$.
3. $-7.7\,p^2x^3$.
4. $-2.9\,axy$.
5. $6(x+y)^5$.
6. $8(a-b)^4$.
7. $-32(2a+b)^2$.
8. $-.1(x-y)^3$.
9. $\dfrac{9\,x^2}{16\,y^2}$.
10. $\dfrac{27\,x^3}{64\,y^3}$.
11. $-5\,a^3b + 15\,a^2b^2$.
12. $-1.08\,x + 4.05$.
13. $-\frac{3}{2}\,ax^2 + \frac{1}{2}\,ay^2$.

14. $12\,x^2 - 17\,xy + 6\,y^2$.
15. $4\,a^4 - 13\,a^2b^2 + 3\,b^4$.
16. $.09\,a^4 - .12\,a^2b + .04\,b^2$.
17. $6\,x^4 - 23\,x^3 + 15\,x^2 + 2\,x - 6$.
18. $10\,a^4 + 13\,a^3b - 27\,a^2b^2 - 2\,ab^3 + 8\,b^4$.
19. $a^4 + a^2x^2 + x^4$.
20. $2\,x^4 + 5\,x^3 - 8\,x^2 + 11\,x - 20$.
21. $12\,x^4 - x^3 - 27\,x^2 - 3\,x + 10$.
22. $6\,x^5 + 5\,x^4y - 16\,x^3y^2 + 14\,x^2y^3 - 6\,xy^4 + y^5$.
23. $x^5 - x^4 - 8\,x^3 + 14\,x^2 + 15\,x - 25$.
24. $16\,x^4 - 48\,x^3y + 108\,x^2y^2 - 108\,xy^3 + 81\,y^4$.
25. $x^2 + ax + bx + ab$.
26. $ax + bx + ay + by$.
27. $x^2 + 2\,xy + y^2 + (a-b)x + (a-b)y - ab$.
28. $x^2 - a^2 + 2\,ab - b^2$.
29. $2\,x^5 - 5\,x^4 - 2\,x^3 + 9\,x^2 - 7\,x + 3$.
30. $3\,x^4y - 10\,x^3y^2 + 4\,x^2y^3 + 6\,xy^4 + y^5$.
31. $x^5 - 5\,x^4y + 10\,x^3y^2 - 10\,x^2y^3 + 5\,xy^4 - y^5$.
32. $4\,x^5 + 9\,x^4 - 16\,x^3 + 22\,x^2 - 21\,x + 6$.
33. $x^6 - x^5 - 7\,x^4 + 3\,x^3 + 17\,x^2 - 5\,x - 20$.
34. $x^6 - 6\,x^4y + 9\,x^2y^2 - y^6$.
35. $a^4 + a^2b^2 + b^4$.
36. $16\,x^4 + 36\,x^2y^2 + 81\,y^4$.
37. $x^7 - 9\,x^5y^2 + 7\,x^4y^3 + 13\,x^3y^4 - 19\,x^2y^5 + 8\,xy^6 - y^7$.
38. $-x^5 + 2\,ax^4 + 8\,a^2x^3 - 16\,a^3x^2 - 16\,a^4x + 32\,a^5$.
39. $a^3 + b^3 + x^3 + 3\,ab^2 + 3\,a^2b$.
40. $a^2b^2 + c^2d^2 - a^2c^2 - b^2d^2$.

EXERCISE 50, Page 96

1. -420.
2. 2.
3. $-2\,a$.
13. -330.
14. 1.6.
4. $5\,xy$.
5. $-10\,a$.
6. $3\,x^2$.
15. $-3\,a + 2\,ab + b$.
16. $3\,x^2 + 2\,x - 5$.
7. $-16\,b$.
8. $40\,a$.
9. $-\frac{1}{2}$.
17. $-8\,x^2 + .4\,x - 30$.
18. $2\,x^3 - x + 1$.
10. $-5(x+y)^3$
11. $7(a-b)^2$
12. -1.5.

19. $3x^3 + 4x^2y + 5xy^2 + 2y^3$.
20. $2x^4 - 3x^2y - 2y^2$.
21. $2x^2 - 5x - 1$.
22. $3x^2 - x - 5$.
23. $2x^3 - 4x^2 - x + 3$.
24. $2x^3 + 3x^2y - 4xy^2 + y^3$.
25. $3a^3 - 4a^2b + 3ab^2 - 2b^3$.
26. $a^2 - ab + b^2$.
27. $x^2 + 2xy + 4y^2$.
28. $x^3 + 2x^2y + 4xy^2 + 8y^3$.

29. $a^4 - 2a^3x + 4a^2x^2 - 8ax^3 + 16x^4$.
30. $x^5 - x^4y + x^3y^2 - x^2y^3 + xy^4 - y^5$.
31. $64x^6 + 16x^4y^2 + 4x^2y^4 + y^6$.
32. $x^2 - x + 1$.
33. $x^2 - xy + y^2$.
34. $x + a + b$.
35. $p - q - x$.
36. $a + m + k$.
37. $a - x + 2y$.
38. $a^2 + b^2 + c^2 - ab - ac - bc$.

EXERCISE 51, Page 97

1. $C = \dfrac{11\,lw}{90}$.   2. 55 yd.   3. 39.6 yd.   4. $C = \dfrac{c}{900}[2h(w + l) + lw]$.

5. $7.64.   7. 660.   9. 38.4

6. $P = \frac{1}{3}abc$.   8. 5 ft.

EXERCISE 53, Page 102

1. $a^2 + 2ax + x^2$.
2. $a^2 + 10a + 25$.
5. $16 + 8a + a^2$.
6. $a^2 - 2ax + x^2$.
25. $16x^6 - 8x^3 + 1$.
26. $25a^2b^2 + 90aby^3 + 81y^6$.
29. $100x^6y^2 + 20x^3y + 1$.
30. $1 - 14xy^5 + 49x^2y^{10}$.
33. $\frac{1}{9}x^2 + \frac{1}{3}xy + \frac{1}{4}y^2$.
34. $\frac{1}{4}a^2 - \frac{2}{3}ab + \frac{4}{9}b^2$.
37. $.04x + .4xy + y^2$.

9. $49 - 14x + x^2$.
10. $81 + 18a + a^2$.
13. $x^2 - 16x + 64$.
14. $c^2 - 2cd + d^2$.

17. $9x^2 - 30x + 25$.
18. $25 + 10ab + a^2b^2$.
21. $9x^2 + 24xy + 16y^2$.
22. $9x^2 - 12xy + 4y^2$.

38. $.04p^2 + .12pq + .09q^2$.
41. $25a^2 + 4a + .16$.
42. $.0004x^2 - .012xy + .09y^2$.
45. $x^2 + 2xy + y^2 + 10x + 10y + 25$.
46. $a^2 + 2ab + b^2 - 8a - 8b + 16$.
49. $x^2 + 2xy + y^2 - 6x - 6y + 9$.
50. $p^2 + 2pq + q^2 + 2py + 2qy + y^2$.

53. $4x^2 + 12xy + 9y^2 - 16x - 24y + 16$.
54. $a^2x^2 + 10ax + 25 - 2axy - 10y + y^2$.

58. 10,404.   60. 106.09.   62. 28.09.   64. 994,009.   66. 9801.
59. 2601.   61. 2704.   63. 998,001. 65. 2401.   67. 9920.16.

EXERCISE 54, Page 104

1. $a^2 - x^2$.
2. $b^2 - y^2$.
5. $x^6 - a^2$.

6. $y^4 - 49$.
9. $a^4 - y^2$.
10. $x^6 - 1$.

13. $x^6 - 81$.
14. $64 - x^4$.
17. $64a^2 - 49$.

18. $9x^6 - 49y^2$.
21. $49a^2b^2c^2 - 1$.
22. $16p^4q^2 - x^2y^2$.

25. $.16\,x^2 - .25\,y^2$.
26. $.09\,x^4 - .25\,y^4$.
30. $a^2 + 2\,ab + b^2 - 25$.
31. $x^2 + 2\,xy + y^2 - 16$.

35. $9\,a^4 + 6\,a^2b + b^2 - 16$.
36. 8096.
37. 9991.
38. 9975.
39. 999,975.
40. 4884.
41. 22,491.

EXERCISE 55, Page 106

1. $x^2 + 5\,x + 6$.
2. $x^2 + 9\,x + 14$.
5. $y^2 + 29\,y + 180$.
6. $x^2 - 7\,x + 10$.
9. $p^2 - 8\,p + 7$.
10. $b^4 - 19\,b^2 + 70$.
37. $y^2 + (c - d)y - cd$.
38. $x^2 + (q - p)x - pq$.

13. $x^4 + 8\,x^2 - 9$.
14. $y^6 + 2\,y^4 - 24$.
17. $x^2 + 6\,x - 135$.
18. $a^4 + 15\,a^2 - 100$.
21. $x^2 - 19\,x + 90$.
22. $x^2 + 4\,x - 21$.

25. $x^2 - 3\,ax - 18\,a^2$.
26. $x^4 - bx^2 - 42\,b^2$.
29. $b^2 - \frac{1}{2}\,b - \frac{1}{16}$.
30. $a^2 - \frac{7}{4}\,a + 3$.
33. $x^2 - .1\,x - .2$.
34. $x^2 + .5\,x + .06$.

40. $a^2 + 2\,ab + b^2 + 9\,a + 9\,b + 14$.
43. $a^2 + 2\,ax + x^2 - 4\,a - 4\,x - 21$.

EXERCISE 57, Page 108

1. 36.
2. 168.
3. 63.
4. 567.
5. $22 - 5\,a$.
6. $2\,a - 6$.
7. $-3\,x - 4$.
8. $-x - 2$.

9. $x^2 - x - 2$.
10. $10\,x^2 + 7\,x - 12$.
11. $17\,x - 12$.
12. $-5\,a - 24$.
13. $24\,a + 20$.
14. $2\,x^2 + 8$.
15. $8\,x$.
16. $2\,a^2 + 8\,b^2$.
17. $8\,ab$.
18. $18\,x^2 + 8$.

19. $-24\,x$.
20. $a^2 + 6\,ab + b^2$.
21. $5\,x^2 + 4\,xy + 20\,y^2$.
22. $a^2 + 14\,ab + b^2$.
23. $12\,a^2 - 132\,ab + 27\,b^2$.
24. $10\,x + 29$.
25. $13\,x - 40$.
26. $4\,a^2 - 24\,a + 61$.
27. $x^3 - 12\,x$.

28. $40\,a - 24\,ab$.
29. $-6\,x^2 + 13\,x - 4$.
30. $x^3 - 7\,x + 6$.
31. $y^2 - 4\,xz + 2\,yz + z^2$.
32. $2\,x + 1$.
33. $2\,x^3 + 4\,x^2 - 2\,x$.
34. $x^2 + 10\,x - 16$.
35. 0.
36. $x^2 - 5\,x + 8$.

37. $27\,x^3 + 54\,x^2 + 36\,x + 8$.
38. $27\,a^3 - 108\,a^2b + 144\,ab^2 - 64\,b^3$.
39. $a^4 + 4\,a^3b + 6\,a^2b^2 + 4\,ab^3 + b^4$.
40. $81\,x^4 - 216\,x^3y + 216\,x^2y^2 - 96\,xy^3 + 16\,y^4$.
42. 18.
43. -15.
44. 24.
45. 50.
46. 47.
48. -23.

EXERCISE 59, Page 110

2. 36.
3. 4.
4. 8.2.
6. 16.
7. 5.
8. $a = 2\,h(l + w) + lw$.
9. 968 sq. ft.
10. $R = \dfrac{2\,h(l + w) + lw}{32}$.
11. 32.
12. $P = \dfrac{lwh}{200}$.
13. 45.
14. 51.
16. 12.

ANSWERS xiii

EXERCISE 60, Page 112

1. $a^2 + 2ab + b^2 + 8a + 8b + 16.$
2. $a^2 + 2ab + b^2 - 6a - 6b + 9.$
5. $9 + 6a + 6b + a^2 + 2ab + b^2.$
6. $25a^2 - 10ax - 10ay + x^2 + 2xy + y^2.$
7. $4a^4 - 4a^3b + 8a^2c + b^2 - 4bc + 4c^2.$
8. $x^2 + 2xy + y^2 - 2ax - 2ay - 2bx - 2by + a^2 + 2ab + b^2.$
9. $x^2 + 2xy + y^2 + 8x + 8y + 15.$
10. $x^2 + 2xy + x^2 + 2x + 2y - 15.$
13. $4x^2 + 12xy + 9y^2 - 4ax - 6ay - 15a^2.$
14. $4x^2 - a^2 + 6ab - 9b^2.$
15. $a^2 + 2ab + b^2 - 9.$
18. $15 - x^2 - 2x.$
20. $a^2 + 2ab + b^2 - 9.$
22. $16 - x^2 - 2ax - a^2$
23. $4x^2 - 9y^2 + 30y - 25.$
26. $a^2 + 2ab + b^2 - 9x^2.$
27. $a^2 - b^2 + 6bx - 9x^2.$
28. $x^2 + 2xy + y^2 - 25a^2.$

35. $72\frac{1}{4}.$
36. $380\frac{1}{4}.$
37. $39,800\frac{1}{4}.$
38. $240\frac{1}{4}.$
39. $2450\frac{1}{4}.$
40. $9900\frac{1}{4}.$
41. $56.25.$
42. $380.25.$
43. $9900.25.$
44. $5625.$
45. $38.025.$
46. $990,025.$
48. $96.04.$
49. $92.16.$
50. $23.04.$
51. $9604.$

EXERCISE 61, Page 114

1. $25a^2 + .4a + .0016.$
2. $.16a^2 - .04a + .0025.$
3. $.0004x^2 - .0012xy + .0009y^2.$
4. $.09a^2 + .024ab + .0016b^2.$
9. $x^2 - .07x + .001.$
10. $a^2 - .58a + .04.$
11. $1,006,009$
12. $100.400.$
13. $99,960,004.$
15. $99,940,009.$
16. $306.25.$
17. $999,975.$
18. $1200.$
19. $292.4.$
21. $16a^2 - x^2 + 2xy - y^2.$
25. $25 - x^2a^2 - 12ay - 9y^2.$
26. $a^2 - 4b^2 + 20bc - 25c^2.$
27. 9991 sq. rd.
28. 9604 sq. rd.
29. $\$35.96.$
30. $\$8075.$

EXERCISE 62, Page 115

1. $5.7.$
2. $.52.$
3. $1.5 - 2x.$
4. $x^2 - 3.5x + 1.5.$
5. $4.155 + .24x - .3x^2.$
6. $-2x^2 - 3.2x - .08.$
7. $x^2 + x.$
8. $4x^2 + 12x + 1.$
9. $2.5x - 5x^2 + .1.$
10. $2.97y^2 + .02yz + .08z^2.$
11. $y^2 + z^2 - 4xz + 2yz.$
12. $.9x^2 - 3.4x + 3.7.$
13. $3x^2 - 10xy + y^2.$
14. $3xy.$
15. $-169x^4.$
16. $c^2 - 2c + 2a.$
17. $x^2y^2 - x^2y - y^2 + y^4.$
18. $4a^2 - ax + bx + my + cy.$

19. 0.
20. $5a^3 + 2a^2x - 11ax^2 + 10x^3$.
21. $6xy$.
22. 4, 8.
23. -12.
24. 0.
25. 12.
26. -12.
27. 5.
28. -18.
29. 16.

30. -1.
31. -26.
32. 4.
33. 1.
34. 5.
35. 29.
36. $8a^2$.
37. $76p^2$.
38. $-2a^2b^2 + 14ab - 5$.
39. 27 ; 9.

EXERCISE 63, Page 116

1. $s = c + pc$.
2. 42.
3. $180.
4. $2.50.
5. $8\frac{1}{3}$ ¢.
6. $s = 2ab + 2ac + 2bc$.

7. 112 sq. in.
8. $12\frac{1}{8}$ sq. ft.
9. $p = \dfrac{25\,lwt}{6}$.
10. 500 lb.
11. 672 lb.

12. 1.2 in.
13. $68\frac{2}{11}$ mi.
14. $d = \dfrac{15\,v}{22}$.
15. $20\frac{5}{11}$; $10\frac{5}{11}$; $40\frac{10}{11}$ mi.
16. $18\frac{3}{4}$, 75 mi.

EXERCISE 65, Page 120

1. 3.
2. 4.
3. 10.
4. 4.
5. 2.
6. 3.
7. -5.

8. 4.
9. 10 ft.
10. 17 in.
11. 4.
12. -2.
13. 3.
14. $3\frac{2}{3}$.

15. 5.
16. -2.
17. 11.
18. $-\frac{4}{5}$.
19. 10.
20. 6.
21. $-\frac{3}{5}$.

22. $\frac{4}{5}$.
23. 2.
24. $2\frac{5}{8}$.
25. $2\frac{1}{4}$.
26. -1.
27. 6.
28. 12.

29. 5.
30. 4.
31. 5.
32. $5\frac{1}{8}$.
33. 5.

EXERCISE 67, Page 122

1. $63, $21.
2. $48, $36.
3. $48, $24, $12.
4. $61, $39.
5. $45, $22.50.
6. 23.22.
7. 38.
8. $9000, $3000.
9. $12,200, $2200.
10. 20, 21, 22.
11. 21, 22.

12. 36, 60.
13. $26, $37, $35.
14. 22, 11, 17.
15. $67, $27.
16. 50, 350, 400.
17. 12.
18. 50.
19. $100.
20. 15.
21. 4×9 ft.
22. 4×5 yd.

23. 6 in.
24. 8 ft.
25. 5 yd.
26. 20×40 ft.
27. 26×34 ft.
28. 43.
29. 8, 9.
30. 5 hr.
31. 6 hr.
33. $4\frac{1}{2}$ hr., 36 mi.
34. 108 mi., 126 mi.

EXERCISE 68, Page 126

1. $x^3 + 9x^2 - 10y^3 + 3c + 2b$.
2. $2a^2 + 3b^2 - 12 + 3b - c$.
3. $a + 2b - 5$.
4. $4x^2 + 100xy + 25y^2$.
5. -3.
6. $2500, 6000.
7. $6x^3 - 13x^2 + 16x - 15$.
8. $4a^2 + 2a - 3$.
9. -4.
10. 9940.09.
11. $16,891.
12. 402.5.
13. 528.
14. 2624.
15. $4a^6 - 25b^4$.
16. 44 in.

17. 8.59 in.
18. $2\frac{2}{3}$ mi.
19. 6
20. $C = \dfrac{lwh}{128}$.
21. $12\frac{3}{16}$.
23. (1) 195.   (5) 247.
   (2) 323.   (6) $224\frac{1}{4}$.
   (3) 285.   (7) $195\frac{3}{4}$.
   (4) 221.   (8) $224\frac{7}{16}$.
25. 15.
27. $i = 12m - t - e$
   $r = \dfrac{12m - t - e}{d}$.
28. 7 %.
29. 16×32 yd.

EXERCISE 69, Page 130

1. 11.
2. 10.
3. 6.
4. 100.
5. 3.
6. $-\frac{4}{13}$.

7. 5.
8. 0.
9. $-.5$.
10. 2.
11. -9.
12. 2.

13. 4.
14. 21.5.
15. -2.
16. .96.
17. 2.17^-.
18. 13.75.

19. 1.33^-.
20. 53.
21. -41.
22. no.
23. no.
24. $5\frac{1}{4}$.

EXERCISE 72, Page 133

1. 13, ···, 17.
2. 19, 21, 23.
3. 6.
4. 4, 5, 6, 7, 8.
5. 21.
7. 8.
8. 125.
9. $37.50.
10. 14, 7.
11. 15 each.
12. 17 each.

13. $32, $32, $32, $14.
14. 80 mi.
15. 30.
16. 20×60.
17. 8×12 in.
18. 36×78 ft.
19. 160×300 ft.
20. 15 yr., 5 yr.
21. 20 yr., 10 yr.
22. 8 lb.
23. $8\frac{1}{4}$ lb.

24. 8 yr., 12 yr.
25. 52, 104, 57, 97.
26. $4.
27. 3 hr. after second starts.
28. 8 hr.
29. $5\frac{4}{5}$ hr.
30. 2 hr. 56 min.
31. 4, 8 mi.
32. A, 24 mi. ; B, 21 mi.

EXERCISE 73, Page 137

1. $3.2\,x^2 - 2.42\,xy - .24\,y^2$.
2. $-7.15\,a^2 - 1.5\,ab - 1.8\,b^2$.
3. $-3.8\,p^2 - .5\,p + 3.85$.
4. $2.6\,x^2 - .5\,x + 2$.
5. 45.
7. $54 - 22\,x$.
8. 6.
9. 0.
10. -18.
11. $D = \dfrac{cdlw}{2700}$.
12. $\$26.43$.
13. $8\,x^4 - 18\,x^3 - 13\,x^2 + 9\,x + 2$.
14. $10, 11, 12$.
15. $x^3 + 3\,x^2 - 2\,x - 1$.
16. $9\tfrac{4}{5}$ sec.

21. (1) $(2\,a^2 + 2\,b^2)x$.
    (2) $(2\,a^2 + 2\,ab)x$.
    (3) $(8\,a^2 + 18\,b^2)x$.
22. (1) $10\tfrac{4}{5}$.   (4) $1\tfrac{1}{2}\,x^2$.
    (2) $13\tfrac{1}{4}$.   (5) $1\tfrac{3}{5}\,a^2b$.
    (3) 9.
23. (1) $9,700$.   (3) $9,200$.
    (2) $6,200$.   (4) $115,800$.
    (5) $3\,a^4 - 5\,a^2b^2 + 2\,b^4$.
24. 2.
25. 7×12 ft.
26. 5.292.
28. $C = \dfrac{lw(p + e)}{1000}$.
29. 28.
30. 7.

EXERCISE 74, Page 141

1. $5(x + y)$
2. $5(a + 3\,b)$.
3. $a(x + y)$.
4. $6(a - 2\,b)$.
10. $18\,x(x - 3)$.
11. $3\,x(1 + x)$.
12. $p(1 + rt)$.

17. $5\,abc(5\,a + 4\,b)$.
18. $a^2b^2c^3(a - bc)$.
21. $7\,a(1 + 2\,a^2)$.
22. $3\,a^2x^2(a - 5\,x)$.
27. $3(a^2 - 2\,ax + 3\,x^2)$.
28. $2\,x(1 + 2\,x - 3\,x^2)$.
30. $11\,a^2b^2c^3(3\,a - 2\,bc + 4\,abc^2)$.

33. $\tfrac{2}{7}(139 + 124 + 150)$
34. 1694.
35. 938.25.
36. $58,190$.
37. $314\tfrac{4}{7}$.
38. $517,000$.

EXERCISE 75, Page 142

15. $(x + a)^2$.
16. $(y + 1)^2$.
19. $(5\,x + y)^2$.
20. $(3\,x + 1)^2$.
23. $(x - 6)^2$.
24. $(3\,a - 5\,b)^2$.

27. $(4\,a - 3\,y)^2$.
28. $(5\,x - 1)^2$.
31. $a^3(b + 2)^2$.
32. $x(y + 1)^2$.
35. $ab(9\,a + 7\,b)^2$.
36. $2\,y(2\,a - 5\,x)^2$.

39. $(x^5 + y)^2$.
41. $(\tfrac{1}{2}\,x + \tfrac{1}{3}\,y)^2$.
43. $(5\,a - 3\,x)^2$.
44. $q(pq - x)^2$.

EXERCISE 76, Page 144

1. $(x + 2)(x - 2)$.
2. $(a + x)(a - x)$.
5. $(5 + a)(5 - a)$.

6. $(y + 7)(y - 7)$.
11. $(3\,ab + 7\,x)(3\,ab - 7\,x)$.
12. $(9\,x + y)(9\,x - y)$.

15. $(3p + 8qr)(3p - 8qr)$.
16. $(2am + 3n)(2am - 3n)$.
20. $(4x^2 + y^2)(2x + y)(2x - y)$.
21. $(2x^3 + a^4)(2x^3 - a^4)$.
24. $(x^2 + 1)(x + 1)(x - 1)$.
26. $(a^4 + x^4)(a^2 + x^2)(a + x)(a - x)$.
29. $2(11 + x)(11 - x)$.

30. $3x(x + 5y^3)(x - 5y^3)$.
35. $(\frac{1}{2}x + \frac{1}{3}y)(\frac{1}{2}x - \frac{1}{3}y)$.
37. $(.3x + .4y)(.3x - .4y)$.
40. $(.9x + .05b)(.9x - .05b)$.
42. 15,200.   45. .36.
43. 34,000.   46. 1.92.
44. 11,200.   47. 584.

EXERCISE 77, Page 146

1. $(x + 1)(x + 2)$.
2. $(a + 1)(a + 3)$.
5. $(a + 3)(a + 4)$.
6. $(p + 2)(p + 8)$.
11. $(y - 7)(y - 9)$.
12. $(y - 3)(y - 21)$.

13. $(x + 5)(x - 3)$.
16. $(y + 10)(y - 8)$.
18. $(x + 2y)(x + 3y)$.
19. $(x + 2)(x - 3)$.
20. $x^3(x + 11)(x - 4)$.
23. $(x + 8y)(x - 2y)$.

25. $(x^3 + 4)(x^3 - 9)$.
26. $(x^2 + 4y^2)(x + 3y)$.
    $(x - 3y)$.
28. $x^2(x - 8)(x + 6)$.
30. $(x + 16y)(x - 3y)$.
32. $(x - 15)(x + 11)$.

EXERCISE 78, Page 147

38. $4(1 + 4a^2)(1 + 2a)(1 - 2a)$.
39. $5(2x + 3pq)(2x - 3pq)$.
40. $6(x^2 - 4xy + 9y^2)$.
41. $c(7 + 2bc)^2$.
42. $3x(x^2 - 10y^4)$.

43. $6(x - 8)(x + 2)$.
44. $x(2x + 11y^2)^2$.
45. $16(a - 3)(a + 2)$.
52. $y(x^6y^4 + z^8)(x^3y^2 + z^4)(x^3y^2 - z^4)$.
53. $x(x + 3)^2(x - 3)^2$.

EXERCISE 80, Page 151

1. $\frac{1}{2}b(x + y)$.
2. $\frac{1}{3}a(p - q)$.
5. $2\pi r(r + h)$.
15. $3(2x - a)^3(3 - 16x^2 + 16ax - 4a^2)$.
16. $(x - y)(2a + b)$.
17. $\frac{1}{2}(x + y)(a + b)$.
19. $(x + y)(a - b - c)$.
28. $(m + 2)x - (n + 4)y + (3 + n)z$.
29. $(1 - a - b)x - (1 - b + a)y - (2 + a - c)z$.

6. $108^2(108 - 1)$.
7. $t(v + \frac{1}{2}gt)$.
8. $.4(x - 2y)$.
20. $(a - b)(a + b + 2)$.
21. $(x - y)(x + y + 1)$.
22. $(a + b)(x - y)$.

11. $.3(a - 4b + 3c)$.
13. $(a + b)(7x + 5y)$.
14. $xy(a + b)(7x + 5y)$.
23. $(x + 3)(y + 2)$.
24. 201.0624.
25. 58,190.
26. .1.

EXERCISE 81, Page 152

1. $(a + .6)^2$
2. $(x - .9)^2$.
5. $(.7 - y)^2$.
6. $(x + y + 3)^2$.

9. $(3p - q + 5)^2$.
10. $(x + y - 6)^2$.
13. $(3x + 3y + 2z)^2$.
14. $(8a - 12 - b)^2$.

15. $(5x - 5y - 12xy)^2$.
16. $(a + b + c)^2$.

EXERCISE 82, Page 153

1. $(3 + .5\,a)(3 - .5\,a)$.
2. $(1.2\,x + .7\,y)(1.2\,x - .7\,y)$.
5. $(.1\,a + 4\,y)(.1\,a - 4\,y)$.
6. $(.9\,x + .05\,s)(.9\,x - .05\,s)$.
12. $(2\,x - 2\,y + 5)(2\,x - 2\,y - 5)$.
20. $(10\,x^2 - 10\,x - 9)(11 - 10\,x^2 + 10\,x)$.
21. $(a + b + x)(a + b - x)$.
22. $(a - b + 2\,x)(a - b - 2\,x)$.
23. $(a + x + y)(a - x - y)$.
38. $(a + b + c + d)(a + b - c - d)$.
39. $(x - 2\,y + 3\,z + 1)(x - 2\,y - 3\,z - 1)$.
40. $(3\,a - 2\,b + 5\,x + 1)(3\,a - 2\,b - 5\,x - 1)$.
41. $(a - 5\,b + 3\,bx - 1)(a - 5\,b - 3\,bx + 1)$.
42. 50.   44. 16.   46. 4053.6.   48. $1257\tfrac{1}{2}$.
43. 100.   45. 360.   47. 4824.4.   49. 8796.4?

14. $(2\,x + 2\,y + 1)(2\,y - 1)$.
15. $(11\,a - 8\,b)(9\,a - 2\,b)$.
16. $y(x^6y^4 + z^8)(x^3y^2 + z^4)(x^3y^2 - z^4)$.
17. $(9x^6 + 4y^2)(3x^3 + 2y)(3x^3 - 2y)$.
18. $x(x^2 + 12\,yz^3)(x^2 - 12\,yz^3)$.
32. $(a^2 + x^2 + y)(a^2 - x^2 - y)$.
33. $(3\,x + y + 1)(3\,x - y - 1)$.
37. $2(z - 1 + z^2)(z - 1 - z^2)$.

EXERCISE 83, Page 155

1. $(x - .2)(x - .3)$.
2. $(a + .6)(a + .2)$.
13. $(x^2 - 8)(x + 1)(x - 1)$.
14. $2\,a(1 - 10\,x)(1 + 3\,x)$.
17. $(a + b - 5)(a + b + 3)$.
19. $(x + a)(x + b)$.
5. $(ax - 6)(ax - 3)$.
6. $(x - 9)(x + 7)$.
9. $(x^2 - 2)(x^2 - 3)$
10. $(a^2 + 4)(a^2 + 3)$
20. $(x + 2\,a)(x - 3\,b)$.
23. $(x - y - 6)(x - y + 3)$.
24. $(5 - a + b)^2$.

EXERCISE 84, Page 155

1. $6(a - 2\,b - c)$.
2. $x(x + .3)(x - .3)$.
5. $a^2(a + 4\,x)^2$.
6. $x(x - 3\,y)(x + 2\,y)$.
13. $x(x - 16)(x - 9)$.
15. $(1 + 2\,a)(1 - 2\,a)(1 + a)(1 - a)$.
16. $x^3(x - 9)(x + 5)$.
30. $(3\,x + 3\,y + 2\,a + 2\,b)(3\,x + 3\,y - 2\,a - 2\,b)$.
31. $.14\,xy(x^2 - 2\,xy + 3\,y^2)$.
18. $(x + 2)(x - 1)(x^2 - x + 2)$.
19. $(x - 1)^2(x + 1)^2$.
23. $(x - .5)(x - .2)$.
26. $(2\,a + 1)^2(2\,a - 1)^2$.
27. $(2\,a - b - 2\,x)^2$.
28. $5(a + b)(x + 2\,y)$.
29. $7(a - b)^2(2 - a + b)$.
32. $\tfrac{1}{2}(x + y)(b + a)$.

EXERCISE 87, Page 161

1. $\frac{4}{4}$.
2. $\frac{3}{4}$.
3. $\frac{11}{14}$.
4. $\frac{2a}{3x}$.
5. $\frac{4x}{5y}$.
6. $\frac{x}{2-3ax}$.
7. $\frac{3xz}{4y^2}$.

8. $\frac{3a}{4b^2}$.
9. $\frac{1}{2a-1}$.
10. $\frac{1}{2a}$.
11. $\frac{1}{a}$.
12. $\frac{x-y}{x+y}$.

13. $\frac{2a}{3x}$.
14. $\frac{2(x+1)}{3}$.
15. $\frac{5}{2(x-y)}$.
16. $\frac{a+b}{2(a-b)}$.
17. $\frac{2}{3x-4y}$.

18. $\frac{2b}{3a}$.
19. $\frac{1}{2x+3y}$.
20. $\frac{7x+8y}{2x^2}$.
21. $\frac{3x}{2(x-1)}$.

EXERCISE 88, Page 163

1. $6\frac{3}{5}$.
2. $13\frac{1}{4}$.
3. $10\frac{11}{17}$.
5. $a-\frac{b}{y}$.
6. $x+\frac{ab}{x}$.
8. $a+\frac{5}{a^2}$.
9. $p^2+\frac{2}{p^2}$.

10. $2x+\frac{7}{5x}$.
11. $5+\frac{5}{4x}$.
13. $x-2+\frac{3}{x}$.
14. $2x^2+3-\frac{5}{2x}$.
16. $b+1-\frac{1}{b}$.
17. $x-2+\frac{1}{x^2}$.

18. $b+\frac{d^2}{a}$.
19. $9x-2-\frac{3}{x^2}$.
21. $2-\frac{x-y}{3a}$.
22. $2x+1-\frac{3a+1}{5x}$.
23. $4a-2-\frac{p+q}{2ab}$.

EXERCISE 89, Page 164

1. $\frac{20}{7}$.
2. $\frac{110}{9}$.
3. $\frac{161}{12}$.
4. $\frac{6+x}{3}$.
6. $\frac{x^2-y^2}{x}$.
7. $\frac{5a+6x}{x}$.

10. $\frac{7x^2-6y^2}{3y}$.
11. $\frac{6y^2-5x^2}{6y^2}$.
15. $\frac{5+6x-2x^2}{2x}$.
16. $\frac{x^2+xy+y^2}{x}$.
18. $\frac{5a^2-b}{2a}$.

19. $\frac{x^2+x-10}{x+4}$.
20. $\frac{26y-25x}{4}$.
21. $\frac{x^2}{x-1}$.
22. $\frac{a^2-2ab+b^2}{4}$.
23. 0.

EXERCISE 90, Page 166

1. 18.

2. 36.

5. $12\,a$.

6. $30\,x^2$.

9. $6\,a^3b^2$.

10. $12\,a^4x^2$.

13. $a^2 - b^2$.

14. $(a - b)^2$.

17. $x^2(x - y)$.

18. $ab(a + b)(a - b)$.

21. $x^2(1 - x)(1 + x)$.

22. $a^3b^2(a + b)(a - b)$.

24. $\frac{14}{10}, \frac{16}{10}, \frac{27}{10}$.

26. $\frac{24\,a}{10\,b}, \frac{7\,b}{10\,b}, \frac{10\,a}{10\,b}$.

28. $\frac{8\,x}{12\,a^2x}, \frac{9\,a}{12\,a^2x}, \frac{12\,a^2}{12\,a^2x}$.

30. $\frac{1}{a(a-1)}, \frac{3\,a}{a(a-1)}$.

32. $\frac{x}{x(4x^2-9)}, \frac{x(2x-3)}{x(4x^2-9)}, \frac{4x^2-9}{x(4x^2-9)}$.

35. $\frac{2(x+2)}{6(x^2-4)}, \frac{15(x-2)}{6(x^2-4)}, \frac{18}{6(x^2-4)}$.

EXERCISE 92, Page 168

1. $\frac{1}{14}$.

2. $\frac{21}{4}$.

3. $\frac{13}{18}$.

4. $\frac{16\,a - 25\,b}{40}$.

5. $\frac{1}{8\,x}$.

6. $\frac{9\,a - 5\,x}{6\,b}$.

7. $\frac{cx + ay}{abc}$.

8. $\frac{14 - 15\,y}{6\,y^2}$.

9. $\frac{21\,ad - 30\,bc}{35\,bd}$.

10. $\frac{10\,a + 3\,b}{12}$.

11. $\frac{19}{6\,x}$.

12. $\frac{8\,x - 9 + 12\,a}{12\,ax}$.

13. $\frac{15\,b - 2c - 6\,a}{6\,abc}$.

14. $\frac{yz + xz + xy}{xyz}$.

15. $\frac{47\,a}{30}$.

16. $\frac{5\,bc + 7\,c + 3}{abc}$.

17. $\frac{bc + ac + ab}{a^2b^2c^2}$.

18. $\frac{2\,x - 1}{x^2 - x - 12}$.

19. $\frac{5\,a - 1}{a^2 + a - 12}$.

20. $\frac{-1}{x^2 - 5\,x + 6}$.

21. $\frac{7\,b - a}{a^2 - b^2}$.

22. $\frac{8\,y - 2\,x}{15}$.

23. $\frac{a^2 - x^2 - 15}{3(a - x)}$.

24. $\frac{b}{a^2 - b^2}$.

25. $\frac{4\,ax}{x^2 - a^2}$.

26. $\frac{3\,a^2 + b^2}{6\,a^2b}$.

27. $\frac{11}{6(x - 1)}$.

28. $\frac{a - 8}{a^2 - 16}$.

29. $\frac{8\,x^2 + 2}{4\,x^2 - 1}$.

30. $\frac{9 - 2\,ax^2 + 28\,a - 12\,ax}{12\,ax^2}$.

31. $\frac{2\,a}{3 - a}$.

32. $\frac{2}{1 - 2x}$.

33. $\frac{3\,x^2 - 2\,x + 6}{x(x - 1)}$.

34. $\frac{11\,x^2 - 10\,x}{x^2 - 4}$.

35. $\frac{x^2 + 5\,x + 10}{(x + 1)(x + 2)(x + 3)}$.

EXERCISE 94, Page 170

1. $\frac{3}{4}$.

2. $\frac{4}{15}$.

3. $\frac{1}{4}$.

4. $\frac{2\,b^2 x}{3\,acy}$.

5. $\frac{4\,a^2 b}{7\,xy}$.

6. 1.

7. $\frac{ab^2 c}{48}$.

8. $\frac{2(x+y)}{a(x-y)}$.

9. $\frac{x+y}{x-y}$.

10. $\frac{ab}{2\,a-1}$.

11. $\frac{x^2+2\,x-3}{x}$.

12. $\frac{x-11}{x-2}$.

13. $\frac{1}{4}$.

14. $(m+1)(m^2+1)$.

15. $\frac{2\,x+3}{3(3\,x-1)}$.

16. $\frac{a-1}{a(x+1)}$.

17. $\frac{3\,x(a+2\,b)^2}{a}$.

18. $\frac{x-1}{x^2}$.

EXERCISE 96, Page 172

1. $\frac{5\,ax}{3}$.

2. $\frac{2\,qy}{3\,bp}$.

3. $\frac{10\,c^2 x}{3\,a^2 y z^3}$.

4. $\frac{3}{x-y}$.

5. $\frac{1}{a+1}$.

6. $\frac{2\,b(y+1)}{y-1}$.

7. $\frac{2\,b-1}{a-3\,y}$.

8. $\frac{x(a+b)}{a(a-b)}$.

9. $\frac{p-4\,q}{p+5\,q}$.

10. $\frac{2\,a(x+1)}{x-1}$.

11. $\frac{(a+2)(a+1)}{4}$.

12. $\frac{a^2}{b}$.

13. $\frac{b(x+y)}{2\,ax}$.

14. $\frac{a(x-2)}{(a-b)(x-1)}$.

15. $\frac{a}{cx}$.

16. $\frac{12(x+y)}{y}$.

17. $\frac{7\,y(a+3\,b)}{3\,b(a+b)}$.

18. $\frac{10\,ayz^3}{x(x-3\,y)}$.

EXERCISE 97, Page 172

1. $\frac{ab}{3(a+2\,b)}$.

2. $x+3\,y-\frac{a}{x}$.

3. $\frac{x^3+x^2-x+5}{x}$.

4. $\frac{1}{16}$.

5. $\frac{2\,a}{7\,b}$.

6. $\frac{1}{6(a-1)}$.

7. $\frac{(a+3)(a-4)}{ab^3}$.

8. $2\,a^2 b^2 c^2$.

9. $\frac{2}{a^2-1}$.

10. $\frac{2(y+3)}{27\,x}$.

11. $\frac{-7\,a-4\,b}{4\,b}$.

12. $\frac{b(x-y)}{2\,a}$.

13. $\frac{a(x+1)}{2(x-1)}$.

14. $\frac{2\,a-5\,b}{a^2-b^2}$.

15. $\frac{x-y}{2}$.

16. $\frac{a^2+b^2}{ab(a-3\,b)}$.

17. $\frac{a(a+x)}{a-x}$.

18. $\frac{x}{x+2\,y}$.

19. No.

20. Yes.

21. Yes.

22. 1.

EXERCISE 99, Page 174

1. $\dfrac{a+b+c}{3}$.

2. 71.

3. $\dfrac{p+q+r+s}{4}$.

4. $57\frac{1}{4}$ sec.

5. $\dfrac{ab}{43560}$.

6. $\dfrac{cd}{160}$; $\dfrac{ef}{14520}$; $\dfrac{pq}{2640}$.

7. 1.72^{+} A.

8. $P = \dfrac{bc}{a}$.

9. $31\frac{1}{2}$.

10. $C = \dfrac{hm}{d}$.

11. 31.

13. \$317.50.

14. 4 yr.

15. 4 %.

16. $\frac{3}{20}$.

17. $\dfrac{12}{35}$; $\dfrac{2x}{x^2-y^2}$.

18. $\dfrac{3}{130}$; $\dfrac{2y}{y^2-x^2}$.

19. 2.162^{+}.

20. 4608.

21. 860.

EXERCISE 100, Page 178

1. $-\frac{1}{4}$; $-\frac{1}{3}$; $\frac{3}{4}$; $-\frac{10}{3}$; $\frac{1}{4}$.

2. $\dfrac{x+3}{x-3}$.

3. $\dfrac{a+b+c}{a-b+c}$.

4. $\dfrac{2x-3y}{x^2}$.

5. $\dfrac{x-a-b}{x+a+b}$.

6. $\dfrac{x+3}{x+2}$.

21. $17a^3 + 51a^2 + 3a + 9$.

22. $\dfrac{10a-10}{8a^2-8}$.

7. $\dfrac{x-y}{a+b}$.

8. $\dfrac{x+y+p+q}{x-p+q-y}$.

9. $\dfrac{y-2x}{y+2x}$.

10. $\dfrac{3+m}{4-m}$.

11. $\dfrac{3+x}{3-x}$.

12. -1.

13. $\dfrac{-1}{a+b+c}$.

14. $\dfrac{1}{2a}$.

15. $\dfrac{y-x}{a+b}$.

17. $4(x^2-y^2)$.

18. $3(x^3+2x^2+4x+8)$.

19. $6a^2 + ab - b^2$.

20. $5x^2 - 16x + 3$.

23. $\dfrac{2x-8}{(x-2)(x-3)(x-4)}$.

EXERCISE 101, Page 180

1. $x^2 - x + 1 + \dfrac{1}{x+1}$.

2. $a^3 - a^2 + a - 1 + \dfrac{9}{a+1}$.

3. $r^2 - 3r + 9 - \dfrac{54}{r+3}$.

4. $3a + \dfrac{6ab}{3a^2-2b}$.

5. $2p + q + \dfrac{3q^2}{2p-q}$.

6. $x^2 - 1 + \dfrac{2}{x^2+1}$.

7. $x^3 + x^2 + 2x + 3 + \dfrac{5x+3}{x^2-x-1}$.

8. $2x^2 - 2x + \dfrac{2x+7}{x^2+x+1}$.

9. $x^2 - x + 2 - \dfrac{3x-3}{x^2+x-1}$.

10. $x + 2y - \dfrac{4y^2+1}{x+y}$.

11. $3x^2 + 9 - \dfrac{13x+1}{x^2-3}$.

12. $x^2 - 1 + \dfrac{1-a}{x-1}$.

13. $x^2 - 1 - \dfrac{x-1}{x^3+2}$.

16. $\dfrac{x^2}{x-1}$.

17. $\dfrac{x^3-2x}{x-1}$.

24. $\dfrac{15x^3-14x^2+5x-5}{3x-1}$.

27. No.　28. Yes.

18. $\dfrac{8x^2-y}{2x+1}$.

19. $\dfrac{a^2+1}{a-2}$.

20. $\dfrac{x^2+xy}{x+a}$.

25. $\dfrac{x^3}{1+x}$.

21. $\dfrac{2bc-b^2+c^2-a^2}{2bc}$.

22. $\dfrac{(2+3a)^2}{4}$.

23. $\dfrac{(2ab+1)^2}{4}$.

26. $\dfrac{x^3-1}{x+1}$.

EXERCISE 102, Page 181

1. $\dfrac{a^2-b^2}{a^2b^2}$.

2. $\dfrac{2a-2c}{ac}$.

3. $\dfrac{a^2-6ab+9b^2}{4(a-b)^2}$.

4. $\dfrac{3}{l+2}$.

5. $\dfrac{18x^2-3x+2}{36x^4}$.

6. $\dfrac{y^3-3x^2z^3-6yz^2}{6x^2y^2z}$.

7. 0.

8. $\dfrac{4x-1}{x^2-1}$.

9. $\dfrac{a^2+11a+18}{9-4a^2}$.

10. 0.

11. $\dfrac{5}{1-x^2}$.

12. $\dfrac{a-3b}{a^2-b^2}$.

13. 0.

14. 0.

15. $\dfrac{3xy}{4y^2-x^2}$.

16. $\dfrac{13}{8(1-a^2)}$.

17. $\dfrac{x}{1-x^2}$.

18. 0.

19. $\dfrac{-7}{12x(x+1)}$.

20. $\dfrac{b^3}{(a+b)^3}$.

21. $\dfrac{5x^2y-3y^3}{x(x^2-y^2)}$.

22. $\dfrac{x^2+4x-13}{2(x^2-1)}$.

23. $\dfrac{x^2+90x-9}{6(x^2-9)(x-3)}$.

24. $\dfrac{-1}{(x-3)(x-4)}$.

25. $\dfrac{x^2+2x-1}{x^2-1}$.

26. $\dfrac{17x^2-42x+39}{15(x^2-9)}$.

27. Yes.　28. Yes.

EXERCISE 103, Page 184

1. $\dfrac{3z^3}{10y(a+b)}$.

2. $\dfrac{2y(a+b)}{a-b}$.

3. $\dfrac{a^2}{b(a-b)}$.

4. $x-1$.

5. $a+b$.

6. $\dfrac{1}{x-2y}$.

7. $\dfrac{(4x+1)(x-1)}{6x^2-1}$.

8. x.

9. $\dfrac{a^2(a-y)}{x}$.

10. $\dfrac{6}{l(l-2)}$.

11. $\dfrac{x-y}{5y}$.

12. 1.

13. $\dfrac{1}{3ab}$.

14. $\dfrac{1}{x+y}$.

15. x^2-1.

16. $\tfrac{5}{6}$.

17. $\dfrac{a}{2b}$.

18. $x+y$.

19. $\dfrac{x-1}{x^2(x+1)}$.

20. $-\dfrac{x+2}{3}$.

21. $\dfrac{x^3-x+3}{x+1}$.

22. $a-b-2+\dfrac{b-2}{a-1}$.

23. $\dfrac{a^3-b^3}{ab(a+b)}$.

24. $\dfrac{x}{a(x+1)}$.

25. $\dfrac{3\,a(3\,a+b)}{b^2}$.

26. $\dfrac{3\,a^2b^2}{3\,a+4\,b}$.

27. $\dfrac{ab}{a^2-b^2}$.

28. $7\frac{7}{8}$.

29. 5.

30. $\dfrac{1-10\,x}{2(x^2-9)}$.

31. $\dfrac{2\,x-y}{2}$.

EXERCISE 105, Page 186

1. $D=\dfrac{d(e-s)}{100}$.

2. $T=\dfrac{fh+wy}{100\,c}$.

3. 8.

4. $s=c(1+r)$.

5. \$10.20.

6. \$8.

7. \$3 per doz.

8. $8\frac{1}{4}\cancel{c}$.

9. $4.008+\cancel{c}$.

10. 1425.6.

11. $8\frac{4}{13}$.

EXERCISE 107, Page 189

1. 12.
2. 11.
3. 12.
4. 20.
5. 60.
6. $2\frac{2}{3}$.
7. 5.
8. 3.
9. 8.
10. $-\frac{1}{2}$.
11. 20.
12. $1\frac{1}{2}$.

13. 3.
14. 24.
15. 5.
16. 2.
17. -116.
18. $12\frac{11}{16}$.
19. 2.
20. 3.
21. 2.
22. -1.
23. 7
24. 5.

25. $\frac{1}{5}$.
26. -2.
27. 5.
28. $-1\frac{1}{4}$.
29. -2.
30. $\frac{2}{3}$.
31. 0.
32. 2.
33. 3.
34. 10.
35. $2\frac{4}{15}$.
36. 3.

37. 3.
38. .4.
39. 3.
40. 13.
41. $\frac{4}{10}$.
42. -7.
43. 2.
44. $1\frac{1}{4}$.
45. $-\frac{1}{5}$.
46. 2.
47. $5\frac{1}{3}$.
48. 3.

49. 4.
50. 210.
51. 3.
52. 8.
53. 45.
54. 1.
55. 73.
56. 30.
57. 122 ; 212.
58. 20.
59. $14\frac{4}{11}$.

EXERCISE 109, Page 192

1. $3\,a$.

2. $\dfrac{b}{a}$.

3. $-\dfrac{c}{a}$.

4. $\dfrac{c}{a+b}$.

5. $\dfrac{b}{a-b}$.

6. $\dfrac{a-b}{2\,c}$.

7. $\dfrac{3\,b+2\,d}{2\,a-c}$.

8. $\dfrac{ab}{a-b}$.

9. $\dfrac{a}{b}$.

10. $\dfrac{a^2-2}{2\,a-3}$.

11. $\dfrac{b}{2}$.

12. $\dfrac{5\,c}{12}$.

13. $\dfrac{bc}{ad}$.

14. $\dfrac{1+2\,a}{2}$.

15. $\dfrac{b+2\,p}{2}$.

16. $\dfrac{13\,p}{5}$.

17. $\dfrac{a+b}{p+q}$.

18. $\dfrac{5a+5b}{c+d}$.

19. $\dfrac{c-d}{p+q}$.

20. $\dfrac{3a-c}{3b}$.

21. $\dfrac{27a}{5}$.

22. $\dfrac{7a-3b}{4}$.

23. $\dfrac{q-pr}{ps}$.

24. $\dfrac{ac-b}{a-1}$.

25. $\dfrac{ps-qr}{s-r}$.

26. $m+n$.

27. $\dfrac{bc-ad}{b+c-a-d}$.

28. $\dfrac{a-b}{a+b}$.

29. $\dfrac{3a-c}{15+9b}$.

30. 0.

31. $\frac{1}{2}a$.

32. 1.

33. $\frac{1}{3}a$.

34. $\dfrac{abcd}{ab+bc+ac}$.

35. $\dfrac{a^1}{1+rt}$.

36. $\dfrac{a}{1+rt}$.

37. $\dfrac{a-p}{pr}$.

38. $\dfrac{a-p}{pr}$.

39. $ab-cd$.

40. $\dfrac{p}{a-c}$.

41. $CR, \dfrac{E}{C}$.

42. $\dfrac{3a+bc}{p}$.

43. $\dfrac{3rs+pq}{4l}$.

44. $\dfrac{LW}{l}$.

45. $D-qd$, $\dfrac{D-r}{q}, \dfrac{D-r}{d}$.

46. $\dfrac{2a}{b}$.

47. $\dfrac{s}{\pi r}$.

48. $\dfrac{T}{\pi r}-r$.

EXERCISE 111, Page 195

1. 204.
2. $9600.
3. 33.
4. 30.
5. 80.
6. $238.25.
7. 60.
8. 84.
9. 36.
10. $14,400, $9600.
11. $2520, $1890.
12. 21,105; 36, 90.
13. 27, 28.
14. $10,800, $5400, $1800.
15. 2000, 1000, 500 lb.
16. $16, $32.
17. $4000, $2000, $800.
18. 2266$\frac{2}{3}$ lb.

19. 50, 62.5 lb.
20. $2800, $2000.
21. $4000, $8000, $6000.
22. 95.
23. 95.
24. 96.
25. $4000.
26. 2150 lb.
27. 5.
28. 7.
29. 7.
30. 4.
31. 6 qt.
32. 31 pt.
33. 7 gal.
34. 1 bu.
35. 40 lb.
36. 150 lb.

37. 100 lb.
38. 80 lb.
39. 15 hr.
40. $\frac{8}{11}$ hr., or 16$\frac{4}{11}$ min.
41. 24 hr.
42. 6 hr.
43. 45.
44. 137.
45. $15,000.
46. $20,000.
47. 17 gal.
48. 88.
49. 159.
50. 4.
51. 8.
52. 26, 27, 28.
53. 108 min.

EXERCISE 112, Page 201

8. 5.1 mi.
9. $s=\frac{2}{3}c$.
10. $15.
14. 1080 ft. per sec.

EXERCISE 113, Page 203

1. (1) $3x(x-2y)^2$.
  (2) $4(a+2b)(a-2b)$.
  (3) $(x-2)(x-5)$.
  (4) $x(x^2+a^2)(x+a)(x-a)$.

2. (1) $(ax-4)^2$.
  (2) $3(x^2+abc)(x^2-abc)$.
  (3) $x^2(3x-7y^2)^2$.
  (4) $(16x+9y)(16x-9y)$

5. $\dfrac{x^2+y^2}{2y(x+y)}$.

6. $\dfrac{p^3-p^2+6}{6p^2(p^2-1)}$.

7. $\dfrac{2(x+2)}{x-1}$.

8. $\dfrac{y(x+y)}{2}$.

9. (1) 22,496
  (2) $63\frac{1}{4}$.
  (3) 992,016
  (4) 1,008,016.
  (5) $143\frac{11}{18}$.
  (6) 16,891.

10. $4.47+\%$.
11. 80.

12. $\frac{2}{8}$.
13. 9.
14. 1.
15. $3b-4a$.
16. $\dfrac{a^2-b^2}{a^2+b^2}$.
17. $-\dfrac{l}{2}$.

18. .08.
19. \$4125,
  \$2475.
24. 75.
25. 89.
26. $36\frac{3}{4}$.
27. 10.
28. 12.

EXERCISE 114, Page 205

1. 2.
2. 2.
3. 1.5.
4. .5.
5. $.16\frac{1}{20}$.

6. 2.
7. 1.4.
8. -10.
9. $2.267-$.
10. $1.316-$.

11. $.533+$.
12. $2.518-$.
13. $5.169-$.
14. $-2.690+$.
15. $1.326-$.

16. No.
17. 0.
18. $\frac{1}{16}$.
19. -5.
20. -7.

21. -3.
22. -3.
23. 12.
24. $-\frac{1}{4}$.

EXERCISE 115, Page 206

1. $\dfrac{a}{18}$.

2. $\dfrac{3a-c}{9b+15}$.

3. $\dfrac{1}{r+s}$.

4. $q+2p$.
5. $17a$.
6. $19a^2$.
7. 1.
8. $\dfrac{b}{2a+b}$.

9. $\dfrac{ac^3}{ac-ab+bc}$.
10. 0.
11. $2p-q$.
12. $\dfrac{a^2b-ab^2-3ab^2c}{bc-ac-3bc^3}$.

13. $\dfrac{lm}{p^2}$.
14. $\dfrac{ab(a-b)}{a^3-2ab^2-b^3}$.
15. $\dfrac{Rg}{g-R}$.

16. $\dfrac{Rs}{s-R}$.
17. $\dfrac{pp'}{p+p'}$.
18. $\dfrac{p'f}{p'-f}$.
19. $\dfrac{9C+160}{5}$.
20. $h=\dfrac{T}{2\pi r}-r$.
21. $h=\dfrac{2a}{b_1+b_2}$.
22. $b_1=\dfrac{2a}{h}-b_2$

EXERCISE 117, Page 209

1. 120.
2. 120.
3. 600.
4. \$200.
5. 123.

6. 96.
7. 2200 lb.
8. 860 millions.
9. 90 pt.
10. 240 lb.

11. 60 gal.
12. 17.
13. 11.
14. 8.
15. 11.

16. 60.
17. 10.
18. $4\frac{1}{4}$ lb.
19. $9\frac{33}{49}$ lb.
20. $33\frac{1}{4}$ gal.

21. $1\frac{1}{15}$ gal.
22. $5\frac{4}{5}$ da.
23. $1\frac{1}{4}$ da.
24. 6 da.
25. $28\frac{4}{5}$ min.

26. 36 min.
27. $169\frac{7}{17}$ min.
28. 1100 ft. per sec.
29. 2357 ft. per sec.
30. 104.

31. 3.
32. $1\frac{1}{2}$ gal.
33. 96.
34. 10, 14, 6, 24.
35. 36 lb.

36. $20,000.
37. $22\frac{1}{2}$ min.
38. 6.
39. 15.
40. $\frac{4}{5}$ gal.

EXERCISE 118, Page 214

1. (1) $\frac{1}{4}a(x+y)$.
   (2) $x(x^2+16)(x+4)(x-4)$.
   (3) $x(x+15)(x-9)$.
   (4) $(2x+a+b)(2x-a-b)$.
   (5) $a^2(a+2)^2(a-2)^2$.
   (6) $(x-.4)(x-.3)$
2. (1) $(a+b+x)(a+b-x)$.
   (2) $\pi(a^2-b^2+c^2)$.
   (3) $(a+b)(5x-7y)$.
   (4) $(a+x-2)^2$.

3. 7960 ; 78.54.
4. $\dfrac{x}{x^2-y^2}$.
5. $\frac{2}{3}$.
6. $\dfrac{1}{a}$.
7. 4.
8. $\dfrac{2a+b}{2}$.
9. 59.

10. 60.
11. 247.
12. $\frac{1}{10}$, $\frac{1}{2}$, $\frac{4}{13}$.
13. $\dfrac{q^2}{p}$.
14. .2.
15. .6
16. $W=\dfrac{4\,lwh}{5}$.
    $h=\dfrac{5\,W}{4\,lw}$.

17. 10 ft. 5 in.
18. 108.
20. $r=\dfrac{a-p}{pt}$.

21. $t=\dfrac{a-p}{pr}$.
27. $-2.154-$.
28. $2800.

29. $-5b$.
30. 100°.
31. $1482\frac{2}{9}$°.

32. $F=\dfrac{9\,C+160}{5}$.
33. $442\frac{2}{9}$°.
34. $1995\frac{4}{5}$°.

EXERCISE 119, Page 219

1. 2, 1.
2. 4, 2.
3. 6, −1.
4. 1, −3.

5. 5, 1.
6. −2, 3.
7. $\frac{1}{2}$, −$\frac{1}{4}$.
8. 2, $\frac{1}{2}$.

9. −2, $\frac{1}{3}$.
10. 3, $\frac{1}{3}$.
11. 0, 2.
12. 1, 1.

13. 2, 3.
14. 7, 3.
15. 1, 1.
16. 2, −1.

EXERCISE 120, Page 220

1. 2, 3.
2. 4, 3.
3. 2, 1.
4. 3, −2.
5. −3, 4.
6. −2, −2.

7. 3, 2.
8. 1, −2.
9. 4, 3.
10. 1, 1.
11. −1, −1.

12. 2, −1.
13. 1, 2.
14. $\frac{1}{2}$, −$\frac{1}{4}$.
15. $2\frac{2}{3}$, $\frac{1}{10}$.
16. 3, −2.

17. −1, 1.
18. 2, −2.
19. 3, −1.
20. 3, −2.
21. −1, 1.

22. −3, 0.
23. 3, −$\frac{1}{4}$.
24. 5, 4.
25. −4, 3.
26. 2, 6

EXERCISE 121, Page 222

1. 12, 12.
2. 6, 20.
3. 4, 2.
4. $1\frac{3}{8}$, -4.
5. $-2, 4$.
6. 6, 6.
7. 12, 12.
8. $1, -1$.
9. $\frac{1}{2}, -\frac{1}{3}$.
10. 5, 12.
11. 3, 1.
12. 1, 2.
13. 2, 5.
14. $\frac{1}{4}, \frac{1}{5}$.
15. 2, 9.
16. 2, 9.
17. 5, 3.
18. 6, 6.
19. $6, -2$.
20. 3, 6.
21. 3, 3.
22. $2, -3$.

EXERCISE 122, Page 224

1. $\dfrac{14\,a}{5}, \dfrac{7\,a}{5}$.
2. $2\,a, 2\,b$.
3. $a, 1$.
4. $\dfrac{3\,ab}{a+b}, \dfrac{2\,ab - a^2}{a+b}$.
5. $-b, 2\,a$.
6. $\dfrac{cp - bd}{ap - bm}, \dfrac{cm - ad}{bm - ap}$.

7. $\dfrac{a}{b}, \dfrac{b}{a}$.
8. $\dfrac{a^2 + b^2}{2\,a}, \dfrac{b^2 - a^2}{2\,a}$.
9. $\dfrac{1}{a}, \dfrac{1}{b}$.
10. $p - 2\,q, 2\,p - 4\,q$.
11. $c - d, cd - d^2$.
12. $\dfrac{3\,pq - q^2}{p+q}, \dfrac{3\,pq - p^2}{p+q}$.

13. $\dfrac{cq + br}{aq + bp}, \dfrac{cp - ar}{bp + aq}$.
14. $\dfrac{c - 2\,b}{a + b}, \dfrac{2\,a + c}{a + b}$.
15. $\dfrac{abc - a}{a + b}, \dfrac{abc + b}{a + b}$.
16. $2\,c, 3\,d$.
17. $a - b, a - b$.

EXERCISE 123, Page 225

1. 1, 2, 3.
2. 2, 1, 3.
3. $2, 3, -4$.
4. 3, 4, 7.
5. $2, \frac{1}{3}, \frac{1}{2}$.
6. 1, 1, 1.
7. $\frac{3}{2}, \frac{1}{3}, \frac{5}{4}$.
8. $2, 3\frac{1}{2}, -4$.
9. $-3, 3\frac{1}{2}, -2$.
10. 1, 3, 5.
11. $-14, -5, -5$.
12. $2, 3, -4$.
13. $3, -5, 1$.
14. 2, 2, 2

EXERCISE 125, Page 228

1. 14, 9.
2. $6.20, $18.60.
3. $15\frac{2}{3}, 12\frac{7}{8}$.
4. 35, 145.
5. 15, 9.
6. 45, 50.
7. 12, 21.
8. 920, 80.
9. 33, 42.
10. $67, $27.
11. 158, 460 lb.
12. 19.
13. 12, 9.
14. $3\,\rlap{/}c, 5\,\rlap{/}c$.
15. $3, $2.
16. $1.80, $1.50.
17. 480, 700 lb.
18. 3, 1.
19. 5, 3, 1.
21. $5400, $4600.
22. $67\frac{1}{2}, 72\frac{1}{2}$.
23. $50\,\rlap{/}c, 3\,\rlap{/}c$,
25. $7, $50.
26. $8, $20, $24.
27. $1.52, $1.36.
28. $1.02, $.81, $.70.
29. $46\frac{2}{3}, 53\frac{1}{3}$.
30. 20 lb. @ $20\,\rlap{/}c$,
    40 lb. @ $32\,\rlap{/}c$.
31. 7×5 in.
32. 15×6 ft.
33. $\frac{2}{5}$.
34. $\frac{8}{15}$.
35. $\frac{11}{4}, \frac{7}{8}$.
36. 16, 81.
37. 21, 79.

38. 14, 54.
39. \$3250, \$1800.
40. \$2000 @ 5 %, \$10,000 @ 4 %.
41. 6¢, 5¢.
42. $\frac{13}{16}$.
43. \$15,000 @ 5 %, \$3000 @ 6 %.
44. 57, 43.
45. $83\frac{1}{3}$ lb. @ 18¢, $16\frac{2}{3}$ lb. @ 30¢.
46. 30 × 120 ft.
47. 87, 11.

EXERCISE 126, Page 234

1. 2, .2.
2. .2, .3.
3. − .8, − 2.2.
4. .1, .1.
5. .01, .002.
6. 2, 4.
7. 8.85−, .27−.
8. 1.18+, .24+.
9. .09−, .02−.
10. − 1, 4.
11. $-\frac{1}{2}, \frac{2}{3}$.
12. 9, − 1.
13. 17, 6.
14. 12, 18, − 24.
15. $a + b, a - b, 2a$.
16. 6, 40, 20.
17. $-a + b + c, a - b + c, a + b - c$.
18. $\frac{3a - 2b}{6}, \frac{2a + 3b}{6}, \frac{a + b}{6}$.
19. $m + n, m - n$.
20. $q - p, q + p$.
21. $\frac{c}{a + d}, \frac{-cd}{a + d}$.
22. $\frac{5ab^2c + 7a^2bc}{5b^2 + 3a^2}, \frac{3a^2bc - 7ab^2c}{3a^2 + 5b^2}$.

EXERCISE 127, Page 237

1. $\frac{p}{a - 5b}, \frac{bp}{a - 5b}$.
2. $a + b, a - b$.
3. $\frac{2c}{2a - b}, \frac{c}{2a - b}$.
4. 3, $\frac{2a + 1}{b}$.
5. a, b.
6. $\frac{1}{a+b+c}, \frac{1}{a+b+c}$.
7. $a, -b$.
8. − 1, 1.
9. $\frac{1}{3}, -\frac{1}{2}$.
10. $\frac{2}{3}, \frac{1}{2}$.
11. $\frac{5}{6}, -\frac{3}{4}$.
12. $\frac{5}{3}, -\frac{3}{4}$.
13. $\frac{2n}{1 + n^2}, \frac{2n}{1 - n^2}$.
14. $a, -a$.
15. $1, -\frac{1}{2}, \frac{1}{3}$.
16. $2, -\frac{1}{2}, 1$.

EXERCISE 129, Page 239

1. .0012, .0003.
2. $1\frac{1}{2}, \frac{1}{6}$.
3. 11.4, 10.8.
4. 15¢, 5¢, 5¢.
5. 2,772,000,000, 737,000,000, 1,007,000,000.
6. 550,480,156 lb.
7. .2875, .5025, .3625 in.
8. 984, 700, 555 ft.
9. 40 lb. @ 15¢, 60 lb. @ 30¢.
10. 40 lb. @ 75¢, 60 lb. @ 50¢.
11. 20.82, 1.98.
12. 5.525, 1.475.
13. 14, 9.
14. 12, \$60.
15. 90 mi.
16. $\frac{5}{8}$.
17. 13, 8.
18. 68, 50.
19. 49.
20. 23.
21. 64.
22. 151.
23. $5\frac{1}{3}$, 17.

24. 24 da., 48 da.
25. $14\frac{2}{17}$, $18\frac{6}{13}$, $34\frac{7}{9}$ da.
26. gold, $2\frac{1}{3}$ lb.; silver, $18\frac{3}{9}$ lb.
27. aluminum, 35 lb.; iron, 45 lb.
28. copper, $51\frac{1}{4}$ lb.; tin, $48\frac{3}{4}$ lb.
29. 480 mi.
30. 11, 36.

31. 27.
32. 80 lb. @ 25 $\not c$.
    120 lb. @ 50 $\not c$.
33. $11\frac{3}{4}$ cream, $8\frac{1}{4}$ milk.
34. 38.
35. $70, $100.
36. 50 $\not c$, 30 $\not c$.
37. 24, 16.
38. $\dfrac{a+b}{2}$, $\dfrac{a-b}{2}$.

39. $\dfrac{ce-bf}{ae-bd}$, $\dfrac{cd-af}{bd-ae}$.
40. $\dfrac{rs-qr}{p-q}$, $\dfrac{rs-pr}{q-p}$.
41. $\dfrac{d-bc}{a-b}$, $\dfrac{d-ac}{b-a}$.

EXERCISE 132, Page 250

1. (2, 1).
2. (1, − 1).
3. (2, 3).
4. (− 2, 1).
5. (− 1, 3).
6. (− 1, 1).
7. (− 3, − 3).
8. (− 3, 4).
9. (− 4, − 2).
10. (0, 0).
11. (− 2, − 3).
12. 3, − 2).
13. (− 2, 1), (3, 2), (0, − 3).

EXERCISE 133, Page 250

3. $7.50.
4. $3.75.
6. $4.80.
8. $24.35.
9. $1\frac{111}{481}$.
10. 19.7⁻ ft.
11. $K = .62\, m$.
13. $r = \dfrac{100\,(12\,m-e)}{d}$.

EXERCISE 134, Page 252

1. (1) $6(y+1)(y-2)$.
   (2) $bx^3(x^2+1)(x+1)(x-1)$.
   (3) $x(4a+b)^2$.
   (4) $(x-13)(x+2)$.
   (5) $5x(x-2y)^2$.
   (6) $(2x+1)^2$.
2. 2.
3. $\dfrac{18x^2-3x+2}{36x^4}$.
4. 8, 12.
5. 10, 15.
6. 7 gal.
9. $c = \dfrac{a}{R-ad}$.
10. 4,602.

11. (1) $6(1+4abx)(1-4abx)$.
    (2) $(y+14)(y-9)$.
    (3) $(3axy+1)^2$.
    (4) $(\frac{2}{3}a+4b)(\frac{2}{3}a-4b)$.
    (5) $(\frac{2}{3}a-3b)^2$.
    (6) $(3a-x)^2$.
12. $\dfrac{a(1-a)}{1+a}$.
13. $\dfrac{(a+3)(a-4)}{a^2}$.
14. $\frac{1}{3}$, $\frac{1}{5}$.
15. 4, 1, − 2.
16. 24 lb. @ 20 $\not c$.
    48 lb. @ 32 $\not c$.
18. 7, 29.

19. $g = \dfrac{v^2}{2h}$.

20. $1,445\frac{1}{2}$.

21. $\dfrac{2pq}{p-q}$.

22. $\dfrac{12a}{9a^2-4}$.

23. $26,010$.

24. $2, 1$.

25. 7.

27. $19-, 28+$.

28. $N = \dfrac{12,000H}{SCA}$.

29. $-3.38-$.

30. $\dfrac{bq-dp}{bc-ad}$, $\dfrac{aq-cp}{ad-bc}$.

EXERCISE 137, Page 257

1. $6x^3 - 22x^2 + 10x + 10$.
2. (1) $\frac{1}{4}a(h-k)$.
   (2) $(x^8+1)(x^4+1)(x^2+1)(x+1)(x-1)$.
   (3) $(2x+4y+3a)(2x+4y-3a)$.
   (4) $228 \times 228 \times 227$.
   (5) $(.6x+1)^2$.
   (6) $(x+3)(x-2)(x^2-x+6)$.

3. 69.75.
4. 0.
5. $1\frac{2}{3}, 2\frac{3}{4}$.
6. $\dfrac{p-q}{p}, \dfrac{p+q}{q}$.
7. 45.

9. $\frac{1}{8}$.

10. $r = \dfrac{4h^2+s^2}{8h}$.

11. $\frac{1}{2}x^4 - \frac{5}{12}ax^3 + \frac{3}{4}a^2x^2 + \frac{2}{3}a^4$.

12. (1) $(x-a+b)(x+a-b)$.
    (3) $(11y-x)(7x-y)$.
    (5) $(a+b)^2(5x-3y)$.

13. $2 - 3x + 3x^2 - 3x^3 + \dfrac{3x^4}{1+x}$.

14. $\dfrac{x+3}{x+2}$.

15. $4b + 3a$.

16. $5, 7$.

18. 8800.

19. $R = \dfrac{drW}{2\pi lP}$.

20. 2.

21. $.115+$.

22. $1.6x^2 - 2xy + 2.4y^2$.

23. $.712+$.

24. -3.

25. $\dfrac{-5cr}{r+2c}$.

27. 10×4 ft.

28. q, p.

30. $\dfrac{x^2+y^2}{5xy(x+y)}$.

EXERCISE 139, Page 263

1. $3a + b$.
2. $2x + 5y$.
3. $9ab + 5c$.
4. $t + u$.
5. $x^2 - 2x + 1$.
6. $1 - a - a^2$.
7. $3x^2 - 2x + 1$.
8. $5 + 3x + x^2$.
9. $n^3 - 2n^2 + 3$.
10. $2x^3 + 3x^2 - 2x - 3$.
11. $4m^3 - 5m - 1$.
12. $2n^3 - 3n^2 + 4n - 5$.
13. $3x^3 + 4x^2y - 4xy^2 - 3y^3$.
14. $1 + x + 2x^2 - x^3$.
15. $7x^3 - 3x^2 - 4x - 2$.

EXERCISE 140, Page 264

1. 85.
2. 51.
3. 325.
4. 427.
5. 581.
6. 753.
7. $6,012$.
8. 90.08.
9. 14.114.
10. 3.2105.
11. $.17071$.
12. $1,230.321$.
13. $2.646-$.
14. $3.317-$.
15. $3.536-$.
16. $3.503+$.
17. $1.826-$.
18. $1.453-$.
19. $.949-$.
20. $2.278+$.
21. $2.582-$.
22. $1.275-$.
23. $.342-$.
24. $21.954+$.
25. $.221+$.
26. $1.003+$.
27. $6.008-$.
28. $2.229-$.

EXERCISE 142, Page 265]

2. 17.
3. 53 in.
4. 113.
5. 78.102+ ft.

6. 250.3⁻ yd.
7. .892-.
8. 5.39−, 5.83+, 8.54+.
9. 515.5+ mi.

11. 6.7+ mi.
12. 16.4+ mi.
13. 5.2⁻ mi.

EXERCISE 143, Page 267

1. $a + b + 1.$
2. $p + 3q + 3.$
3. $x - 2y - 3.$
4. $2a + x - 4.$
5. $x + m + 3.$
6. $x + \dfrac{p}{2}.$
7. $x - \frac{1}{4}y.$
8. $5 - \frac{1}{2}x.$

9. $a^2 + a + \frac{1}{4}.$
10. $x^2 - x + \frac{1}{4}.$
11. $x^2 + x - \frac{1}{4}.$
12. $x^2 + xy - \frac{1}{2}y^2.$
13. $b^2 + 2b - \frac{1}{4}.$
14. $6 - \frac{1}{3}a + \frac{1}{4}a^2.$
15. 4.618+.
16. 2.633+.
17. 3.588+.

18. .158+.
19. − 1.822+.
20. 1.931+.
21. 1.368+.
22. 2.726+.
23. 1.434+.
24. .453+.
25. .891+.
27. .490⁻.

28. 12.655+.
29. 13.097⁻.
30. 19.102+.
31. 8.291+·
32. 4.888⁻.
33. 27.0⁻.
34. 434.54⁻.

EXERCISE 145, Page 269

2. 8.
3. 9.14+.
10. 756.
11. 9.7979+ sq. mi. ;
6270.67⁻ A.

4. 8.54⁻.
5. 11.03⁻.
12. 6.55⁻ sq. mi.
13. 18.742+ sq. mi.
14. 188.9+ ft.

6. 7.
7. 10.61+.

8. 120 ft.
9. 61.48+.
15. 582.8⁻ ft. per sec.
16. 3.46⁻.
17. 2.33.-

EXERCISE 147, Page 272

1. $6\sqrt{6}.$
2. $\frac{1}{4}\sqrt{6}.$
3. $-2\sqrt{10}.$
13. $2a\sqrt[4]{8ax^2}.$
14. $5b^2\sqrt[3]{2a^2}.$
15. $3\sqrt{11a}.$
16. $4a^2x\sqrt{a}.$
17. $2a^2x\sqrt{2a}.$
18. $10a^3\sqrt{2a}.$
19. $7x^2y^3\sqrt{3x}.$
20. $-6x^7y^5\sqrt{7x}.$
21. $-3ax\sqrt[3]{3x^2}.$
22. $a(x-y)\sqrt{x-y}.$

4. $-6\sqrt{7}.$
5. $2\sqrt{5}.$
6. $4\sqrt{2}.$
23. $7x(a+1)^2\sqrt{x(a+1)}.$
24. $\dfrac{4ac^2}{x^2}\sqrt{3an}.$
25. $\dfrac{x^2z^5}{a}\sqrt{7xz}.$
26. $3ac^2d^2\sqrt{abd}.$
27. $2axy^2\sqrt{2xy}.$
28. $2ax^2y\sqrt{6ax}.$
29. $(a+b)\sqrt{x}.$
30. $(a-2b)\sqrt{y}.$

7. $2\sqrt[3]{6}.$
8. $2\sqrt[3]{3}.$
9. $3\sqrt[3]{2}.$
31. $2(a+b)\sqrt{2x}.$
32. $(x+y)\sqrt{x-y}.$
33. $2(a+b)\sqrt{2x}.$
34. $4(x+y)\sqrt{x-y}.$
35. $(x-y)\sqrt{2}.$
36. 3.4641+ ; 5.19615+ ;
8.66025+; 15.58845+.

10. $\sqrt[3]{9}.$
11. $-10\sqrt{3}.$
12. $2\sqrt[4]{3}.$

EXERCISE 149, Page 274

1. $17\sqrt{2}$.
2. $\sqrt{2}$.
3. $11\sqrt{3}$.
4. $-5\sqrt{x}$.
5. $7x\sqrt{2x}$.
6. $(a+ab)\sqrt{ab}$.
7. $(x^2-y^2)\sqrt{xy}$.
8. $12a\sqrt{2}$.
9. $2\sqrt[3]{2}$.
10. $4\sqrt[3]{3}$.
11. $18\sqrt[3]{a}$.
12. 0.
13. $8\sqrt{5}$.
14. $15\sqrt{2}$.
15. $8\sqrt{3}$.
16. $4\sqrt{5}$.

17. $(3x-2x^2y+2y^2)\sqrt{2y}$.
18. $(3a+5b-c)\sqrt{abc}$.
19. $5\sqrt[3]{3}-2\sqrt[3]{2}$.
20. $-47x\sqrt{x}$
21. $4\sqrt{b}$.

22. $10\sqrt[3]{2c}$.
23. 0.
24. $4ac\sqrt{b}$.
25. $6b\sqrt[3]{2a}-6b\sqrt[3]{a}$.
26. $8\sqrt{2}$.

27. $8\sqrt{3}$.
28. $11\sqrt{3}-10\sqrt{2}$.
29. $12\sqrt[3]{2}-21\sqrt{3}$.
30. $37ab\sqrt{3a}$.
32. 5.6568^+.

EXERCISE 151, Page 276

1. $15\sqrt{3}$.
2. 18.
3. $ab^2\sqrt{a}$.
4. $30\sqrt{10}$.
5. $8\sqrt{6}$.
6. $6a^2\sqrt{b}$.

7. $x^3\sqrt{6}$.
8. $6y$.
9. $\dfrac{x}{2}\sqrt[3]{x^2}$.
10. $\dfrac{9ab}{20}\sqrt{6b}$.

11. $30\sqrt{21}$.
12. $7\sqrt{5}$.
13. $\frac{1}{12}$.
14. $\frac{4}{9}\sqrt[3]{3}$.
15. $60\sqrt{2}$.
16. $12\sqrt{10}$.

17. 2.
18. abc.
19. $a^2bc\sqrt{ab}$.
20. 1.
21. $8a$.
22. 5.92^-.

EXERCISE 153, Page 277

2. 7.071^+ ;
   17.677^+ in.
3. 187.453^- yd.
5. $4\frac{1}{2}^-$ in.

6. 4.9^+.
7. 2.8^+.
8. $4\frac{1}{4}^-$ in.
9. 2.8^-.

11. 616 sq. in.
12. 6.16 sq. ft.
13. $1257\frac{1}{7}$ sq. yd.

15. 1.78^+ in.
16. 5.64^+ yd.
17. 117.7^+ ft.

EXERCISE 155, Page 280

1. $\frac{1}{2}\sqrt{30}$.
2. $\frac{5}{7}\sqrt{7}$.
3. $\frac{2}{3}\sqrt{15}$.
4. $\frac{2}{5}\sqrt{30}$.
5. $\frac{2}{3}\sqrt{15}$.
6. $\frac{1}{5}\sqrt{15}$.
7. $\frac{1}{4}\sqrt{10}$.
8. $\frac{1}{10}\sqrt{35}$.

9. $\dfrac{1}{a^2}\sqrt{5a}$.
10. $\dfrac{1}{2b^2}\sqrt{10b}$.
11. $2\sqrt[3]{14}$.
12. $2\sqrt[3]{25}$.
13. $\sqrt[3]{180}$.
14. $15\sqrt[3]{4}$.

15. $\dfrac{1}{a}\sqrt[3]{2ab}$.
16. $\frac{1}{3}\sqrt{3xy}$.
17. $\frac{1}{5}\sqrt{10ab}$.
18. $\dfrac{3}{2a}\sqrt{2a}$.
19. $\dfrac{2}{x}\sqrt{x}$.

20. $\frac{5}{12}\sqrt{6}$.
21. $\dfrac{1}{2a}\sqrt{6ax}$.
22. $2a\sqrt[3]{5}$.
23. $\frac{4}{3}\sqrt[4]{8}$.
24. $\frac{3}{2}\sqrt[3]{6}$.
25. $\dfrac{3}{4x}\sqrt{14a}$.

26. $\frac{4}{5}\sqrt{6a}$.
27. $\dfrac{1}{a+b}\sqrt{5(a+b)}$.

28. $\dfrac{1}{(x+1)^2}\sqrt{3(x+1)}$.
29. $\frac{1}{6}\sqrt{6}$.
30. $\frac{10}{7}\sqrt{8}$.

EXERCISE 156, Page 281

1. $\frac{1}{4}\sqrt{6}$. 2. $5\sqrt{3}$. 3. $2\sqrt{ax}$. 4. $-8\sqrt{6}$. 5. $4\sqrt{2}$. 6. $30\sqrt{3}$.
7. $-11\sqrt{2}-8\sqrt{3}$. 14. $12\sqrt{6}+10\sqrt{5}$. 21. $(5a+b)\sqrt{x}$.
8. $8\sqrt{5}$. 15. $ab\sqrt{3a}$. 22. $5(x+2y)\sqrt{a}$.
9. $7\sqrt{15}$. 16. $12\sqrt[3]{2}-21\sqrt{3}$. 23. $(2b-a)\sqrt{b}$.
10. 0. 17. $\frac{3}{2}\sqrt[3]{2}-\frac{7}{4}\sqrt{2}$. 24. $(b-2a)\sqrt{5}$.
11. $6\sqrt{3}-6\sqrt{2}$. 18. $\frac{11}{5}\sqrt{15}$. 25. $13.880+$.
12. $5\sqrt{7}$. 19. $8\sqrt[3]{2}-9\sqrt[3]{3}$.
13. $-5\sqrt{3}$. 20. $4(2a+b)\sqrt{5}$.

EXERCISE 157, Page 282

1. $(x-y)\sqrt{2}$. 9. $(a-b)^2$. 16. $2-4\sqrt{2}$.
2. $(a+b)\sqrt{10}$. 10. $2\sqrt{4x^2-y^2}$. 17. $-6-2\sqrt{6}$.
5. $15(x-y)\sqrt{x+y}$. 13. $6+3\sqrt{2}-\sqrt{15}$. 18. $7\sqrt{6}-12$.
6. $6\sqrt{ab}$. 14. $2\sqrt{6}-4\sqrt{3}+8\sqrt{5}$. 19. $2\sqrt{15}-6$.
20. $-282-72\sqrt{10}$. 22. $x(\sqrt{3}+\sqrt{2}-\sqrt{5})$.
21. $30\sqrt{6}+54\sqrt{5}-34$. 25. $a-b-\sqrt{a^2-b^2}+\sqrt{a^2-ab}$.

EXERCISE 158, Page 284

1. $\frac{1}{2}\sqrt{2}$. 6. $\frac{1}{4}\sqrt{2}$. 11. $\frac{1}{10}\sqrt{6}$. 15. $\dfrac{5\sqrt{2x}}{6x}$. 19. $\dfrac{\sqrt{35}}{10}$.
2. $\frac{2}{3}\sqrt{3}$. 7. $\frac{3}{35}\sqrt{7}$. 12. $\frac{1}{b}\sqrt{ab}$.
3. $\frac{5}{7}\sqrt{7}$. 8. $\dfrac{a\sqrt{x}}{bx}$. 16. $\dfrac{\sqrt{21x}}{3x}$. 20. $\dfrac{a\sqrt{xy}}{by}$.
4. $\dfrac{2\sqrt{x}}{x}$. 9. $\dfrac{ab\sqrt{y}}{y}$. 13. $\dfrac{2\sqrt{xy}}{3y}$. 17. $\frac{5}{11}\sqrt{11}$. 21. $.378^-$.
5. $\dfrac{a\sqrt{b}}{b}$. 10. $\frac{1}{5}\sqrt{10}$. 14. $\dfrac{\sqrt{5x}}{5x}$. 18. $\dfrac{3\sqrt{3ax}}{5x}$.

EXERCISE 159, Page 284

1. $3abc^2\sqrt{2ac}$. 4. $\frac{1}{5}\sqrt[3]{25}$. 7. $\frac{4}{5}\sqrt{5}$. 9. $(a+2b)\sqrt{3a-6b}$.
2. $2a\sqrt{5a}$. 5. $\frac{3}{25}\sqrt{15}$. 8. $\dfrac{a}{3}\sqrt{3a}$. 10. $3(a+b)\sqrt{2}$.
3. $\frac{3}{7}\sqrt{7}$. 6. $\frac{1}{4}\sqrt{14}$. 11. 16.
12. $22+25\sqrt{10}$. 15. $(x-y)^2\sqrt{x+y}$. 18. $24\sqrt{5}$.
13. 3. 16. $(2b-a)\sqrt{x}$. 19. $(11+3a)\sqrt{a}$.
14. $ab(a\sqrt{a}+b\sqrt{b})$. 17. $12a+35b-41\sqrt{ab}$. 20. $3\sqrt{2}+3\sqrt{3}$.
21. $\sqrt{3}+12\sqrt{2}$. 24. 12.73^-. 26. 35.36^-. 28. 576.
22. $26\sqrt{2}$. 25. $4.24+$. 27. 432. 29. 1260.
23. $\frac{3}{4}\sqrt{3}$.

EXERCISE 161, Page 286

1. $1039\frac{1}{4}$ cu. in.
2. 3.03^+ gal.
3. $r = \sqrt{\dfrac{v}{\pi h}}.$
4. 8.5^+ in.
5. 40.41^+ in.
6. $r = \dfrac{1}{2}\sqrt{\dfrac{s}{\pi}}.$

7. 2.82^+ in.
8. 17.84^- yd.
9. $r = \sqrt[3]{\dfrac{3\,v}{4\,\pi}}.$
11. $e = \sqrt{\dfrac{s}{6}}$, 14.7.
12. $a = \dfrac{11\,c}{8}.$

13. \$5.50.
14. \$22.
15. $c = \dfrac{8\,a}{11}.$
16. \$14.55.

17. \$2181.82.
18. $a = \dfrac{25\,c}{18}.$
20. $a = \dfrac{26\,c}{19}.$

EXERCISE 162, Page 290

1. $\pm 4.$
2. $\pm 2.$
3. $\pm \frac{1}{3}.$
4. $\pm 2.$
5. $\pm \frac{1}{2}\sqrt{2}.$
6. $\pm 5.$

7. $\pm 1.$
8. $\pm \frac{4}{5}.$
9. $\pm a.$
10. $\sqrt{\dfrac{b-c}{a}}.$

11. $\pm \dfrac{c}{2}.$
12. $\pm(a+b).$
13. $\pm 1.$
14. $\pm 14.$
15. $\pm .3.$

16. $\pm .7.$
17. $r = \sqrt{\dfrac{7\,a}{22}}.$
18. $t = \sqrt{\dfrac{2\,s}{g}}.$

19. 6.
20. 8×32 rd.
21. 6×24 rd.

EXERCISE 163, Page 292

13. $2, -8.$
14. $2, -12.$
15. $2, -4.$
16. $5, -1.$
17. $10, -2.$
18. $6, -1.$

19. $-3, -8.$
20. $1, -\frac{7}{4}.$
21. $2, -\frac{4}{5}.$
22. $-1, \frac{7}{2}.$
23. $2, -\frac{13}{3}.$
24. $-\frac{4}{5}, \frac{4}{5}.$

25. $1, -\frac{4}{5}.$
26. $3, 6.$
27. $3, 4.$
28. $\frac{1}{3}, -\frac{4}{5}.$
29. $3, -1.$
30. $3, -\frac{1}{3}.$

31. $4, -\frac{4}{5}.$
32. $1, -\frac{10}{3}.$
33. $-2, -\frac{1}{3}.$
34. $1, -\frac{10}{3}.$
35. $\pm 1.$
36. $5, -\frac{13}{4}.$

37. $3, -\frac{4}{5}.$
38. No.
39. Yes.
40. Yes.

EXERCISE 164, Page 294

1. $2\,a, -6\,a.$
2. $3\,b, -7\,b.$
3. $2\,c, -5\,c.$
4. $2\,a, 3\,a.$
5. $3\,b, 4\,b.$

6. $ab, -6\,ab.$
7. $\dfrac{3\,b}{2}, -\dfrac{4\,b}{3}.$
8. $\dfrac{5\,cd}{3}, -3\,cd.$

9. $\dfrac{-p \pm \sqrt{p^2 - 4\,q}}{2}.$
10. $3\,m, 6\,m.$
11. $3\,a, -2\,a.$
12. $5\,b, -3\,b.$

13. $\dfrac{1}{a}, -\dfrac{3}{2\,a}.$
14. $\dfrac{a}{c}, \dfrac{3\,a}{7\,c}.$
15. $\dfrac{1}{a}, -\dfrac{6}{a}.$

16. $\dfrac{b}{2\,a}, -\dfrac{3\,b}{2\,a}.$
17. $\dfrac{-v \pm \sqrt{v^2 + 2\,gs}}{g}.$
18. $\dfrac{-v \pm \sqrt{v^2 + 2\,gs}}{g}.$

EXERCISE 165, Page 295

1. $1, -6$.
2. $1, -\frac{3}{4}$.
3. $\frac{1}{2}, -\frac{4}{3}$.
4. $\frac{3}{4}, -\frac{1}{4}$.
5. $\frac{5}{3}, -\frac{7}{2}$.
6. $\frac{1}{3}, \frac{3}{4}$.
7. $\frac{2f}{3}, -\frac{f}{2}$.
8. $\frac{7}{4a}, -\frac{3}{a}$.
9. $3, -5$.
10. $2, -9$.
11. $5, -3$.
12. $5, 2$.
13. $3, 4$.
14. $\frac{1}{2}, -3$.

EXERCISE 167, Page 296

1. $9, -5$.
2. $3, -4$.
3. $\pm 8, \pm 9$.
4. $3, -1\frac{3}{4}$.
5. 50×150 ft.
6. $2, 3$.
7. $5, 2$.
8. $5, 6$.
9. $5, 6, 7$.
10. $7, 8$.
11. $2, 5$.
12. 5.
13. 8.
14. $6, 7, 8$.
16. 4×9 yd.
17. 18 ft.
18. 8×10 rd.
19. 4.14^{+} rd.
20. 20%.
21. 20 rd.
22. 1 rd.
23. 10 rd.
24. $\frac{5}{7}$.
25. $8, 9$.
26. 3 in.
27. 6×12 ft.
28. 24×12 ft.

EXERCISE 168, Page 298

1. $15 - 5a$.
2.
   (1) $x(x+11)(x-13)$.
   (2) $5x^2(x+5)(x-5)$.
   (3) $(1-4ax)^2$.
   (4) $(x-3a)(x+2a)$.
   (5) $(9+a^2)(3+a)(3-a)$.
   (6) $x(x^2+x-3)$.
3. $\dfrac{6-4x}{(x-1)(x-2)(x-3)}$.
4. $2b^2 - 3b + \frac{1}{2}$.
5. $6, 7, 8$.
7. $\left(19 - \dfrac{x}{4}\right)\sqrt[3]{x}$.
8. $\frac{1}{2}, \frac{1}{2}, -1$.
9. $l = \dfrac{gt^2}{\pi^2}$.
10. $\dfrac{-d \pm \sqrt{d^2 - 4cf}}{2c}$.
11. $999,400.09$.
12. $\dfrac{9(x-2y)}{a}$.
13. 4.393^{+}.
14. 7×15 ft.
15. $0, 3$.
16. $a^8bx^2\sqrt{axy}$.
17. $\dfrac{3b}{2}$.
18. s, r.
19. $l = \sqrt{6\,Rd - 3\,d^2}$.

EXERCISE 169, Page 301

1. $1.207^{+}, -.207^{+}$.
2. $1.866^{+}, .134^{-}$.
3. $.911^{-}, -.244^{+}$.
4. $1.618^{+}, -.618^{+}$
5. $3.833^{+}, .667^{-}$.
6. $2.823^{-}, .177^{+}$.
7. $2.439^{-}, .228^{-}$.
8. $.2, .3$.
9. $.7, .15$.
10. $1.2, -.6$.
11. $.34, -.5$.
12. $-.85, -.24$.
13. $.24, .35$.
14. $3.5^{+}, .5^{-}$.
15. $2.3^{+}, .2^{+}$.
16. $1.0^{-}, -8.6^{+}$.
17. $5.4^{-}, -1.5^{-}$.
18. $3.2^{+}, -.4^{-}$.
19. $3.5^{-}, .4^{-}$.
20. $17.90^{-}, .30^{+}$.
21. $.11^{-}, -3.43^{+}$.
22. $.29^{-}, -4.59$.
23. $13.72^{+}, -.45^{+}$.
24. $.719^{+}, .181^{-}$.
25. $3.086^{+}, .014^{-}$.
26. $.315^{+}, -.475^{+}$.
27. $.393^{-}, -.236^{+}$.
28. $.537^{-}, -13.037^{-}$.
29. $.402^{-}, -22.402^{-}$.

EXERCISE 170, Page 302

1. ± 1, ± 3.
2. ± 1, ± 4.
3. ± 2, ± 3.
4. ± 2, ± 4.
5. $\pm \frac{3}{2}$, ± 1.
6. ± 1, $\pm \frac{1}{2}$.
7. 2, 1.
8. ± 3, $\pm \sqrt{2}$.
9. $\pm 3\sqrt{3}$, $\pm 2\sqrt{2}$.
10. ± 2, $\pm \sqrt{3}$.
11. $\pm \sqrt{2}$, $\pm \sqrt{3}$.
12. $\pm 1.62^-$, $\pm .62^-$.
13. $\pm 1.31^-$, $\pm .54^+$.
14. $\pm 1.58^+$, $\pm .71^-$.
15. $\pm 1.37^-$, $\pm .37^-$.
16. No.
17. Yes.
18. Yes.

EXERCISE 171, Page 304

1. $\begin{cases} 1, -13. \\ 2, 16. \end{cases}$
2. $\begin{cases} 4, -1. \\ \frac{1}{2}, -2. \end{cases}$
3. $\begin{cases} 1, 2. \\ 0, -3. \end{cases}$
4. $\begin{cases} 2, -26. \\ 1, 15. \end{cases}$
5. $\begin{cases} 4, -\frac{3}{2}. \\ 1, -\frac{12}{5}. \end{cases}$
6. $\begin{cases} -2, -\frac{24}{25}. \\ 1, -\frac{7}{25}. \end{cases}$
7. $\begin{cases} 3, -\frac{16}{9}. \\ -1, \frac{17}{9}. \end{cases}$
8. $\begin{cases} -3, 7. \\ -\frac{1}{3}, 4. \end{cases}$
9. $\begin{cases} \frac{7}{5}, 2. \\ \frac{7}{5}, 1. \end{cases}$
10. $\begin{cases} \frac{9}{11}, -\frac{12}{13}. \\ \frac{18}{11}, \frac{9}{13}. \end{cases}$
11. $\begin{cases} 2, -\frac{7}{5}. \\ 1, -\frac{24}{15}. \end{cases}$
12. $\begin{cases} 3, \frac{17}{4}. \\ 2, 3. \end{cases}$
13. $\begin{cases} 8, -12. \\ 6, -4. \end{cases}$
14. $\begin{cases} 12, -7\frac{1}{5}. \\ 5, -8. \end{cases}$
15. $\begin{cases} 10, -42. \\ 6, -10, \end{cases}$
16. No.
17. Yes.
18. Yes.
19. No.
20. $\begin{cases} -3.3, 2. \\ 2, -3.3. \end{cases}$

EXERCISE 173, Page 306

1. 7, 3.
2. 3.5, 2.2.
3. 3.8, 2.5.
4. 3.62$^+$, 1.82$^+$.
5. 4, 28.
6. $2\frac{1}{2}$, $3\frac{1}{5}$.
7. 5, 13.
8. 3, 8.
9. 7 × 12 ft.
10. 4.2 × 2.7 ft.
11. 40 × 30 yd., 20 × 60 yd.
12. 78 × 36 ft.
13. 7 × 11 rd.
14. 20 × 40 ft.
15. 30 mi.
16. 36.
17. 9.
18. 20.
19. $\frac{9}{12}$.
20. $\frac{4}{5}$.
21. $\frac{2}{3}$.
22. 3.
23. 24, 16.
24. $4\frac{1}{2}$, $7\frac{1}{2}$.
25. 14, 3.
26. 110 ft.
27. 15.

EXERCISE 174, Page 309

1. 257.6 ft.
2. 353.556 ft.
3. 2129$^+$ ft.
4. $t = \sqrt{\dfrac{2s}{g}}$.
5. 7.88 sec.
6. 17.62$^+$ sec.
7. 27.
8. 11.3, 14.3$^+$.
   $$d = 4 + 4\sqrt{\dfrac{L}{l}}.$$
9. $a = \left(\dfrac{1+g}{1-d}\right)c$.
10. \$12.
11. \$4.
12. \$3600.
13. $c = \left(\dfrac{1-d}{1+g}\right)a_o$
14. 7.92$^-$ ¢.
16. $r = \sqrt{\dfrac{a}{\pi}}$.

EXERCISE 175, Page 313

16. 19.6+ sq. in.   18. 4″, 5″, 6″.   21. 2¼ sec.   23. 66+ ft.
17. 3+ in.   20. 196.   22. 11.3⁻ mi.

EXERCISE 176, Page 315

2. (1) 8.1.   (2) −13¼.   (3) 10.9.
   (4) 43.3.   (5) − 64.3.   (6) − 9.18.
   (7) 19¼.   (8) − 1.6.   (9) − 150.
   (10) 290.   (11) − $6 x^3 y^2$.   (12) ps^2.

3. $a^2 + 19\,ab + 3\,b^2$.   5. $17 x^3 + 3 x^2 y + 3 x y^2 − y^3$.
4. $(a + b − 3 c)y^3$.   6. $(a−r)x^2 + 2\,bxy + (p − 3 c)y^2$.
7. $− 8.7 p^2 − 13.2$.   9. $a^4 − 7\,a^2 b^2 + b^4$.   11. $4 a^2 − 2\,ax + x^2$.
8. $5 x^4 − 21 x^2 y^2 + 4 y^4$.   10. $3 x^2 + 2 x − 1$.   12. 1333.   13. 54.
14. (1) $x^2(4 p + 1)(4 p − 1)$.   (2) $(4\,ab − 1)^2$.
    (3) $(5\,ab + 13 x)(5\,ab − 13 x)$.   (4) $x(2 x − 3 y^2)^2$.
    (5) $(x + 7)(x − 11)$.   (6) $(3 a + b + 2 x)(3 a + b − 2 x)$
    (7) $3(r^2 + s^2)(3 r^2 − 3 s^2 + 2)$.   (8) $x(x − 3 y)(x + 2 y)$

15. 155.55.

16. $\dfrac{3 x + 5 a}{x(3 x − 5 a)}$.

17. $\dfrac{15 x^3 − 45 x^2 − 7 x + 21}{x^3 − 3 x^2 + 5 x − 15}$.

18. $3 m + p + \dfrac{4 p^2}{3 m − p}$.

19. $\dfrac{6 a^3 − 13 a^2 + 5 a − 7}{2 a − 1}$.

20. $\dfrac{− 3 + 16 x − 5 x^2}{9 x^2 − 4}$.

21. $\dfrac{r + 3 s}{r − 4 s}$.

23. 3.16⁻.
24. .84−, .86+.

25. $\dfrac{3 b − a}{21 + 9 c}$.

26. $\dfrac{3 p q^2 r + 7 p^2 q r}{3 q^2 + 5 p^2}, \dfrac{5 p^2 q r − 7 p q^2 r}{5 p^2 + 3 q^2}$.

27. $q = \dfrac{R p}{p − R}$.

28. $d = \dfrac{R c − a e}{a e c}$.

29. No.   30. 40.527.
31. 18.8−.
32. $(5 x − 3 y − 4)\sqrt{xy}$.
33. $2(x − 4 y)\sqrt{x − y}$.
34. $(a − b)\sqrt{a^2 − 3\,ab − 4 b^2}$.
35. $x^2 − 3 x + \frac{1}{4}$.
36. 4.458 +.   37. 5, 7.
38. .42−, −2.07+.
39. ±2.1+, ± .7−.
40. Yes.   41. $\begin{cases} 1,\ 11. \\ 6,\ 1. \end{cases}$
42. $r = −\dfrac{h}{2} \pm \sqrt{\dfrac{h^2}{4} + \dfrac{T}{2\pi}}$.
45. $x^5 + x − 1$.   46. $x^2 − 3$.
47. 38.
48. $a + b,\ 3 a − 5 b + 4 c + 7 d$.

49. $− 7 − 3 x + 2 y − z,\ − a − 3 x + 2 y − z,\ b − 3 x + 2 y − z$.
50. $− 2 a + 3 b,\ − 5$.   61. $\frac{1}{2},\ −\frac{1}{3}$.
59. $(x − 3 y)\sqrt{x^2 + 2 xy − 3 y^2}$.   62. 15,064, 17,480.

63. $(4x + .3)^2$, $(x^2 + .09)(x + .3)(x - .3)$.

64. 400, 800, 1600 lb.

66. $q + 2p$.   **68.** $\frac{18}{7}, -\frac{4}{7}, \frac{10}{7}$.

69. \$12,000 @ 5 %, \$8000 @ 7 %.

71. $x^2 + \frac{1}{2}x - \frac{3}{2}$.

73. $y + 2cy - d$.

77. $\begin{cases} 5, 2. \\ 3, 6. \end{cases}$

78. $x - 2$.

79. $(x + 2)(x - 2)^2$.

80. 12.68^-.

81. $500 - cx$, $5 - \dfrac{cx}{100}$.

83. $\frac{3}{2}$.

84. $(72b - 12a)$ cents.

86. $\dfrac{ap}{l}, \dfrac{br}{p}$.

88. $\dfrac{ab}{m}$.

89. $7(a - b)\sqrt{a - b}$.

91. 1.001,105.

94. $1.13^+, -.53^+$.

96. $x - \dfrac{y}{3} + \dfrac{2y^2}{3x}$.

98. $\dfrac{4}{(a + 1)(a^2 + 4b^4)(a + 2b^2)}$.

99. $\dfrac{xy}{p}$.

100. 16 lb.

101. 124, 61, **15.**

102. 2, 5.

103. $x - 2a + \dfrac{b}{3}$.

104. 114.

105. $\begin{cases} 1, \frac{12}{7}. \\ -3, \frac{4}{7}. \end{cases}$

107. Gold, 76 lb.; silver, 30 lb.

EXERCISE 177, Page 324

1. x^{a+5}.

2. $-15 x^{n+3}$.

3. $3 z^{2x-5}$.

4. $-6 a^{2n+2}b^{3n}$.

5. $30 a^{7x}b^{4x}c^{5x}$.

6. $28 x^{5a}y^{2b}z^{8c}$.

7. $24 a^{3x-2}b^{n+2}$.

8. $15 x^{4n+1}$.

9. $100 a^{9x}b^{12y}$.

10. $3 x^{n+5} - 5 x^{n+4} + 4 x^{n+3}$.

11. $3 x^{4n+2} - 5 x^{4n+1} + 4 x^{4n}$.

12. $-20 a^{4x}b^{3y} + 15 a^{3x}b^{2y}$.

13. $15 x^{n+2} + 16 x^{n+1} - 15 x^n$.

14. $x^{6n} - 16 y^{4n}$.

15. $24 a^{2x+5} - 31 a^{2x+4} + 10 a^{2x+3}$.

16. $2 a^{2n+4} + a^{2n+3} - 12 a^{2n+2} + 9 a^{2n+1}$.

17. $x^{5a} - x^{4a} - 2x^{3a} + 2 x^{2a} + x^a - 1$.

18. $2 a^{n+1}$.

19. $2 a^{3x}$.

20. $3 a^{2n+3}b^n$.

21. $a^{2x}b^y$.

22. $24 x^{3n} - 15 x^{2n} - 3 x^n$.

23. $4 x^{2n} + 3x^n$.

24. $2 x^{n+2} - 3 x^{n+1} + 5 x^n$.

25. $3 a^{2x} + 2 a^x + 1$.

26. $3 x^{3n}(2 x^{2n} - 1)$.

27. $4 a^{2x}b^y(3 a^x b^y - 2)$.

28. $(x^n + 3)^2$.

29. $(2 x^n - 3)^2$.

30. $(x^n - 3)(x^n - 1)$.

31. $(x^{2n} - 3)^2$.

32. $(x^n - 2 y^n)^2$.

33. $(x^a + 1)(x^a - 1)$.

34. $(x^{3p} + 1)(x^{3p} - 1)$.

35. $(3 x^n + 2y)(3 x^n - 2 y)$.

EXERCISE 178, Page 326

1. $2x^2+11x+12.$
2. $2x^2-11x+12.$
3. $2x^2-5x-12.$
4. $2x^2+5x-12.$
5. $6a^2+19a+15.$
6. $6a^2-a-15.$
7. $5x^2+34x-7.$
8. $3x^2+xy-24y^2.$
9. $12a^4-11a^2b-5b^2.$
10. $\frac{3}{4}x^2+\frac{7}{8}x+\frac{1}{4}.$
11. $2a^2+.1ab-.06b^2.$
12. $\frac{3}{4}x^2+2ax-\frac{3}{4}a^2.$

EXERCISE 179, Page 327

1. $(2x+1)(x+1).$
2. $(3x-2)(x-4).$
3. $(2x+1)(x+2).$
4. $(3x+1)(x+3).$
5. $(3x+5)(2x-1).$
6. $(x+3)(2x-1).$
7. $2x(x+4)(3x-2).$
8. $x^2(3x+2)(x-2).$
9. $(4a-5)(2a+3).$
10. $(2x+5)(x-2).$
11. $(4x+1)(3x-2).$
12. $(4x-1)(x+3).$
13. $(5x-1)(x+5).$
14. $3x(x-2)(3x+1).$
15. $2y(3x+2)(x-1).$
16. $(.3a+4)(.2a+5).$
17. $(4x-9y)(3x+7y).$
18. $(8a-9b)(4a+5b).$
19. $(2x+3)(x+1)(2x-3)(x-1).$
20. $(3x+2)(x+4)(3x-2)(x-4).$
21. $(.5p-.4)(.3p+.1).$
22. $2x(6x-y^2)(2x+9y^2).$
23. $(5a+4b)(5a-4b)(a^2+b^2).$
24. $(8x^2-9y^2)(2x^2+y^2).$
25. $(4.4a-.3b)(a-.5b).$
26. $(5a+4b)(a+b)(5a-4b(a-b).$
27. $(5+4x)(4-5x).$
28. $(5-3xy)(1+7xy).$
29. $(a+b+8)(a+b-3).$
30. $(3x-3y-2z)(x-y+3z).$

EXERCISE 180, Page 328

1. $\dfrac{x+2}{2x-5}.$

2. $\dfrac{x+2y}{2x+3y}.$

3. $\dfrac{2x+y}{2x-y}.$

4. $\dfrac{3x+4a}{x(x+a)}.$

5. $\dfrac{23a-b}{(2a-3b)(a-2b)(4a+b)}.$

6. $\dfrac{10x-3x^2-1}{(3x-4)(2x+5)(3x-1)}.$

7. $\dfrac{5x(x+3)}{(2x+1)(2x-1)(x+1)}.$

8. $\dfrac{33p+36q}{(2p-q)(p+2q)(4p+7q)}.$

9. $\dfrac{9(2x-3y)}{4(3x-2y)}.$

10. $\dfrac{(2x+1)^2}{2(x+1)^2}.$

11. $\dfrac{2x-1}{3x+2}.$

12. $1.$

13. $6.$

14. $\frac{7}{8}$ or $1.$

EXERCISE 181, Page 330

1. 2, 3.
2. 2, −2.
3. 2, −1.
4. −1, −7.
5. 1, $\frac{1}{2}$.
6. 2, $\frac{1}{3}$.
7. $\frac{1}{3}$, −2.
8. 3, −2.
9. 3, $\frac{1}{3}$.
10. $\frac{1}{4}$, −$\frac{1}{3}$.

11. $\frac{1}{2}$, −$\frac{1}{4}$.
12. $\frac{1}{4}$, −3.
13. $\frac{10}{3}$, −$\frac{2}{3}$.
14. $3a$, $4a$.
15. $\frac{7a}{3}$, −a.
16. $2b$, −$\frac{4b}{5}$.
17. $\frac{2}{a}$, −$\frac{4}{5a}$.

18. $\frac{1}{a}$, −$\frac{3}{2a}$.
19. $\frac{2}{3a}$, −$\frac{4}{a}$.
20. ±1, ±3.
21. ±1, ±$\frac{3}{2}$.
22. ±1, $\sqrt{2}$.
23. ±1, ±$\frac{1}{3}$.
24. ±1, ±$\frac{5}{3}$.
25. 3, −$\frac{3}{4}$.

26. 3, −$\frac{3}{4}$.
27. 5. −$\frac{1}{4}$.
28. 3, −1.
29. 8, $4\frac{1}{2}$.
30. 9, −5.
31. .4, −10.
32. 3, −4.
33. 0, 9.
34. $\frac{3}{4}$, −2.

EXERCISE 182, Page 332

17. 8.
18. 4.
19. 9.
20. 16.
21. 8.
22. 32.
23. 20.
24. $6\frac{3}{4}$.
25. $\sqrt{2}$.
26. $8\sqrt{x}$.
27. $9\sqrt[3]{a}$.
28. 0.
29. $7\sqrt{x-y}$.
30. x.

31. x^2.
32. a^2.
33. $5(p−q)$.
34. $\sqrt{6}$.
35. $5\sqrt[3]{y^2}$.
36. $20\sqrt{2}$.
37. $16(a−b)^2$.
38. 66.
39. $16\sqrt{2x}$.
40. $−(a−b)\sqrt{2}$.
41. $(x−y)\sqrt{x+y}$.
42. $4(a−b)^{3/2}$.
43. $\frac{1}{x}\sqrt{x}$.

44. $\frac{5}{3a}\sqrt{a}$.
45. $\frac{7}{5a}\sqrt{5a}$.
46. $\frac{1}{b}\sqrt{ab}$.
47. $\frac{1}{y}\sqrt{xy}$.
48. $\frac{b}{ax}\sqrt{ax}$.
49. $\frac{7}{10}\sqrt{5}$.
50. $\frac{1}{bc}\sqrt{abc}$.

EXERCISE 183, Page 334

5. (1) $\frac{1}{4}$.
   (2) $\frac{5}{14}$.
   (3) $\frac{3}{4}$.
   (4) $\frac{1}{4}$.
6. $\frac{7}{3}$.
7. $\frac{7}{4}a$.
8. −$\frac{1}{3}$.
9. $1\frac{1}{2}$.

10. $\frac{bc}{a}$.
11. 3.
12. 2, −$\frac{1}{4}$.
13. 5, −4.
14. 4.
15. 5.
16. 480, 720.

17. 300, 400, 500.
18. $6040, $9060, $24160.
19. $25, $105, $5, $5.
20. $5200, $8600.
21. $5070, $7745, $9966.

www.ingramcontent.com/pod-product-compliance
Lightning Source LLC
Chambersburg PA
CBHW030450210326
41597CB00013B/610